D0025151

THE IMAGE EMPIRE

A HISTORY OF BROADCASTING IN THE UNITED STATES

VOLUME I — A TOWER IN BABEL

VOLUME II — THE GOLDEN WEB

VOLUME III — THE IMAGE EMPIRE

THE IMAGE EMPIRE

A History of Broadcasting in the United States

Volume **III**—from 1953

ERIK BARNOUW

New York OXFORD UNIVERSITY PRESS 1970

Library of Congress Catalogue Card Number: 66-22258

10 9 8 7 6 5

Printed in the United States of America
on acid free paper

CONTENTS

THE IMAGE EMPIRE

INTRODUCTION

Then 'twas the Roman, now 'tis I.

A. E. HOUSMAN[1]

During the 1950's, in a frontier atmosphere of enterprise and sharp struggle, an American television system took shape. But even as it did so, its pioneers pushed beyond American borders and became programmers to scores of other nations. In its first decade United States television was already a world phenomenon. Since American radio had for some time had international ramifications, American images and sounds were radiating from transmitter towers throughout the globe.

They were called "entertainment" or "news" or "education" but were always more. They were a reflection of a growing United States involvement in the lives of other nations—an involvement of imperial scope. The role of broadcasters in this American expansion and in the era that produced it is the subject matter of *The Image Empire,* the last of three volumes comprising this study.

A television-radio system is like a nervous system. It sorts and distributes information, igniting memories. It can speed or slow the pulse of society. The impulses it transmits can stir the juices of emotion, and can trigger action. As in the case of a central nervous system, aberrations can deeply disturb the body politic. These complex roles lend urgency to the study of television and radio and the forces and mechanisms that guide and control them.

1. "On Wenlock Edge" from *A Shropshire Lad,* authorized edition, in *Collected Poems of A. E. Housman.* Copyright 1939, 1940, © 1959 by Holt, Rinehart and Winston, Inc. Copyright © 1967, 1968 by Robert E. Symons. Used by permission of Holt, Rinehart and Winston, Inc.

The first of our three volumes, *A Tower in Babel,* brought our story to 1933; it told of separate transmitter towers, competing for attention in a chaos of voices. The second volume, *The Golden Web,* covering the years 1933-53, saw the national networks become a dominant force and then shift attention—their own and that of the nation—from radio to the fantastic promise of television. At this moment the story of *The Image Empire* begins.

1 / ENTERPRISE

"Papa, what is the moon supposed to advertise?"

CARL SANDBURG, *The People, Yes*

Television has brought back murder into the home, where it belongs.

ALFRED HITCHCOCK

A coast-to-coast television boom was coming. As Dwight D. Eisenhower prepared to take over the United States presidency in January 1953, that much was clear. Every sign pointed to the coming boom. Cities where television was in operation were experiencing a local chain explosion. As sets entered homes, theater attendance, night-club revenue, and taxi receipts dropped. So did radio listening. Advertisers were taking note and revising plans and budgets.

Only manufacturing restrictions relating to the Korean War had held back the boom. A "freeze" in new station licenses had been in effect from 1948 to 1952. Many cities—including such places as Denver, Little Rock, and Austin—had not had television during this period. Under these conditions *network* television could have only limited meaning; besides, coast-to-coast networks had not been possible until late in 1951, when AT&T television cables and microwave relays were first able to span the continent.

But now peace was in the air. An Eisenhower flight to Korea before he took office had spurred peace expectations. Restrictions on war materials were meanwhile being lifted. In the closing months of 1952 a number of new stations received a go-ahead. Among the first was KTBC-TV, Austin, licensed to Mrs. Lyndon B. Johnson, wife of the U. S. Senator from Texas; before it even reached the air, advertising sales were such that *Broadcasting* magazine reported: "AUSTIN'S BRINGING IN A GUSHER."[1] Hundreds of additional applicants clamored for a go-ahead. Sponsors, shifting from

1. *Broadcasting*, October 27, 1952. The station went on the air Thanksgiving Day.

war to consumer goods, were more than ready, lured by tales of fabulous advertising successes—such as the Hazel Bishop story. This lipstick maker, doing a $50,000 annual business, had taken up television in 1950; solely through television, its sales had zoomed to $4,500,000 in 1952, and continued to climb sharply.[2]

But if a boom was certain, its shape was not. The very certainty of the boom was bringing to television migrations from other fields—film, theater, newspaper, magazine, radio. Each had felt tremors; in each there was a search for escape routes, a look toward new frontiers. But each inevitably had its own ideas on the destiny of television and the direction it should take.

One of the largest migrations was from the film world. It was in a turmoil that stemmed partly from the rise of television, but also from other events. As a result of an anti-trust conviction in the U. S. Supreme Court, the "big five" among the studios—Paramount, RKO, Twentieth Century-Fox, Warner Brothers, Loew's (including MGM)—were being forced to sell thousands of theaters and to end block booking—in other words, to give up the weapons by which they had controlled the film industry.[3] The result was an almost panicky retrenchment. Studios were being shuttered and production lots sold off. Most long-term contracts with artists and technicians were being canceled. As for the television monster, the big studios were determined not to assist its growth in any way; they would withhold their products from it. To defeat the monster, they would totally ignore it.

But artists and technicians, fired by the hundreds, were not ignoring it. Except for the big stars—who were as aloof as the studio heads—they drifted toward the enemy. They were especially nudged in that direction by a phenomenon named Lucy.

I Love Lucy, one of the few filmed programs among network leaders, began the year 1953 in spectacular style. Lucille Ball, co-star with her husband Desi Arnaz, was pregnant; for months this had provided the comedy plotting as the nation watched Lucille—or Lucy—grow larger. A program about the big day was filmed when the event became imminent.

2. *Advertising Age*, March 29, 1954.
3. *U. S. v. Paramount et al.*, 334 U.S. 131 (1948), was followed by several consent decrees which settled details for disposing of the theaters. The case also involved Columbia Pictures, Universal, and United Artists, which had all engaged in block booking; but they did not own theaters at the time of the decision and were therefore less drastically affected.

Then on January 19, 1953, Desiderio Alberto Arnaz IV, 8½ pounds, was born on the exact day of the Lucy-has-her-baby telecast. The event found 68.8 per cent of television sets tuned to *I Love Lucy*[4] and was headline news even in competition with the Eisenhower inaugural, which came the following morning.

Hollywood had not thought much of *I Love Lucy* as film art. It was produced like a live program, in one session before a studio audience in tiered bleachers, and was only filmed for possible re-use income. But it had turned two minor Hollywood names into household deities, and other actors took note. Many began to work on pilot films. Some of these were produced in fringe studios; a few—for the sake of low cost—were made on location abroad, like the spy series *Foreign Intrigue,* shot against Swedish and French backgrounds. Many other artists decided to hurry to New York, which had been the main radio center and seemed certain to be the main television center. Most network programming—about 80 per cent—was *live,* but perhaps film talents would be usable. An influx of Hollywood writers, directors, actors, cameramen, set-dressers, designers, special-effects men, dancers descended on New York production offices. They hoped—they felt sure—the new medium was for them.

But the Broadway world was also ready. With its own activity shrinking since the 1920's, Broadway was always in crisis. This was to some extent a result of the rise of radio—which theater people had largely ignored. But they were not likely to ignore television. This seemed different to them, and they had been prominent in television experimentation since the 1930's. Theater leaders like Oscar Hammerstein II, Richard Rodgers, Elmer Rice, and the Theater Guild partners were all busy with television projects or negotiations. In television they saw a new theater renaissance, made to their measure. They too brought special abilities and attitudes—differing sharply from those of the Hollywood migration.

Equally different was another invasion—from broadcasting itself. In the shrinking world of radio programming, phonograph records were replacing a horde of artists. Many were panicky. Network television divisions tended to recruit executives from theater or film, and they were inclined to shunt radio people aside. "It's a different medium," was the endlessly repeated—often condescending—advice. But radio people had some advantages. Television was taking over the business structure of

4. According to Trendex, a rating system based on telephone calls. *Broadcasting,* January 21, 1953.

radio, with its agencies, sponsors, time periods, and time-buying practices. Some sponsors wanted to adapt their radio successes to television. The transition had been made successfully by series of various sorts, including variety (the Arthur Godfrey and Kate Smith programs), quiz (*You Bet Your Life, Two for the Money*), panel (*Meet the Press, Information Please*), drama (*Suspense, Aldrich Family*).

Still another invasion was composed of animated-film people. In the big-studio era, before the anti-trust debacle, each "major" had sheltered an animation unit. In the great retrenchment these units had been reduced, set adrift, or disbanded. A number of brilliant artists—John Hubley, Ernest Pintoff, and others—had set up small, independent enterprises, many in New York. Living mainly on commercials, they fought for a place in network programming.

There were still other groups—less numerous, equally clamorous. Theater newsreels, like animation units, were suffering from the retrenchment of the film world, and their personnel gravitated toward television. As newspaper and magazine revenue fell, editors, writers, and photographers drifted in the same direction. So did educators, who saw in television a new day in learning, a school without walls. Many ministers looked on it as a latter-day pulpit; some, as a revival tent. Sports promoters were making it a new wrestling and boxing arena. Opera people were staking a claim.

Watching the maelstrom, preparing their own moves, were advertising men, public relations advisers, copyright lawyers, talent agents, and numerous other middlemen.

Television in 1953 was a chaotic scramble for position and ultimate control. The outcome was in doubt.

Beyond the uncertainties involved in these struggles lay another and more crucial uncertainty—a political one. In January 1953 it overhung all other struggles. It had to do with Senator Joseph R. McCarthy of Wisconsin, the political Goliath of the hour, and with the yearning for a David to smite him.

MC CARTHY AND THE GENERAL

Senator Joseph McCarthy has been described as the most brilliant demagogue ever bred on American shores. No politician, wrote Richard H.

Rovere, had "surer, swifter access to the dark places of the American mind."[1]

As the new administration took office, McCarthy was looked on as leader of a coming purge. The fields of film, radio, and television programming had already experienced a purge based on blacklists, but the idea was still spreading. McCarthy, it was reported, had plans to cleanse the Federal Communications Commission; also some of its licensees; also that part of the State Department concerned with communicating with people of other lands—the international information division, including the Voice of America. Thus he planned moves in the agencies most concerned with American broadcasting and the environment in which it operated, at home and abroad.

The blacklist mania, though antedating McCarthy,[2] had become "McCarthyism"—a term that brought a chill to some and an almost religious glow to others. A "loyal underground" of McCarthy followers in government agencies was feeding his dossiers with tidbits on anyone of suspicious habits who might be one of "them." Dog-eared lists of such people were in circulation. McCarthy had become the center of a confident, active coterie of some seventy people—"a heady mixture," said *Look*, of millionaires, beautiful women, broadcasting stars, hard-working political bosses, clergy, and "hot-shot investigators." Two broadcasting figures, the commentators Fulton Lewis, Jr., of the Mutual Broadcasting System and George Sokolsky of the American Broadcasting Company, were close to the inner circle.[3] The group spoke constantly of "communists," "traitors," "fellow travelers," "dupes," "pinkos," "undesirables," "subversives," and seemed to use the terms interchangeably.

McCarthyism was really a linking of various political vendettas, old and new. Many southerners—and quite a few northerners—had always regarded an interest in Negro rights as communistic. Admirers of Franco considered opposition to the Spanish dictator a communist symptom; this alone had landed broadcasting artists on blacklists. More recently opposition to nuclear-weapons tests had come to be held pro-Russian and thus communistic. Another theme, of growing importance since World War II, concerned China. The powerful "China Lobby" had virtually made support for the Chiang Kai-shek regime a test of loyalty to America,

1. Rovere, *Senator Joe McCarthy*, p. 3.
2. For the development of blacklists in broadcasting see *The Golden Web*, pp. 174–81, 246–57, 261–83.
3. *Look*, December 1953.

and opposition to him a communist trait. Significantly, the publications *Red Channels* and *Counterattack*, which struck hard at broadcasters critical of Chiang Kai-shek, were financed by the importer Alfred Kohlberg, who was also a China Lobby leader and McCarthy backer. He was credited with having stimulated McCarthy's violent attacks on General George C. Marshall for attempting to mediate between Chiang Kai-shek and the Chinese communists—an attempt that McCarthy denounced as treason.[4]

It was widely rumored that Eisenhower hated McCarthy, and this reassured some people. Many of the prominent men who flocked to join the Eisenhower crusade—to clean up the "mess in Washington"—tended to assume that McCarthy had facts up his sleeve and would in time substantiate his charges; yet many did not care for his methods. They were sure that Eisenhower would, sooner or later, put McCarthy in his place.

Eisenhower had had a dramatic chance to do this during the campaign, but had not taken it. General Marshall had been Eisenhower's mentor, and Eisenhower revered him. In the course of a scheduled campaign tour of Wisconsin, Eisenhower planned a brief defense of his wartime chief, in words penned for the purpose by Emmet John Hughes, an intermittent Eisenhower speech-writer. But Republican Party chiefs, afraid of a party split, dissuaded him from carrying out the plan. Still, confidence that Eisenhower would in time stem the McCarthy steamroller, and restore an atmosphere of sanity, was widespread.

The first omens, at the start of the new administration, were not favorable.

Eisenhower, on taking office, had an immediate opportunity to appoint a member of the Federal Communications Commission, to fill a vacancy. He appointed a McCarthy protégé, John C. Doerfer—apparently hoping to placate McCarthy. Doerfer promptly made McCarthyist moves. Some stations had nettled McCarthy by not carrying his campaign speeches, and the Senator had hinted that this was a communist symptom. As commissioner, Doerfer set out to substantiate this—an effort that occupied him and the commission for several years.

Late in 1953 another FCC vacancy occurred. Again Eisenhower appointed a McCarthy friend—this time Robert E. Lee, whom *Look* listed among investigators of the McCarthy inner circle. A former FBI agent,

4. For Kohlberg's relation to *Red Channels* and *Counterattack*, see *The Golden Web*, pp. 253–4; for the Kohlberg-McCarthy relationship, see *The Reporter*, April 29, 1952.

he was said to have provided information on which McCarthy based his charges of subversives in the State Department and the Voice of America. Lee acknowledged admiration for McCarthy, though saying he was not "beholden" to him.[5]

Meanwhile, at the State Department, the McCarthy drive was bringing quick developments. John Foster Dulles, the new Secretary of State, at once accepted as his director of "personnel and security" Scott McLeod, another FBI man and McCarthy adherent. Emmet John Hughes found him a "loudly aggressive superpatriot."[6]

These "security" devotees, McCarthy-linked, were a sharp contrast to others arriving on the Washington scene. Among leaders of various fields drawn into the Eisenhower administration was the gentlemanly Dr. Robert L. Johnson, president of Temple University. He had no political ambitions but had long admired General Eisenhower, and had urged him to run; when asked to enter the State Department as head of its international information activities, including the Voice of America and overseas libraries and exhibits, he felt he could not refuse to join the crusade. He brought with him as right-hand man his associate Martin Merson, who had studied federal reorganization problems for the incoming administration. Arriving in Washington, they found security officer Scott McLeod already on the job, and assured him of cooperation. Both Johnson and Merson assumed there must be a basis for McCarthy's charges of widespread communist infiltration in their division; awaiting documentation, they were prepared to act on it.

Their division had 9000 employees, including the staff of the Voice of America. On some of them, fragmentary reports began to be provided by McLeod, but none indicated disloyal activity. To Dr. Johnson's astonishment, not a single case of disloyalty was found. Eventually six employees were dismissed on the lesser charge of being "security risks"—a term used to cover such problems as alcoholism or suspected homosexuality. Even some of these cases seemed doubtful to Johnson and Merson, so that they became appalled at the injustice done by the sweeping charges.

But meanwhile, throughout these proceedings, the two executives were continually badgered by two young McCarthy aides, Roy M. Cohn and G. David Schine. They came with lists of "undesirables" who had to be

5. *Look*, December 1, 1953, March 20, 1956; Washington *Post*, October 7, 1953.
6. Hughes, *The Ordeal of Power*, p. 84.

fired right away to avoid trouble with "Joe," and lists of others to be hired in their place. No information was provided—just lists. The certainty of budget cuts by congressional committees—if the desired actions were not taken—was often mentioned. Cohn and Schine made casual references to chats with Vice President Richard Nixon and other administration leaders. At lunch they would send for a telephone and call McCarthy. When Dr. Johnson suggested that he would like to discuss the personnel matters directly with McCarthy, Cohn and Schine doubted that the Senator would see him until he had cleansed his division of undesirables. When Dr. Johnson sought assurance of White House backing against the Cohn-Schine harassments—he had been promised direct access to the President—he could not get beyond C. D. Jackson, former publisher of *Fortune,* who said he could not possibly take the matter up with the President, because Eisenhower was determined not to offend members of Congress. Johnson continued efforts to confront McCarthy directly, and was told that perhaps George Sokolsky, the ABC commentator and Hearst columnist, could help him. Sokolsky, over a dinner, felt he could help the State Department official to see Senator McCarthy, but suggested another preliminary meeting to talk it all over. This second meeting was held in the apartment of David Schine's parents in the Waldorf Towers in New York, where the commentator lectured the official about undesirables and read him names out of a little book.

The threatened budget cuts came, and Dr. Johnson had to eliminate 1500 employees. He did so by abolishing some activities altogether. This at once brought furious protests from Cohn and Schine, with demands to restore a number of employees. It seemed that Johnson had inadvertently eliminated portions of the "loyal underground."

Meanwhile Secretary of State John Foster Dulles, apparently still appeasing McCarthy, had sent word to Dr. Johnson that all books and materials by "communists, fellow-travelers, et cetera" should be removed from State Department libraries throughout the world. When Johnson asked for an official list of authors who should be considered "communists" or "fellow-travelers" or "et cetera," no one could provide one, although unofficial lists were available from McCarthy aides. These listed names like Franklin P. Adams, Sherwood Anderson, Brooks Atkinson, W. H. Auden, Stephen Vincent Benét, Louis Bromfield, Van Wyck Brooks, Henry Steele Commager, Elmer Davis, Bernard De Voto, John Dewey, Theodore Dreiser, Edna Ferber, Archibald MacLeish, Quentin

Reynolds, Arthur Schlesinger, Jr., Carl Van Doren, Mark Van Doren, Edmund Wilson, and others. There was even Foster Rhea Dulles, cousin of the Secretary of State. (When the Secretary learned this, he was quite annoyed.) McCarthy was meanwhile proclaiming to the world that there were thirty thousand volumes by "communists" in the overseas libraries—a calculation presumably based on the unofficial lists of dangerous authors. McCarthy meanwhile issued statements denouncing Dr. Johnson—without reply from the White House. Johnson was forbidden to reply.

The field of music was subjected to a similar flurry. In Vienna an exhibition of photographs of leading American composers, arranged by the information division, had to be canceled at the last moment because twenty-three of the composers had been "questioned" by the State Department security office.[7]

Throughout this period the President repeatedly made it clear to intimates that he did not like McCarthy. He went so far as to say that he could not imagine anyone feeling clean "after shaking hands with that fellow." Yet the President shunned occasions for repudiating McCarthy. It would demean the presidency, said Eisenhower, to attack McCarthy directly. "I just will not—I *refuse*—to get into the gutter with that guy."[8]

Still the chief hope of stopping McCarthy remained Eisenhower himself. The struggle for the President's mind continued.

This hope—this struggle—produced in 1953 one of the strangest telecasts of the year, a program which in sixty minutes compressed a panorama of its time.

TV DINNER

It all began when the Anti-Defamation League of B'nai B'rith considered plans for its fortieth anniversary. It hoped, through a celebration, to further its aim of combating discrimination.

One idea was an attention-getting award to a prominent person. An inquiry was sent to President Dwight D. Eisenhower. Would he accept, at a banquet to be staged in Washington, the League's Democratic Legacy Award for his contributions to civil rights? The answer came back: he would.

The situation was curious. Many leaders of the organization felt that

7. Merson, "My Education in Government," *The Reporter*, October 7, 1954.
8. Hughes, *The Ordeal of Power*, pp. 66, 124.

Eisenhower had done almost nothing for civil rights. His inaction since assuming the presidency seemed to them deplorable. And this was, in fact, one reason for the award. They hoped in this way to further his interest in civil rights.

The President's acceptance created instant television possibilities, which the Anti-Defamation League began to develop. It invited Oscar Hammerstein II and Richard Rodgers to stage entertainment for a *Dinner With the President*, scheduled for November 23 at the Mayflower Hotel—to which an array of national leaders would be invited, and which would be offered to television.

Hammerstein and Rodgers were at this moment among the most admired leaders in entertainment. They had behind them a series of fantastic theatrical successes, two of which—*South Pacific* and *The King and I*—were still running. Rodgers had recently composed the score for a highly acclaimed NBC-TV film series, *Victory at Sea*. Hammerstein was president of the Authors League of America, which represented writers in various fields including television and radio. In this role he had forcefully and courageously protested network blacklist practices. But above all, Hammerstein and Rodgers had the golden touch. Their acceptance, added to that of the President, created a bandwagon effect.

Performers who quickly agreed to take part included Desi Arnaz, Lucille Ball, Eddie Fisher, Jane Froman, Rex Harrison, Helen Hayes, Ethel Merman, Lilli Palmer, Thelma Ritter, Jackie Robinson, William Warfield.

Meanwhile a dizzying list of guest acceptances came in. On hand would be Chief Justice Earl Warren of the U. S. Supreme Court and five other Supreme Court Justices; five cabinet members, including Secretary of State John Foster Dulles and Attorney General Herbert Brownell; the Chairman of the Joint Chiefs of Staff, Admiral Arthur W. Radford; also such assorted dignitaries as J. Edgar Hoover and Bernard Baruch; and governors, senators, ambassadors, publishers, judges.

The promoters took care to invite top leaders of each network. This resulted in unusual coverage: CBS-TV agreed to carry the live telecast, NBC-TV and ABC-TV to carry a kinescope film later the same evening.[1] The small Dumont network also agreed to carry a kinescope, and the NBC and MBS radio networks decided to carry a portion of the event.

1. A kinescope was a film of the program photographed from a television tube during the live telecast. It was the standard means for securing supplementary coverage.

Thus national attention was focused on a dramatic moment. In a sense, the entire effort was directed at one man, Dwight D. Eisenhower. Hope of success lay in the extraordinarily public nature of the assault on his convictions and conscience.

What could be done to create from these components a civil rights event—one that might help stem the drift toward vilification and hatred? This was the problem facing Hammerstein and Rodgers.

The White House had its own problem. Eisenhower had no doubt agreed to the proposed award for the same reason it was offered—an uneasiness over the drift to McCarthyism, a sense of need for leadership. Some aides saw the occasion as a moment for the hoped-for commitment. Speech-writers began drafting an acceptance speech.

This was always—as Emmet John Hughes has described—a difficult process. Rhetoric made Eisenhower supremely uneasy. He would scowl at almost any fine phrase. He liked to have drafts read aloud to him, and would listen attentively without interruption. Then, without any general comment or judgment, he would seize on a particular phrase. "Right. Now here you say . . . that *sounds* good . . . but what does it mean? . . ."

If the point could just be brought down to the *individual,* he would plead. "A carpenter or a farmer or a bricklayer or a mechanic . . . the *individual,* that's what counts." He would say this fervently, but usually without resolving the writing problem.[2]

The draft wound through many revisions and was not finally approved until the day of the telecast. As the banquet began, the text was being mimeographed for distribution to the press. The President was said to be still unhappy about it, but his copy lay on the banquet lectern.

The show—with the nation's power structure as studio audience—began in witty revue style. It opened with a shot of the Statue of Liberty—which then moved and turned out to be Thelma Ritter, who complained that she got pretty weary holding that torch aloft. It was hard work, she said; she got snowed on and rained on, and got discouraged. "Sometimes I wonder if people really appreciate Liberty." Interrupted by a boat whistle, she turned her head, glancing down. "Welcome stranger. . . . Take a left at the Battery, you can't miss it."

Other parts became explicit. Hammerstein and Rodgers, as hosts, said they welcomed the opportunity for theater people to speak for democ-

2. Hughes, *The Ordeal of Power,* pp. 24, 53.

racy. In the theater, said Hammerstein, "nobody tells us with whom we may associate . . . no one tells us what to write." He was speaking quite specifically of the theater, rather than of television or radio—a point that must have been clear to network executives present.

The script made the guests part of the action. Television cameras stood ready to provide planned close-ups across the sea of diners. At the dais Lilli Palmer said:

> PALMER: A thoughtful Greek philosopher said long ago that a democracy was necessarily limited to the number of people who could be reached by a single human voice. A challenging definition—answered in the twentieth century by the technology of television and radio, pioneered in this democracy by such men as . . .
>
> CUT TO: SARNOFF
>
> . . . David Sarnoff of the Radio Corporation of America . . .
>
> CUT TO: GOLDENSON
>
> . . . Leonard Goldenson of the American Broadcasting Company . . .
>
> CUT TO: PALEY
>
> . . . and William S. Paley of the Columbia Broadcasting system.[3]

There was also, at one point, a planned close-up of J. Edgar Hoover, but it was spoiled. Someone had moved a table lamp, which got the close-up instead. Occasional close-ups of Eisenhower showed him watching with riveted attention.

At some time during the performance Eisenhower made an extraordinary decision. It apparently happened during the final sequence—a magnificent cameo by Helen Hayes. Dressed as Harriet Beecher Stowe, author of *Uncle Tom's Cabin,* she told of a visit to the White House—to Abraham Lincoln. She began:

> HARRIET: One hour with our President has lifted my spirits and endowed me with new strength. . . .

He had showed her, she said, a great hope. There had always been tyrants, but the war had given hope that people would no longer accept them or the kind of world they were always determined to make.

> HARRIET: Yes, that is our hope. . . . Our danger is this: when the conflict is over and war-weariness has set in, we may be tempted to forget, to slip back into the old ways. Then, and then only, will

3. *Dinner With the President,* all television networks, November 23, 1953.

our sons have died in vain. . . . A day will come when this, our little life, will be ended. All will be gone. All who raged, all who threatened; the weaklings who yielded; the men who, like our President, stood and bore infamy and scorn for the truth. Yes, life will be over, but eternity will never efface from our souls whether we did well or ill, fought bravely or failed like cowards, whether at the end we could say, with truth, "I have fought the good fight —I have kept the faith." (*She pauses, then adds softly*) For mine eyes have seen the glory. . . .

Orch (sneak in): Battle Hymn of the Republic

DISSOLVE FROM HAYES TO: FULL SHOT, LINCOLN MEMORIAL.[4]

From the Lincoln Memorial the viewer was brought back to the banquet tables. President Eisenhower came to the lectern to receive the Democratic Legacy Award. He did not look at the prepared acceptance speech.

EISENHOWER: Ladies and gentlemen, for many years I have been served by able staffs—in war and in peace. I have a staff now of which I am intensely proud. . . .

Presidential press secretary James Hagerty, sitting next to a representative of the Anti-Defamation League, murmured: "My God, what now?"[5]

EISENHOWER: They are always anxious that I do well, no matter where I appear, and tonight was no exception. I have been briefed—and briefed—and briefed.

A roar of laughter—and applause—swept through the Mayflower ballroom and into television homes. As it subsided he explained that he had listened to hours of suggested comments on civil liberties. But he realized that his audience was very knowledgeable about such matters—*habeas corpus* and so on—so he preferred not to follow the suggestions. A President had to learn to say no.

Then he floundered a bit, as he often did, in a way that made his aides squirm. He said he was proud to be an American.

EISENHOWER: Now—why are we proud? Are we proud because we have the richest acres in the world? I have heard that the Nile valley is one of the richest places in the world. Now that's a great nation, but do you want to give up your citizenship for that of a nation that has merely richer ground—richer minerals underneath its soil?

4. The speech was from the play *Harriet,* by Florence Ryerson and Olin Clements.
5. Interview, Nathan Belth.

He granted that many nations—presumably including Egypt—might be older, harder-working. Still—"we love America"—because here, a man could walk upright. Then he said the performance had taken his thoughts back home, to the old days. He felt on safer ground now, and lunged ahead. His voice warmed.

> EISENHOWER: I was raised in a little town of which most of you have never heard, but in the West it's a famous place! It's called Abilene, Kansas. We had as marshal for a long time a man named Wild Bill Hickok. If you don't know about him, read your westerns more. Now that town had a code, and I was raised as a a boy to prize that code. It was—meet any one face to face with whom you disagree. You could not sneak up on him from behind —do any damage to him—without suffering the penalty of an outraged citizenry. If you met him face to face and took the same risks he did, you could get away with almost anything, as long as the bullet was in the front. And today, although none of you have the great fortune, I think, of being from Abilene, Kansas, you live after all by that same code, in your ideals and in the respect you give to certain qualities. In this country, if someone dislikes you or accuses you, he must come up in front. He cannot hide behind the shadow. He cannot assassinate you or your character from behind, without suffering the penalties of an outraged citizenry.

The President's ad-libbed speech may have reached some 38 million television viewers and 20 million radio listeners.[6] And, although the right to be shot in front rather than from behind was not a recognized civil right, editorials throughout the country discussed the subject—along with the right to face an accuser. Civil rights advocates took heart.

McCarthyism had actually made a fleeting appearance on the broadcast itself. To write the script for the Rodgers-Hammerstein show, the Anti-Defamation League had engaged William N. Robson, former CBS writer-director noted for his work on *Columbia Workshop* and other series. When CBS learned this, its spokesman protested, pointing out that Robson was listed in *Red Channels*. "We don't use him."[7] CBS made no

6. *Memorandum Submitted to the American Public Relations Association*, pp. 1–12. The memorandum summarized audience research data, costs, and results. Broadcast time was free, and performers gave their services. B'nai B'rith spent about $18,500, mainly for hotel, travel, and publicity; it gained 20,000 new members in the month following the telecast.
7. Interview, Nathan Belth.

allegations against Robson—merely that he was "on the list" and that his name among program credits would bring protests and probably trouble.[8]

Because Robson, a decade earlier, had written the script for its thirtieth anniversary, the Anti-Defamation League insisted on keeping him, and finally had its way. He wrote the banquet broadcast in collaboration with Milton Geiger. Robson was informed of the backstage maneuvers, and therefore found it curious that top CBS executives greeted him like an old friend at the banquet. "Bill, like old times!"[9]

Viewers and listeners were left unaware of all this. President Eisenhower, ad libbing, ran over his allotted time, and a number of credits were omitted from the air show. Because some artists thought other factors had caused the omission, and to give credit for work done, the Anti-Defamation League listed the program credits in a trade press advertisement.

The script for *Dinner With the President* had a form reminiscent of radio documentaries. It used several narrators, who linked statements by other voices and brief dramatic vignettes. At the banquet the narrators were shown at lecterns. Switches were made to other performers, on a stage or at other lecterns. In the Mayflower ballroom, where almost every shot involved a halo of celebrity faces, the form had special value. But it was already felt to be a form outmoded by television—a last gasp of an old style of documentary.

If *Dinner With the President* gave heart in some quarters, there was no sign that blacklisters felt deterred. At networks and advertising agencies talent-hiring continued to be watched by security consultants armed with lists. Government agencies were increasingly in the McCarthy grip. Raymond Swing, chief commentator at the Voice of America, described its demoralization as "so thick you could feel it." He himself, having weathered a McCarthy harassment, decided to leave; he considered the agency crippled beyond recovery.[10] Dr. Robert Johnson left, followed by Martin Merson, who charged that "mutual espionage" had become an American intramural sport and had made the United States government—and especially the Voice of America—a "butt of worldwide ridicule and contempt." A later administrator found it a "shambles.[11]

8. For the organization of blacklists and related economic pressures, see *The Golden Web*, pp. 253–83.
9. Robson, *Reminiscences*, pp. 25–6.
10. Swing, *Good Evening!*, pp. 279–80.
11. *The Reporter*, October 7, 1954; Button, *Interview*, p. 8.

At the FCC the purge had launched strange events. A station owner particularly objectionable to McCarthy because of failure to carry his speeches was Edward Lamb, who had stations in the Midwest and South. Lamb was a lawyer with various other business interests including television and radio—which had made him rich. A former Republican turned Democrat under the spell of Franklin D. Roosevelt, Lamb had become an important Democratic Party contributor, and was often mentioned as a possible treasurer of the Democratic National Committee. His standing made him an attractive McCarthy target. The trade press began to carry reports that Senator Joseph McCarthy and associates were "gunning" for Lamb. Commissioner Doerfer was said to be in charge of "the Lamb case."[12]

As a lawyer, Lamb had represented labor leaders and handled civil liberties cases. He had also visited the Soviet Union and published an analysis of what he considered the successes and failures of its economic system. He probably seemed an easy target.

As a licensee Edward Lamb had signed affidavits that he was not and never had been a communist. In March 1954, while awaiting FCC action on pending licenses and renewals, he received instead an FCC notice charging that he had been a Communist Party member. The implication was that he had committed perjury and that his licenses would be voided. The FCC released the accusing letter to the press—apparently before mailing it.

Lamb reacted with a vigor that was perhaps unexpected. He placed advertisements in the New York *Times* and many other publications, offering $10,000 reward to anyone who could disprove a single one of his non-communist affidavits.[13] Meanwhile he demanded a public hearing by the full FCC and asked for a bill of particulars on its charges.

The FCC declined, saying this would be premature. Meanwhile its investigators apparently began an intense hunt for particulars—of various kinds. A former secretary of Edward Lamb was asked if she could tell of any "girl trouble" he had had. Finally the FCC brought to Washington, and placed on the witness stand, a lady who said she had known Lamb as a communist in Columbus, at meetings where gin was drunk and caviar eaten. But then she took it all back, said she had never known

12. Based on accounts in *Broadcasting,* beginning December 8, 1952; Lamb, *Trial by Battle,* pp. 3–21; other sources as mentioned.
13. New York *Times,* May 6, 1954.

Lamb, had never been a communist, and had only been trying to be helpful to the government. She testified:

> The FCC lawyers told me it was my duty to testify because Lamb's radio station could beam atom bombs from foreign countries and also beam in enemy broadcasts.

The case against Lamb collapsed in confusion and his licenses were eventually renewed. But the case stirred questions. What was going on at the FCC?

The search for subversion had its lighter moments. It was revealed that Lucille Ball of *I Love Lucy*—No. 1 in Trendex ratings—had registered as a Communist Party member in 1936. CBS and the sponsor, Philip Morris, were dismayed; they feared "trouble." The House committee on un-American activities began an inquiry. But a few days later Representative Donald Jackson of California, a member of the committee, announced that he had "cleared" her. *Broadcasting* reported: "LUCILLE BALL CLEARED OF COMMUNIST ASSOCIATION." It was explained that she *had* registered with the Communist Party, but only to please her grandfather.[14] The case was quickly forgotten.

FIVE SETS, ONE MURDER

While wars of McCarthyism rumbled on many levels, studio struggles between the migrations continued. Shifts in style—and in fortune—often came suddenly.

Quiz and variety programs were beginning to have a settled look. Some, like the Ed Sullivan series, were already old favorites. "Ed Sullivan," said Oscar Levant, "will last as long as other people have talent."[1] But the forms of drama, newscast, and documentary were in flux.

In drama, in spite of *I Love Lucy*, the dominance of live production was expected to continue. David Sarnoff of RCA was said to be determined that it should; so was William Paley of CBS.

The fact that *local* schedules were using a lot of film—mostly old westerns and gangster films, not made by the major Hollywood studios—was not considered significant. In radio, local programming had always had similar dependence on recordings of all sorts; yet throughout most

14. *Broadcasting*, September 21, 1953.
1. Levant, *The Unimportance of Being Oscar*, p. 232.

of radio history, network schedules had remained live.[2] This was expected to be the pattern in television.

In drama, at the start of 1953, the radio migration was holding its own. Its main contribution to network schedules was the *episodic series*—a form derived from radio. It featured continuing characters, with a separate plot for each program. Once the formula was established, many writers could contribute. Most episodic series used mystery-crime or comedy formulas. Many were radio transplants: *Mr. and Mrs. North, Big Town, The Big Story, The Aldrich Family, My Friend Irma, Beulah.* New series included *Treasury Men in Action, Mr. Peepers.* Almost all were live—at least, they began as live series.

I Love Lucy, the most notable exception, was not seen as a threat to the live tradition. Its form was that of a live series. By adding the cost of film, it had merely acquired a speculative re-use value.

A typical episodic series was *Man Against Crime,* starring Ralph Bellamy. It had begun in 1949 and ran until 1954, often high in ratings. In the radio tradition, it was produced by an advertising agency, William Esty; the program staff worked from the Esty office. Free-lance writers came and went; fifty different writers contributed to the series. The live-production costs ran $10,000 to $15,000 per program; a writer usually got $500 to $700.

While following a radio pattern, the series faced television problems. In radio the length of a play could be gauged by counting words—it usually ran 140 to 150 words per minute. Television timing, because of action intervals, was more treacherous. From one rehearsal to another, the length varied wildly.

On *Man Against Crime* the problem was solved by requiring writers to include a "search scene" near the end of each program. The hero-investigator would search a room for a special clue. A signal would tell Ralph Bellamy how long to search. If time was short, he could go straight to the desk where the clue was hidden; if there was need to stall, he could first tour the room, look under sofa cushions, and even take time to rip them open.[3]

The CBS studio in the Grand Central Terminal building where *Man Against Crime* was produced was under unceasing pressure. Here the cast had only one full rehearsal with cameras and lights. Earlier rehears-

2. See *The Golden Web,* pp. 109, 163–6.
3. Reisman, *Interview,* p. 40.

als were "dry runs" in offices or rented ballrooms. During the studio rehearsal, work might be in progress on sets for other programs.

Man Against Crime was sponsored by Camel cigarettes. This affected both writing and direction. Mimeographed instructions told writers:

> Do not have the heavy or any disreputable person smoking a cigarette. Do not associate the smoking of cigarettes with undesirable scenes or situations plot-wise.[4]

Cigarettes had to be smoked gracefully, never puffed nervously. A cigarette was never given to a character to "calm his nerves," since this might suggest a narcotic effect. Writers received numerous plot instructions:

> It has been found that we retain audience interest best when our story is concerned with murder. Therefore, although other crimes may be introduced, somebody must be murdered, preferably early, with the threat of more violence to come.

The hero, said the instructions, "MUST be menaced early and often." Violence, if on-camera, was very briefly staged; one good blow or shot might suffice. Physical struggle was hardly feasible amid flimsy sets.

Although "other crimes" could be used as plot elements, arson was not one of them. Fires were not to be mentioned because they might remind a viewer of fires caused by cigarettes.

No one could cough on *Man Against Crime*.

Romance, or the possibility of it, was as essential as violence. A plot had to include "at least one attractive woman." A passing romance for the hero was encouraged, "but don't let it stop the forward motion of the story."

Doctors could be shown only in "the most commendable light." There were rumors of a coming report on health effects of smoking—a report of this sort had appeared in the United Kingdom—and this made the sponsor increasingly nervous about antagonizing doctors. Since doctors tended to take a dim view of fictional doctors, it seemed best to avoid doctors. On *Man Against Crime* it was usually someone other than a doctor who said, "He's dead." It generally took only a moment.

Before a writer was hired, his name had to be checked by phone with a designated agency division for a "yes" or "no." Lists of "yes" people were later confirmed, without explanation, by memorandum. Memoranda

4. This and following quotations are from *Blueprint for "Man Against Crime,"* pp. 1–3.

on "no" people were avoided, or used cryptic terms favored in security matters. One memo said: "Watch out for the following writers . . . [list of names] . . . skip all these characters—they can't 'write.'"[5]

The writer had to limit action to five sets, one of which had to be the "fashionable" Manhattan apartment from which the hero-investigator worked. Before the middle commercial, action had to "rise to a cliff-hanger." Costume changes were difficult and unwelcome. Between scenes Ralph Bellamy was always rushing from set to set. Transitions between scenes were sometimes eased by use of a film clip, as of traffic or a subway train.

With episodic series proliferating, groups rehearsing throughout the city, and artists and technicians converging for the briefest of studio run-throughs, the pressures on all were brutal. Possibilities for error were huge. For these reasons the sponsors of *Man Against Crime* decided, after three years of live production, and in spite of the steeper costs that would result, to "go to film." They moved to a studio in the Bronx built by Thomas Edison in 1904, near the Botanical Gardens off Bedford Park Boulevard. Here actors came across books with Mary Pickford's name in them, and relics of Richard Barthelmess, Thomas Meighan, King Baggott. It was a moment's wondrous contact with another, not unsimilar, era of crude beginnings.[6] In the Bronx, costs went to $20,000 to $25,000 per program for three days of shooting—two in the studio, one on location.

But if film was called for, there were better places than the Bronx. The move really foreshadowed the doom of New York's episodic programs. During 1953 filmed series from other sources leaped into prominence. *Dragnet,* a series on the Los Angeles Police Department launched in 1952, suddenly achieved the impossible: in November 1953 it momentarily passed *I Love Lucy* in the Nielsen ratings,[7] and the two series began a neck-and-neck race for leadership. *Dragnet* had a highly mobile style, with many outdoor sequences for which Hollywood was made to order. Meanwhile other Hollywood-produced episodic series were erupting on the air: *The Life of Riley* with William Bendix, *Topper* with Leo G. Carroll, *Private Secretary* with Ann Sothern, *Make Room for Daddy* with Danny Thomas—and dozens of others. Some were sold for network use while others were *syndicated*—sold on a station-by-station

5. Esty agency memorandum.
6. Bellamy, *Reminiscences,* p. 10.
7. *Broadcasting,* November 30, 1953. Nielsen ratings were based on mechanisms inserted in a sampling of television sets, keeping a record of stations tuned.

basis. Desilu, the company formed by Desi Arnaz and Lucille Ball, was a leading force in the eruption, but there were others, including Ziv Television Productions, a radio syndicate branching aggressively into television; Hal Roach Productions, heir to a theatrical short-film tradition; Screen Gems, short-film offspring of Columbia Pictures; and most significantly, Revue Productions, subsidiary of the talent agency MCA.[8]

It was not usual for a talent agent to make films on the side. Normally, talent guilds would have blocked the practice, because it involved a clear conflict of interests. MCA as agent was supposed to get the best possible terms for an artist; MCA as producer had an opposite incentive. But in a time of Hollywood panic, the readiness of MCA to finance production and provide employment had been welcomed, and soon mass-produced episodic series from MCA-Revue were pouring across the United States—and elsewhere.

A new element was entering the situation. In 1953 Japan launched a commercial television system. A commercial system was also being planned for the United Kingdom to supplement its noncommercial British Broadcasting Corporation. Television had begun in Mexico, Brazil, Argentina, Cuba, and was about to begin in a dozen other countries. All might be markets. All were suddenly reasons for American producers to plunge into filmed programming, instead of the dead-end risks of live production. By 1954 the live episodic series was obsolete. A chapter of New York television had ended.

But the episodic series had been only one element of live television drama in New York. Even while it slipped into oblivion, another kind of live drama was stirring into vigorous life. Amid the turbulence, it suddenly produced a short, surprising, brilliant "golden age."

ANTHOLOGY

Why did the years 1953–55 produce, under chaotic conditions in primitive television studios, an outburst of television drama that was remembered decades later and left an imprint on theater and film?

To an extent the chaos may have had a creative impact. In its midst something new was developing; many an artist had an exhilarating sense of finding and sharpening new tools.

To a surprising extent, the anthology series grew in an atmosphere

8. For previous MCA history, see *The Golden Web*, p. 172.

different from that surrounding the episodic series. Surprising—because the same networks, studios, and technicians were involved.

The episodic series was conceived for rigid control. It invited writers to compose, to a defined formula, scripts for specified actors and often for a particular sponsor, who was inclined to think of a play as a setting for his commercials. Where a star had an unusual, individual brilliance, effervescent writing might develop. But too often the formula invitation attracted the hack, and turned talented writers into journeymen.

The anthology series started from a different premise. The play was the thing. Actors were chosen to fit the play, not vice versa. The anthology series said to the writer: "Write us a play." There were no specifications as to mood, characters, plot, style, or locale—at least, not at first.

Length was specified. And the play had to be produceable as a live program in a gymnasium-sized studio. Except for such technical requirements, *Philco Television Playhouse* and other anthology series began as carte-blanche invitations to writers—and writers responded.

High fees were not the lure. During the most fertile period of the anthologies—1953–55—writers generally received $1200 to $2500 for a one-hour script.

Whereas the episodic series had emerged from a radio tradition, the anthology series emerged from a theater tradition. From the start, artists from the theater were active in the anthology series.

The theater, in the years after World War II, comprised not only Broadway but also the "off-Broadway" world. As the Broadway theater shrank, tending to become a privileged-class diversion, many actors and writers tried to keep their talents alive in small, peripheral theaters, often run cooperatively. Many of these had a marginal economic life but produced experimentation and ferment—not unlike that produced during the Depression by the Federal Theater and its offshoots, such as the Mercury Theater. Related to the off-Broadway phenomenon were workshops like Actors Studio, where even well-established stars gathered regularly to exercise and stretch their talents. At Actors Studio the great magnet was the leadership of Lee Strasberg—to Kim Stanley, "the most dedicated thing the American theater has to offer."

Kim Stanley grew up in Texas, did some stock-company acting, then came to New York and worked for two years as a waitress—and occasionally as a model—before becoming involved in an off-Broadway company that did things like Cocteau's *The Infernal Machine* and E. E.

Cummings's *him*. "We didn't really completely understand the plays, but we were very enthusiastic and wanted to." The group was run as a co-operative. "The person who sold lemonade—which was me, the first eight months I got there—had just as much vote as the head of the theater." She and others broke away to form another group at the Cherry Lane Theater. Meanwhile, in 1950, she was accepted at Actors Studio. It was above all a place to work constantly—for the training acquired in some countries via repertory. She was no beauty. She knew she would not make it "that way." Constant study was an escape from the humiliation of "going the rounds" to casting offices, but it was also an expression of conviction: she would make it, if at all, through study and discipline. When opportunities came, she was ready. She got started on Broadway—the first really good part was in the 1953 *Picnic*—and meanwhile began to appear before nationwide audiences on *Philco Television Playhouse* and *Goodyear Television Playhouse*. It was a time of discovery for audiences as well as for Kim Stanley and others.[1]

Paul Newman had studied economics under a football scholarship at Kenyon College, went into the sporting goods business in Cleveland, fled to the Yale Drama School and then to Broadway, where he likewise landed a part in *Picnic*—which ran fourteen months. He too began to study at Actors Studio and to appear in anthology series, as did Kim Stanley's understudy, Joanne Woodward—later Mrs. Paul Newman. From other groups and workshops, similar threads converged. Starting from the off-Broadway American Negro Theater, Sidney Poitier kept coming back to Paul Mann's Actors Workshop to flex his talents. He got parts in several Broadway shows and films, but a *Philco Television Playhouse* appearance —in *A Man Is Ten Feet Tall*, by Robert Alan Aurthur—was the turning point in his career.

The Philco and Goodyear series were broadcast Sunday evenings, not conflicting with Broadway. Each program got seven or eight days of glorious, intensive rehearsal—"Those were marvelous days," Joanne Woodward later recalled dreamily—under young directors like Delbert Mann and Arthur Penn, who were themselves finding their way.[2]

One reason they were marvelous days was that a producer, not a committee, was in control. Fred Coe, producer of the Philco and Goodyear

1. Stanley, *Reminiscences*, pp. 4–12.
2. Newman, *Reminiscences*, pp. 2–6; Woodward, *Reminiscences*, pp. 14–18; Ewers, *Sidney Poitier: the Long Journey*, pp. 77–9.

series, had established himself in television with experimental productions for the Theater Guild in NBC's Studio 3G in the late 1940's. By the time NBC decided to put him in charge of a Sunday evening anthology series, he commanded so much respect that he was given his head. When Philco and Goodyear became alternate-week sponsors, the situation did not at once change. Television drama was still new to them and their advertising agencies, and they were, for the moment, not inclined to interfere with a going thing. Coe chose directors, dealt with writers. Under his encouragement the talents of scores of writers and directors struck fire.

During 1953 programs of this sort included, in addition to the Goodyear and Philco series, *Kraft Television Theater, Studio One, Robert Montgomery Presents, U. S. Steel Hour, Revlon Theater, Medallion Theater.* There was also *Omnibus,* set up by a Ford Foundation grant but carried on under commercial sponsorship—a 90-minute series of diverse elements, using scripts of various lengths. These series were later joined by *Motorola Playhouse, The Elgin Hour, Matinee Theater,* and *Playhouse 90.* Thus a sizable "open market" attracted the writer. What had been a trickle of scripts in 1952 became a deluge a year later. The various series were flooded with submissions—synopses and completed scripts. Writers came and went. Brilliant scripts arrived from surprising sources.

An advertising writer for women's garments, Reginald Rose, came to script editor Florence Britton at *Studio One* with a script titled *The Remarkable Incident at Carson Corners,* followed soon afterwards by *Thunder on Sycamore Street,* and suddenly found he had launched a new career. A thin-faced ex-paratrooper, Rod Serling, working at a Cincinnati radio station, sent a script to *Kraft Television Theater;* flying to New York for a script conference he nervously dropped his suitcase in the J. Walter Thompson office and scattered socks and underwear but sold *You Be the Bad Guy* and, soon afterwards, other scripts, including *Patterns.* An ex-novelist, David Davidson, gave *U. S. Steel Hour* its premiere script, *P. O. W.* A veteran astronomy lecturer from Hayden Planetarium, Kenneth Heuer, and a writer of children's books, George Selden, turned up at *Studio One* with *The Genie of Sutton Place.*

A landmark in the history of anthology series, and an inspiration to many writers, was Paddy Chayefsky's *Marty,* broadcast on *Goodyear Television Playhouse* on May 24, 1953. The role of Marty, a young butcher in the Bronx, was played by the unknown Rod Steiger, and the direction was by Delbert Mann.

Marty had a deceptive simplicity. "I tried to write the dialogue as if it had been wire-tapped," said Chayefsky. The talk had an infectiously natural rhythm. Marty says, "You want to go you should go." At the same time the play provided a crowded tapestry of metropolitan life: the Waverly Ballroom, the RKO Chester, an all-night beanery. The play opened in a butcher shop.

YOUNG MOTHER: Marty, I want a nice fat pullet, about four pounds. I hear your kid brother got married last Sunday.
MARTY: Yeah, it was a very nice affair, Missus Canduso.
YOUNG MOTHER: Marty, you ought to be ashamed. All your kid brothers and sisters, married and have children. When you gonna get married?

Pressures of this sort are also a factor in the talk on Saturdays, when the unattached young men gather with much male camaraderie—ritualistic, self-protective.

ANGIE: Well, what do you feel like doing tonight?
MARTY: I don't know, Angie. What do you feel like doing?

At home the same pressure is applied. The mother is a widow.

MOTHER: So what are you gonna do tonight, Marty?
MARTY: I don't know, Ma. I'm all knocked out. I may just hang arounna house.
The mother nods a couple of times. There is a moment of silence. Then . . .
MOTHER: Why don't you go to the Waverly Ballroom?
This gives Marty pause. He looks up.
MARTY: What?
MOTHER: I say, why don't you go to the Waverly Ballroom? It's loaded with tomatoes.
Marty regards his mother for a moment.
MARTY: It's loaded with what?
MOTHER: Tomatoes.
MARTY: (*Snorts*) Ha! Who told you about the Waverly Ballroom?
MOTHER: Thomas, he told me it was a very nice place.
MARTY: Oh, Thomas. Ma, it's just a big dance hall, and that's all it is. I been there a hundred times. Loaded with tomatoes. Boy, you're funny, Ma.

MOTHER: Marty, I don't want you hang arounna house tonight. I want you to take a shave and go out and dance.

Marty is short and stocky. He calls himself "a fat little man." Girls have brushed him off often enough at the RKO Chester and elsewhere to make him cautious. He thinks he will stay home and watch Sid Caesar on television. How he finally puts on his blue suit and goes back to the Waverly; picks up a skinny schoolteacher whom his friends consider a "dog" and who doesn't impress his mother either; how their opinions halt his romance temporarily but how he finally pursues it—this is the story of *Marty*. When he does pursue it, we are not sure whether he is drawn mainly by his liking for the girl, or a feeling of identification with her (she too has been hurt), or whether he is escaping from the ritual of:

MARTY: I don't know, Angie. What do you feel like doing?

Beneath the simple story line we sense other dimensions. Chayefsky himself, in discussing *Marty*, speaks of an oedipal element in his home situation, the persistence of adolescence in grown men, and a latent homosexual ingredient in the Saturday gatherings.[3] Such elements are never explicit in the play—we can be aware of them or not, as we choose—but they give the work a psychological and sociological resonance characteristic of Chayefsky.

The choice of heroes and heroines defying Hollywood standards of beauty was a central aspect of Chayefsky's work, and important to its success. He sensed that television, with its potential for intimacy, offered opportunities for such a revolt, and the response bore him out. Rod Steiger could scarcely believe the impact of his own performance. "People from all over the country and all different walks of life, from different races and religions and creeds, sent me letters. The immense power of that medium!"[4]

Chayefsky's work had wide influence on anthology drama—its successes and failures. During 1953–55 the riches came in profusion: The Philco-Goodyear series offered Chayefsky's *Holiday Song* (1953), *The Mother* (1954), *Bachelor's Party* (1955), and *A Catered Affair* (1955); Robert Alan Aurthur's *Man on a Mountaintop* (1954) and *A Man is Ten Feet Tall* (1955); Horton Foote's *A Young Lady of Property* (1953); Gore Vidal's *Visit to a Small Planet* (1955). The *Studio One* series came up

3. Chayefsky, *Television Plays*, pp. 173–7.
4. Steiger, *Reminiscences*, p. 2.

with Reginald Rose's *Thunder on Sycamore Street* (1954) and *Twelve Angry Men* (1954). The Kraft series had Serling's *Patterns* (1955). All these had a subsequent history.[5]

The structure of these plays related to circumstances under which they were produced. Such problems as costume changes and aging were unwelcome. This encouraged plays of tight structure, attacking a story close to its climax—very different from the loose, multi-scene structure of films.

Ingenuity could ease the limitations. An actress could start a play wearing three dresses and peel them off en route between scenes. Color lighting could make painted wrinkles invisible in one scene, emphasize them in the next. A studio was sometimes crammed with small sets and fragments—telephone booths, park benches, street corners. Some producers—especially on *Studio One*—felt compelled to prove they could stage whole floods, as in Robert Anderson's *The Flood,* or sink an ocean liner, as in *A Night to Remember,* adapted by George Roy Hill and John Whedon from the book by Walter Lord. But amid such technical wizardry an actor was likely to become a zombie, and in the end it emphasized what others could do better. The Philco-Goodyear directors were not inclined to think of live production as a limiting factor; it merely influenced the kind of drama to be explored. They found its niche in compact rather than panoramic stories, in psychological rather than physical confrontations.

Close-ups became all-important. A Marty-Clare scene in an all-night cafeteria was played almost wholly in close-up. The human face became the stage on which drama was played.

To this close-up drama, live television brought an element that had almost vanished from film—one which few viewers noticed consciously but which undoubtedly exercised a hypnotic influence.

Film had long been dominated by its own kinds of time, made by splices in the editing room. The final tempo and rhythm were generally created not by an actor, nor by actors interacting, but by an editor and the director working with him. The possibilities of this kind of control

5. All appeared in print: see Chayefsky, *Television Plays;* Rose, *Six Television Plays;* Serling, *Patterns;* Vidal (ed.), *Best Television Plays.* Feature films of like title were made from *Marty, Bachelor's Party, A Catered Affair, Twelve Angry Men, Patterns; A Man is Ten Feet Tall* became the film *Edge of the City.* Broadway plays were derived from *Twelve Angry Men, Visit to a Small Planet;* elements of *Holiday Song* turned up in *The Tenth Man.*

had led to production methods in which films were shot in fragments—often as short as two or three seconds. A feature film might consist of seven or eight hundred shots. The manipulation of "film time" offered creative pleasures so beguiling to film makers that they had virtually abolished "real time" from the screen. Its appearance in long stretches of television drama gave a sense of the rediscovery of reality—especially for people whose only drama had been film.

When a play involved a number of sets, these were generally arranged around the periphery of the studio. In the middle the cameras—three or more—wheeled noiselessly from set to set, each tended by its cameraman, and each trailing behind it the long, black umbilical cord leading to the control room. Each camera had a turret of several lenses, so that its cameraman could, in a second or two, switch from close-up to medium or long shot. Each cameraman had in front of him, mounted on the camera, a list of the shots he would do during the play. Over earphones he got constant supplementary instructions from the control room. "A little tighter, Joe." A light on his camera told him—and others—when he was on the air. When it went off, he changed position and lens for his next shot. One camera was likely to be on a boom, which could take camera and cameraman sailing to a position above the action; this required an extra operator.

If a play had only one set, it might be in the middle of the studio so that cameras could roam on all sides. Some directors favored shots through specially planned apertures. A hinged picture might swing open to permit a through-the-wall camera angle, then close quickly for shots from the other side. Timing mishaps were part of the lore of live production.

Live television, like Ibsen theater, drove drama indoors. "Outdoor" sets looked artificial and tended to be avoided. In contrast, film was irresistibly drawn to outdoor drama and physical action.

Paddy Chayefsky's own prefaces to his plays superbly articulated the feelings of many anthology writers. He wrote: "There is far more exciting drama in the reasons why a man gets married than in why he murders someone." He was just becoming aware, he wrote, of the "marvelous world of the ordinary."[6]

That this "marvelous world" fascinated millions is abundantly clear from statistics. These plays—akin to genre paintings—held consistently

6. Chayefsky, *Television Plays*, p. 178.

high ratings. But one group hated them: the advertising profession. The reasons are not mysterious.

Most advertisers were selling magic. Their commercials posed the same problems that Chayefsky drama dealt with: people who feared failure in love and in business. But in the commercials there was always a solution as clear-cut as the snap of a finger: the problem could be solved by a new pill, deodorant, toothpaste, shampoo, shaving lotion, hair tonic, car, girdle, coffee, muffin recipe, or floor wax. The solution always had finality.

Chayefsky and other anthology writers took these same problems and made them complicated. They were forever suggesting that a problem might stem from childhood and be involved with feelings toward a mother or father. All this was often convincing—that was the trouble. It made the commercial seem fraudulent.

And then these non-beautiful heroes and heroines—they seemed a form of sabotage, as did the locales. Every manufacturer was trying to "upgrade" the American people and their buying habits. People were being urged to "move up to Chrysler." Commercials showed cars and muffins and women to make the mouth water. A dazzling decor—in drama or commercial—could show what it meant to rise in the world. But the "marvelous world of the ordinary" seemed to challenge everything that advertising stood for.

A classic expression of this feeling was contained in a letter written in 1954 to the playwright Elmer Rice, who had suggested a series of telecasts based on *Street Scene*. Since he had won a Pulitzer Prize with this play, the advertising agency apparently felt he deserved to know the objections. Chief among them, he was told, were the setting and "lower class social level" of the characters. The letter went on:

> We know of no advertiser or advertising agency of any importance in this country who would knowingly allow the products which he is trying to advertise to the public to become associated with the squalor . . . and general "down" character . . . of *Street Scene*. . . .
>
> On the contrary, it is the general policy of advertisers to glamorize their products, the people who buy them, and the whole American social and economic scene. . . . The American consuming public as presented by the advertising industry today is middle class, not lower class; happy in general, not miserable and frustrated. . . .[7]

7. Rice, "The Biography of a Play," *Theatre Arts*, November 1959.

Quite aside from the revulsion against "down" settings and people, advertisers were beginning to feel uneasy about political implications. Lower class settings had a way of bringing economic problems to mind. And some writers kept edging into dangerous areas. In 1954, and increasingly in 1955, sponsors and their agencies began to demand drastic revisions and to take control of script problems. The result was a fascinating series of disputes and explosions.

Most were settled behind closed doors. A writer who brought a script quarrel to public attention risked his livelihood. But a few cases leaked out. They throw indirect light on pressures at work in the nation and the industry.

Reginald Rose's *Thunder on Sycamore Street,* broadcast on the Westinghouse *Studio One* series over CBS-TV, directed by Franklin Schaffner, was derived from an incident that took place in suburban Cicero, Illinois. A group of residents, disturbed to find a Negro family moving into their neighborhood, organized to get the family out. The resulting events received newsreel coverage.

Rose presented to *Studio One* a meticulous outline of a drama based on these events. He proposed that three small suburban houses be built side by side in the studio—virtually identical, with a neat patch of lawn for each. Action would move from one to another, showing the evolution of vigilante activity against the new family in the third house—the Negro family. In the end, a mild-mannered man in the second house would turn against his neighbors and take his stand with the new family.

As usual, Rose's plan was shrewdly organized, and enthusiastically received by the staff. It was approved with one proviso. Network, agency, and sponsor were all firm about it. The black family would have to be changed to "something else." A Negro as beleaguered protagonist of a television drama was declared unthinkable. It would, they said, appall southern viewers.

Rose first considered the demand a mortal blow, but persuaded himself otherwise. Wasn't *vigilantism,* after all, the essential theme? He agreed to make the neighbor an "ex-convict." The problem seemed to have been settled smoothly.

But Rose, to minimize the setback, adopted an ingenious strategy. The audience would not be allowed to know, throughout most of the play, why the new neighbor was unwanted. It would only be aware of the determination to get rid of him.

This evasive strategy turned the play into an extraordinary social Rorschach test. Comments indicated that viewers filled in the missing information according to their own predilections. Some at once assumed he was a communist; others, that he was a Puerto Rican, atheist, Jew, Catholic, Russian, or Oriental. The information that he was an ex-convict, mentioned with utmost brevity in the final act, was accepted as a logical supplementary detail. The sponsors found, with some uneasiness, that they had presented precisely the kind of disturbing drama they had tried to avoid.[8]

This helps to explain the agitation surrounding a later *U. S. Steel Hour* program, produced by the Theater Guild. In 1955 the press carried stories of three lynchings in Mississippi. One victim, a fourteen-year-old boy named Emmett Till, had whistled at a white woman, which prompted two men to take him, shoot him, and dump him in the Tallahatchie River. A trial was held. In the face of overwhelming evidence—and international indignation—a local jury acquitted the defendants. However, the community later showed coolness toward them.

Rod Serling was interested in the phenomenon of a community closing ranks against outside pressures. He felt the Mississippi community had been saying, "They're bastards, but they're *our* bastards." He suggested to the Theater Guild a play on this theme. Being a realist and eager for acceptance, he dropped the idea of a Negro victim and changed it to an old pawnbroker. The killer was a neurotic malcontent lashing out at a scapegoat for his own shortcomings. The play, titled *Noon on Doomsday,* was accepted and scheduled.

The script might have been produced with only minor revisions had not a reporter, after a conversation with Serling, mentioned in a newspaper story that the play had been suggested by the Till case. Then "all hell broke loose." Serling became involved in interminable meetings in which some thirty high-level executives of the Theater Guild, the Batten, Barton, Durstine & Osborn advertising agency, and the U. S. Steel Corporation became involved in script revisions. Everything frightened them. White Citizens Councils were said to be threatening boycotts against U. S. Steel, and Serling was assured this was no idle threat. (They had done it, he was told, against Ford and Philip Morris.) The southern location had to be changed. An unspecified location was not good enough; it had to be New England; to *prove* it was New England, it

8. Rose, *Six Television Plays,* pp. 107–8.

had to open with a shot of a white church spire. Anything that might conceivably suggest the South had to be changed. Once this sleuthing began, almost anything seemed southern. A Coca Cola bottle had to be removed from the set. An executive busied himself restoring the missing g's in present participles. Plot details were changed every day, and the final show was an absurdity in a total vacuum.[9]

Another incident caused international repercussions. General Motors sponsored the anthology series *General Motors Theater,* seen over the Canadian Broadcasting Corporation television network; it sometimes re-used scripts from United States anthology series and sometimes used Canadian originals. In 1954 it scheduled a script titled *The Legend of the Baskets,* by Canadian writer Ted Allan, about a Mexican peasant who turned out beautiful baskets at a leisurely pace, at low price. Approached by American industrialists with a plan to mass-produce the baskets (he would supervise and get rich), he turned it down because it would destroy his artistic satisfaction. The script had caused no alarm at lower echelons but during rehearsals an executive, representing the sponsor, decided the script was an attack on the capitalist system and had to be canceled. An earlier script was repeated. This censorship of Canadian television by an American sponsor caused something of an uproar in the Canadian press.[10]

Such harassment inevitably doomed the anthology series, which in 1954–55 began a rapid decline. It had a few further moments of impressiveness, as in Serling's *Requiem for a Heavyweight,* a 1956 *Playhouse 90* offering, but most leading anthology talents were turning elsewhere. Paddy Chayefsky, in the preface to his collected television scripts, published in 1955, wrote: "I came out of the legitimate theater, and I want to go back again." He added: "I will not be able to calculate the debt I owe to television. . . ." It had been a brief but glorious dramatic moment. It had been, thought Kim Stanley years later, "the best of all possible worlds." Horton Foote, looking back, thought: "You just have to shut it out of your heart somehow."[11]

Writers, directors, actors turned to Broadway but also in other directions. The year 1955 saw the appearance of the motion picture version of *Marty,* again directed by Delbert Mann. It won four Oscars, including

9. *The Relation of the Writer to Television,* pp. 10–12.
10. *Variety,* March 10, 1954.
11. Chayefsky, *Television Plays,* p. xiv; New York *Times,* February 23, 1964; January 31, 1965.

the Best Picture of the Year award, and got the Grand Prix at the Cannes film festival. A worldwide success, it had been produced independently on a low budget. The deflation in big-studio production during the Hollywood panic and the resulting shortage of features encouraged such projects. The triumph of *Marty* drew a parade of anthology talent into similar ventures, for which backing was suddenly available. Besides Delbert Mann, such directors as Arthur Penn, Sidney Lumet, and John Frankenheimer moved into a new arena.

Ironically, while sponsors were beginning to exert a chokehold on anthology drama, feature films were winning new freedom. In 1952 the U. S. Supreme Court had reached the belated conclusion that film was a part of the press and, as such, endowed with a constitutionally guaranteed freedom.[12] This threw doubt on the legality of the antediluvian state and local censor boards which, under continuing litigation, seemed headed for oblivion. During the 1950's a number of these boards disappeared—with mixed results. Films of the sort once reserved for stag parties took over many theaters, and flourished. At the same time, independent production became the magnet for any director who wanted to deal with a subject without compromise. Film, when at its best, gained strength even as television was losing it.

If artists were deserting the live anthology series, so were sponsors. In 1955 Philco switched to a Hollywood-produced series, and others followed. A few New York anthology series hung on, but with a sense of doom. The *U. S. Steel Hour,* a survivor, relied increasingly on commissioned adaptations—a policy that tended to shut out new talent. Unsolicited work ceased to be a major source of material.

The death of the live anthology was Hollywood's gain; the trend was to film. The anthology form survived to some extent on film, but was eclipsed by filmed episodic series of upbeat decor, preferred by most sponsors. Identification with a continuing, attractive actor had merchandizing advantages, and some actors were willing to do commercials. Above all, the series *formula* offered security: each program was a variation of an approved ritual. Solutions, as in commercials, could be clear-cut.

The years 1954–55, which saw the fall of the anthology, also marked

12. *Burstyn* v. *Wilson,* 343 U.S. 495 (1952). It reversed a doctrine enunciated in *Mutual Film Company* v. *Industrial Commission of Ohio,* 236 U.S. 230 (1915), to the effect that film was a mere spectacle and not an "organ of public opinion."

an increase in blacklist pressures. A group calling itself Aware, Inc., formed in the closing days of 1953, declared war on the "communist conspiracy in the entertainment world."[13] One of its leaders was Vincent W. Hartnett, one-time *Gangbusters* assistant who had written the introduction to *Red Channels* and meanwhile built up his own dossiers on writers, directors, and performers. He had associates in various broadcast guilds.

A favorite Aware dictum held that it was time for all Americans to "stand up and be counted." Those not ready to stand up were to be considered "pro-communist." Aware, Inc., began to issue bulletins listing people it regarded as suspect. A listing usually cut off radio and television employment like a guillotine. One reason was that sponsors received notes and phone calls from Laurence Johnson, supermarket executive in Syracuse, New York, calling attention to Aware listings. His point was that supermarkets should not, as a matter of patriotism, carry products advertised through subversive performers.[14]

Hartnett carried on constant research for additions to his lists. His approach is suggested by a letter written in December 1954 to actor Leslie Barrett, a member of AFTRA—the American Federation of Television and Radio Artists.

> Dear Mr. Barrett:
>
> In preparing a book on the Left Theater, I came across certain information regarding you. A photograph of the 1952 New York May Day Parade shows you marching.
>
> It is always possible that people have in good faith supported certain causes and come to realize that their support was misplaced. Therefore I am writing you to ascertain if there has been any change in your position.
>
> You are, of course, under no obligation to reply to this letter. As a matter of fact, I am under no obligation to write you. However, my aim is to be scrupulously fair and to establish the facts. If I do not hear from you, I must conclude that your marching in the 1952 May Day Parade is still an accurate index of your position and sympathies.
>
> I am enclosing a 3-cent stamp for a reply.
>
> Very truly yours,
> Vincent Hartnett

13. *Broadcasting*, December 7, 1953.
14. See *The Golden Web*, pp. 273–7, for details on Laurence Johnson.

Barrett realized the influence of Aware, Inc., listings. His lawyer wrote an immediate reply, assuring Hartnett that Barrett had never in his life marched in a May Day Parade and had no communist sympathy or leaning. Another letter came.

> Dear Mr. Barrett:
>
> To my surprise I received today a letter dated December 13 from Mr. Klein, a lawyer. . . .
>
> I was according you a privilege of commenting on certain information in my possession, and I hoped for the courtesy of a reply from you, not from Mr. Klein. As things stand at this point, I have not received from you any reply to my December 9th letter. . . .
>
> Parenthetically, is this the same Harvey L. Klein who is listed as having signed Communist Party nominating petitions in 1939-40? . . .
>
> Frankly, I am disappointed. . . . In my previous experience in similar cases, people who had nothing to hide did not pull a lawyer into the discussion. They simply and candidly denied or affirmed the evidence. I hope you will be equally candid and direct. You will find me most sympathetic and understanding.
>
> Sincerely yours,
> Vincent Hartnett

Barrett himself replied at once, repeating the denial. At his lawyer's request he included a denial of the innuendo about Klein. Hartnett wrote him another letter.

> Dear Mr. Barrett:
>
> I am happy to receive your personal warranty. . . .
>
> I appreciate your writing to me, and I hope you incurred no expense by the unnecessary move of calling in a lawyer. This only muddied the waters.
>
> Frankly, two people in radio and tv who knew you thought the man pictured in the May Day Parade photo was you. Research to establish a positive identification of the man is continuing.
>
> Sincerely yours,
> Vincent Hartnett[15]

15. The correspondence was read at an AFTRA meeting, was entered in the minutes, and later became evidence in the case of *Faulk* v. *Aware et al.* Quoted, Nizer, *The Jury Returns*, pp. 226–9.

To say that such tactics caused indignation among artists would be an understatement. But criticism of "anti-communist" warfare was no longer considered safe—it "helped the communists." Dan Petrie, a director for *Studio One, U. S. Steel Hour,* and other anthology series, received a phone call at two o'clock one morning. He and his wife were awakened from sound sleep.

"Hello?"

"Hello. This is Dan Petrie?"

"Yes."

"You the director?"

"Yes. Who is this?"

"Never mind who I am. Do you have a wife who's tall and blond?"

"Yes, I do. Now who's calling? What are you calling at this hour for?"

"I just want to give you a little piece of advice, Mr. Petrie. You better tell your wife to be careful about how she talks about the blacklist at cocktail parties."

(CLICK)[16]

Amid such maneuvers, the dying New York anthology series was not likely to become a rallying point against blacklists, vigilantism, McCarthyism—especially in the absence of any sign of support from the executive level.

If there was to be resistance, it would have to come from another sector of the program world. And it did.

THIS IS THE NEWS

Television news, at the start of 1953, was an unpromising child. It was the schizophrenic offspring of the theater newsreel and the radio newscast, and was confused as to its role and future course.

Each parent had known glory. The theater newsreel had a tradition going back almost to the turn of the century. Radio news had had its own brilliant days, especially before and during World War II, when a number of its on-the-air voices had become world famous. Some of

16. Petrie, *Reminiscences,* p. 40. He later directed the film *Raisin in the Sun* and other features.

these remained active, although others had vanished—victims, in some instances, of McCarthyist blacklists.

As network news units tackled television in the post-war period, they found the problems staggering. Much of their own experience seemed suddenly irrelevant. They had news-gathering operations and on-the-air voices—both of which now seemed to be considered secondary. Turning for help to the newsreels, they entered an alliance that was uneasy from the start.

NBC, as early as 1945, had hired newsreel veteran Paul Alley, of the Hearst-MGM News of the Day, to put together news telecasts. Given a radio-sized budget, he had to begin on a shoestring. He obtained his first film footage free from the Signal Corps Pictorial Center on Long Island. He himself wrote narration, picked music from disk recordings of the NBC Thesaurus,[1] and hired another newsreel man, David Klein, at $10 a night for part-time editing. A small budget increase allowed Klein to become a full-time assistant. Another staff member came to them from the Office of War Information, bringing with him an $8000 Mitchell camera which he was sure the government no longer needed. This equipment was the mainstay of NBC-TV news operations for two years.

When in 1947 NBC decided to take up television newscasts at a more resolute level, it did so by means of a curious contract. It commissioned Jerry Fairbanks Productions, a producer of theatrical shorts and industrial films, to provide film for NBC newscasts. A year later the network switched to Fox Movietone. When CBS at about the same time launched a television news series featuring Douglas Edwards, it made a similar contract with Telenews, a unit related to the Hearst-MGM News of the Day and later absorbed by it. The rationale for these contracts was that the networks had news sources and voices and just needed help with pictures. An additional reason for the arrangements was that the networks wanted to avoid—at least, postpone—dealing with various film unions. The newsfilm contracts continued to play a part in network news telecasts for a number of years while the networks gradually came to terms with the unions and developed film staffs of their own—first NBC, then CBS, then ABC. The expanding staffs came largely from the theatrical newsreels, which shrank as their moment in history passed.[2]

By 1953 the showpieces of television network news were the NBC-TV

1. Recorded music syndicated to stations for local use.
2. Interviews, Robert Butterfield, Paul Alley, David Klein.

Camel News Caravan, featuring the breezy, boutonniered John Cameron Swayze and sponsored by Camel cigarettes, and the CBS-TV *Television News with Douglas Edwards,* which acquired Oldsmobile as sponsor. Both were 15-minute early-evening newscasts. Each relied heavily on ex-newsreel men and followed many newsreel traditions. Each was likely to include six to eight film items. Film from distant places came by airplane. The programs also made use of cable pickups of correspondents in other cities; these pickups grew slowly in importance.

Like the newsreels in their heyday, the networks tried to maintain film crews in a few principal news centers—such as New York, Washington, Chicago, Los Angeles, London, Paris. Many other major cities were covered by "stringers"—often cameramen attached to foreign film units, who could do some shooting on the side, or retired cameramen who liked to keep busy. The stringer was paid for footage used; in important spots he might also get a retainer. He received new film to replace film sent in, whether used or not. When sending a fragment of a reel, he received a full reel to replace it; this alone was enough to encourage submissions.

A camera crew of the sort used in principal centers consisted of two or three men using several hundred pounds of equipment—cameras, magazines, tripods, batteries, lights. In 1953 NBC-TV's *Camel News Caravan* used mainly 35mm equipment; CBS-TV's *Television News with Douglas Edwards* used 16mm equipment and was looked on as second-rate because of this. The equipment was more maneuverable, but considered not quite professional.

Each of these series managed to fill its time with interesting and diverting material, and viewers were not particularly aware of shortcomings. But for presenting "the" news, crews and stringers in a few dozen cities were ludicrously inadequate in number, while the tools of their trade were often irrelevant. Except for catastrophes of some duration—fires, floods, wars—the crews were usually "covering" predictable events, many of which had been staged for the purpose—press conferences, submarine christenings, cornerstone layings, beauty contests, horse races, campaign speeches, ribbon-cuttings, dam dedications, air force demonstrations, high society debuts, award banquets.

A favorite pronouncement of the day was that television had added a "new dimension" to newscasting. The truth of this concealed a more serious fact: the camera, as arbiter of news value, had introduced a drastic curtailment of the scope of news. The notion that a picture was worth a thousand words meant, in practice, that footage of Atlantic City beauty winners, shot at some expense, was considered more valuable than a thousand words from Eric Sevareid on the mounting tensions of Southeast Asia. Analysis, a staple of radio news in its finest days and the basis for the fame of Swing, Murrow, Kaltenborn, Shirer, and others, was being pushed aside as non-visual.

Near the end of each *Camel News Caravan* telecast came a moment when John Cameron Swayze exclaimed with unbounded enthusiasm: "Now let's go hopscotching the world for headlines!" What followed was a grab-bag of items that had unhappily taken place without benefit of camera. Each event had to be dispatched, it seemed, in one sentence. Then Swayze would say: "That's the story, folks. Glad we could get together!" *Broadcasting* considered him "the best dressed TV news commentator . . . whose suave handling of the news matches perfectly his handsome face and impeccable garb."[3]

While all this introduced a distortion of values, a more serious problem was the news management involved in staged events. Behind every planned event was a planner—and a government or business purpose. Television dependence on such events gave the planner considerable leverage on news content. This might be especially serious in the case of foreign news, often based on tenuous sources. A crisis in Iran, for example, might be presented—apparently authoritatively—through a 40-second filmed statement made at a State Department press conference. The extent to which this might color—and limit—American conceptions of distant events was hardly guessed at this time.

Aside from these influences, *Camel News Caravan* had a few special distortions of its own. Introduced at the request of the sponsor, they were considered minor aspects of good manners rather than of news corruption. No news personage could be shown smoking a cigar—except Winston Churchill, whose world role gave him a special dispensation from Winston-Salem. Shots of "no smoking" signs were forbidden.

In addition to *Camel News Caravan*, NBC-TV had its early-morning *Today* series, partly newscast and partly variety program, which made some use of newsfilm. Footage from NBC's own crews and stringers was sometimes supplemented with film from other sources—purchased or donated. This included footage from American public-relations firms serving foreign governments and registered with the Justice Department

3. *Broadcasting*, March 2, 1953.

under the Foreign Agents Registration Act. In newscasts, as in theater newsreels, such footage usually appeared without identification of its source. The law required identification of any material intended to influence the public "with respect to the political and public interests, policies, or relations of a government of a foreign country," but it was assumed by the networks that this did not apply to film footage if they did not use accompanying written commentary provided by the public relations firm—that is, the foreign agent. This assumption had not been challenged by the Justice Department, although film, through its selection of detail, could serve as effective propaganda—all the more so if the source was not identified.[4]

The inherited newsreel addiction to royal panoply was illustrated by an unusual June 1953 telecast, considered a triumph at the time. The date for the coronation of Elizabeth II of England was set months ahead, and NBC-TV at once decided to make it a breakthrough television event. The aim was to get footage on the air within hours after the ceremony—particularly, ahead of CBS. NBC began working with experimenters at the Massachusetts Institute of Technology—the people at the Eastman Company said the plan was not plausible—to build a portable film developer for rapid use on location. The result was a box—about four feet square, two feet high—that could develop 100 feet of film in 25 minutes. Its secret chemical formula had to be mixed on the spot and used fresh; David Klein carried the formula in his wallet. Four such developers were shipped to England by boat and set up in a Quonset hut near an airfield outside London. Here NBC technicians also placed a television set, which they tuned to the BBC coverage of the events in Westminster Abbey. This BBC coverage was photographed from the television tube, and within an hour the developed film was on a chartered DC-6 from which seats had been removed—replaced by bolted-down editing equipment. Editing went on as the plane sped westward via the great circle route, skirted Greenland, refueled at Gander, landed at Boston. From there the film was put on the NBC-TV network, beating CBS-TV. It was hailed as a coup, although it was actually BBC coverage, reprocessed.[5] Ironically ABC-TV, which had not entered the race, won it. BBC pictures reached

4. For the development of this problem see Cater and Pincus, "The Foreign Legion of U. S. Public Relations," *The Reporter*, December 22, 1960. The law was the Foreign Agents Registration Act of 1938, amended 1942.
5. Interview, David Klein.

Canada before the NBC plane arrived in Boston. ABC-TV, taking a Canadian Broadcasting Corporation telecast via cable, scooped its rivals by minutes.

There was some British indignation over American television commercials exploiting the event. An automobile had been praised for its "royal carriage"; another was called "queen of the road." Juxtaposition of coronation footage with *Today*'s famous chimpanzee, J. Fred Muggs, also brought expressions of outrage.[6]

Dissatisfaction with newsreel superficialities, and their addiction to pseudo-events, existed among both film and broadcasting people. In theaters *The March of Time* films, which appeared from 1935 to 1951, had aimed at something more significant. Avoiding spot news, they had tried to expose underlying issues. An imitator of *The March of Time,* RKO's *This Is America,* had attempted similar things, though in less stentorian manner and on a paler intellectual level. In November 1951, as *The March of Time* was dying, Edward R. Murrow and Fred W. Friendly had launched a television series of similar purpose, *See It Now.* In content, they made a cautious start. The first telecast, for which they spent $3000 on a video line to San Francisco to show the Golden Gate Bridge and Brooklyn Bridge simultaneously on a split screen, was symptomatic. They felt awed by their new medium and needed to spend time exclaiming over the wonder of it. And they had to find out, by trial and error, what could be done with it.

During their first year they were babes in the wood. Knowing nothing about film, they did as newscast producers had done, and turned to a newsreel for help. They contracted with the Hearst-MGM News of the Day for camera work and other technical services on a cost-plus basis, and library footage as needed. Anxious to be professional, they decided on 35mm film. Newsreel men regarded their project with amusement. Palmer Williams, a veteran of the Signal Corps Pictorial Center who had worked on the *Army-Navy Screen Magazine* during World War II, found a group discussing it at Barbetta's restaurant, where off-duty and unemployed newsreel people often gathered. A News of the Day man told Williams about the *See It Now* venture. "They don't know what they're doing! Go and see them—they need help." Williams went to see Friendly, who promptly hired him. Palmer Williams became an indispen-

6. The hubbub seems to have been partly an attempt to block introduction of commercial television in Britain. Dizard, *Television—A World View,* pp. 31–2.

sable element in the Murrow-Friendly operation. "He was as much my teacher," Friendly later recounted, "as was Murrow."

During its first two years *See It Now* examined a number of issues, but not that of rampant McCarthyism. That Murrow, a symbol of courage during World War II, should ignore such a subject troubled many people. Some accused him of having settled into comfortable affluence, to which he answered, "You may be right." He used the same words to those who said he should get behind Senator McCarthy, as some newsmen and executives were doing. It seemed to Friendly that Murrow was harboring his energies for some decisive action, on a battlefield to be chosen with care.[7]

Late in 1953 the action began, abruptly but quietly.

THE FAULT, DEAR BRUTUS . . .

One October day, at lunch time, Murrow handed Friendly a wrinkled clipping. "Here, read this."[1]

It concerned a Lieutenant Milo J. Radulovich, aged twenty-six, a University of Michigan student who was in the Air Force Reserve as a meteorologist, and who had suddenly been asked to resign his commission because his sister and father had been accused—by unidentified accusers—of radical leanings. When Radulovich refused to resign, an Air Force board at Selfridge Field ordered his separation on security grounds. The clipping gave sparse details, but a *See It Now* staff member, Joe Wershba, was sent to Detroit to gather further information. He read the transcript of the Air Force hearing and talked to Radulovich, his family, and their neighbors. At Wershba's urging, Murrow then dispatched a News of the Day camera crew to film their statements. A day later Murrow and Friendly began to look at shipments of film. Lieutenant Radulovich in a filmed statement said that the Air Force had in no way questioned his loyalty but had told him that his father and sister had allegedly read "subversive newspapers" and engaged in activities that were "questionable." These activities had not been specified.

> RADULOVICH: The actual charge against me is that I had maintained a close and continuing relationship with my dad and my sister over the years.

7. Interview, Palmer Williams; Friendly, *Due to Circumstances Beyond Our Control . . .*, pp. xix, 5.

1. The following is based on Friendly, *Due to Circumstances Beyond Our Control . . .*, pp. 5–22; other sources as noted.

In another film sequence the lieutenant's father, a Serbian immigrant, read a letter he said he had written to President Eisenhower. "Mr. President . . . they are doing a bad thing to Milo. . . . He has given all his growing years to his country. . . . I am an old man. I have spent my life in this coal mine and auto furnaces. I ask nothing for myself." He asked only for "justice for my boy."

It became clear that the family and acquaintances of Lieutenant Radulovich were ready and willing to talk, but that no Air Force spokesman, in Detroit or Washington, would say a word. It became a question of whether *See It Now,* being unable to present "both" sides, should drop the case. This would have conformed with the policy of many series. Discussion programs regularly dropped issues unless "both" sides could be presented. But Murrow regarded this as a dubious policy since it allowed one side, by silence, to veto a broadcast discussion.

Murrow decided to proceed and notified CBS management that the title of the next *See It Now* broadcast would be "The Case Against Milo Radulovich, AO589839." He also notified the Air Force at the Pentagon and continued to urge its participation. *See It Now* wanted to do a balanced job of reporting, the Air Force was told. This would be difficult if the Air Force refused to comment, but *See It Now* would in any case do its best. All this suddenly brought a visit by an Air Force general and a lieutenant colonel to the office of Edward R. Murrow—who asked Friendly to join them.

The dialogue was cordial and restrained. The general seemed to consider it unlikely that the broadcast would ever get on the air. The fact that Murrow had once won the Distinguished Service to Airpower award was mentioned. The visitors considered Murrow an Air Force friend, and wanted him to know that. The general concluded: "You have always gotten complete cooperation from us, and we know you won't do anything to alter that." Murrow stared quietly at the general.

Because of the importance they attached to the case, the *See It Now* producers asked CBS to provide newspaper advertising for the telecast, but the management declined. Murrow and Friendly made an unusual decision: they withdrew $1500 from their own bank accounts for an advertisement in the New York *Times.* It did not carry the CBS symbol—the eye. It was signed, "Ed Murrow and Fred Friendly."

The relation of Murrow to CBS management was a very special one. His World War II broadcasts from London and his role in building a

European news staff had been key factors in establishing CBS leadership in the radio news field. During this period William Paley had worked in London as deputy chief of Psychological Warfare, and had come to know Murrow personally and to admire him. Murrow, returning from Europe, became a member of the CBS board of directors. In 1953, in addition to co-producing *See It Now,* he was on the air each evening, Monday to Friday, as a radio newscaster.

As co-producer of *See It Now* he was technically responsible to Sig Mickelson, who headed all CBS television news operations. But the relationship was a pro forma one, involving some discomfort for both Murrow and Mickelson. In practice, Murrow had almost total autonomy. Even Frank Stanton, president of the CBS parent corporation since 1948—when Paley had moved himself up to the chairmanship of the board—was not likely to attempt to limit it. But it was a position of power that Murrow had scarcely tested. In "The Case Against Milo Radulovich, AO589839" he was doing so and was aware of it. Before air time, after a pre-broadcast gulp of scotch, he told Friendly: "I don't know whether we'll get away with this one or not . . . things will never be the same around here after tonight. . . ."

The circumstances under which *See It Now* was assembled and telecast in 1953 were primitive. As with drama, studios planned for radio were proving disastrously inadequate for television needs, and program operations were spilling out of headquarters into makeshift facilities in every corner of town. The "Radulovich" material, shot in Michigan and developed in a laboratory on Ninth Avenue in New York, was assembled in a special *See It Now* cutting room in a loft at 550 Fifth Avenue. The Murrow off-camera narration—but not his final comment, or "tail piece"—was recorded in a radio studio near his office at CBS headquarters at 485 Madison Avenue. This was then mixed via telephone wire with the other sound elements at the Fifth Avenue facility. For the telecast all the material was then taken to the CBS Grand Central studios where many other CBS programs also went on the air. Here there was always anxiety over whether the film and the composite sound track would synchronize properly; sometimes the film slipped "out of sync." The final tail piece by Murrow—on camera—was done live from Grand Central, after the film.

The insane pressure involved in this process always put Murrow on the air in high tension, which communicated itself to all concerned. In "The

Case Against Milo Radulovich, AO589839," the tension was especially felt. All were aware that Murrow was not merely probing the judicial processes of the Air Force and Pentagon—a quixotic venture few broadcasters would have undertaken at this time—but was examining the whole syndrome of McCarthyism with its secret denunciations and guilt by association. They were also aware that the disease was not peculiar to government but had virulently infected the broadcasting industry—including CBS.

Lieutenant Milo Radulovich, in his quiet way, suggested the meaning of the case.

> RADULOVICH: If I am going to be judged by my relatives, are my children going to be asked to denounce me? . . . Are they going to have to explain to their friends why their father's a security risk? . . . This is a chain reaction if the thing is let stand. . . . I see a chain reaction that has no end.

In his tail piece Murrow offered the Air Force facilities for any comments, criticism, or correction it might care to make in regard to the case. He then suggested that the armed forces should be more frank concerning their procedures. He added:

> MURROW: Whatever happens in this whole area of the relationship between the individual and the state, we will do ourselves; it cannot be blamed upon Malenkov, Mao Tse-tung or even our allies. It seems to us—that is, to Fred Friendly and myself—that it is a subject that should be argued about endlessly. . . .[2]

Friendly has described Murrow as "bathed in sweat and smoke" as *See It Now* staff members and engineers crowded to shake his hands. Some had tears in their eyes. The phones began ringing. For days congratulatory telegrams and letters poured in. A few newspaper columnists denounced the program, but there were many paeans of praise. Not a word of comment came from CBS management.

The program had given Murrow a momentum that now would not let him go. In Indianapolis a group of citizens had rented a civic auditorium for a meeting to organize a local chapter of the American Civil Liberties Union. The noted civil liberties lawyer Arthur Garfield Hays was to speak. But local units of the American Legion and other groups became

2. "The Case Against Milo Radulovich, AO589839," *See It Now*, CBS-TV, October 20, 1953.

indignant and managed to work up enough pressure to force cancellation of the hall. A Roman Catholic priest, Father Victor Goosens, then offered his church for the Civil Liberties meeting, and he in turn became an object of indignation.

In November 1953—the same month as *Dinner With the President*—*See It Now* dispatched a Hearst-MGM News of the Day camera crew to Indianapolis to film both the Civil Liberties meeting and an American Legion meeting denouncing the Civil Liberties meeting. The memorable November 24, 1953, program on *See It Now,* "Argument at Indianapolis," was a brilliant, stimulating intercutting of the two meetings. It was also memorable for another—and totally unexpected—reason. The day of the broadcast brought much telephoning between New York and Washington. That night the opening of the *See It Now* program was postponed for an announcement. Murrow introduced Secretary of the Air Force Harold E. Talbott in a statement filmed that day. Talbott said he had reviewed the case of Lieutenant Radulovich and decided he was not a security risk. "I have, therefore, directed that Radulovich be retained in his present status in the United States Air Force."[3]

Although it left the loyalty-security apparatus untouched, the decision was an extraordinary triumph for *See It Now,* of which CBS had reason to be proud. But CBS management seemed more aware of rising friction. There were increasing anti-Murrow pressures on the *See It Now* sponsor, Alcoa. Attacks on CBS were a regular feature of the Hearst newspapers, and particularly a specialty of television columnist Jack O'Brian, who delighted in attacking "Murrow and his partner in port-sided reporting, Mr. Friendly." Some CBS affiliates were also becoming restive. But Murrow was not drawing back. When Raymond Swing, disillusioned with the Voice of America, resigned in protest, Murrow engaged him to write his radio commentaries. Murrow was now allied with Swing in radio broadcasts and with Friendly in television, and was looked to more and more for sanity amid hysteria.

In October 1953 Murrow launched a second television series, *Person to Person*—in a sense, a spin-off of *See It Now.* On two *See It Now* programs Murrow had paid a "television visit" to a celebrity. The celebrity, allowing television cameras to prowl his home, would lead the way, show treasured possessions, and answer questions asked by Murrow from a CBS studio. This became the *Person to Person* formula, and it brought

3. CBS-TV, November 24, 1953.

out a different aspect of the Murrow personality. Here he was the urbane man of the world, intimate of the great. For celebrities the *Person to Person* visits had a public-relations aspect, which seemed to control the kinds of questions used. The series was seldom controversial; it had a *Vogue* and *House Beautiful* appeal, along with a voyeuristic element. It immediately developed a large audience rating—larger than *See It Now*. Some Murrow admirers deplored its superficial, chic quality, but its commercial success clearly strengthened his position at CBS. Asked by the actor John Cassavetes—who, with his wife, had been booked for a *Person to Person* visit—why he did "this kind of show," Murrow answered: "To do the show I want to do, I have to do the show that I don't want to do." Although Murrow probably enjoyed *Person to Person* more than this remark would seem to indicate, it was clearly *See It Now* that dominated his thinking and his life.[4]

It was perhaps inevitable that *See It Now* should eventually take up the subject of Senator McCarthy himself. During 1953 Murrow suggested that the staff gather all available McCarthy footage, and from time to time he and Friendly studied the growing accumulation. McCarthy had long kept opponents off-balance by bewildering shifts. He offered exact numbers for "proved" communists in government agencies, but the numbers kept changing. In television appearances he waved sheafs of paper which he called "documentation," but no one ever learned what the papers contained. New headlines diverted attention from the legerdemain. The *See It Now* producers decided on a simple compilation that would speak for itself. Murrow would add only brief comments.

When the Murrow-Friendly team informed CBS that the March 9, 1954, broadcast would concern Senator Joseph McCarthy and again asked for advertising support, the management again declined. Once more Murrow and Friendly personally paid for an advertisement in the New York *Times*. It said: "Tonight at 10:30 on *See It Now*, a report on Senator Joseph R. McCarthy over Channel 2. Fred W. Friendly and Edward R. Murrow, co-producers."[5]

At the start of the program, in the control room, Friendly found his hand was shaking so hard that when he tried to start his stopwatch, he missed the button completely on the first try. The program ended with words by Murrow.

4. Cassavetes, *Reminiscences*, p. 22. Kendrick, *Prime Time*, pp. 360–61.
5. New York *Times*, March 9, 1954.

MURROW: As a nation we have come into our full inheritance at a tender age. We proclaim ourselves—as indeed we are—the defenders of freedom, what's left of it, but we cannot defend freedom abroad by deserting it at home. The actions of the junior Senator from Wisconsin have caused alarm and dismay amongst our allies abroad and given considerable comfort to our enemies, and whose fault is that? Not really his. He didn't create this situation of fear; he merely exploited it, and rather successfully. Cassius was right: "The fault, dear Brutus, is not in our stars but in ourselves. . . ."

Good night, and good luck.[6]

Murrow and Friendly had not planned *two* programs on McCarthy, but that week they had a crew in Washington photographing a typical McCarthy hearing. As an example of his method of interrogating—bullying—government employees, it was enlightening, and became the next *See It Now* program. It required virtually no comment from Murrow.

If the Murrow programs were damaging, it was partly because he himself had said so little. Senator McCarthy had done most of the talking. He was offered time to reply, but would seem to be replying to himself. Yet he decided to do so. A script was prepared with the help of McCarthy advisers, including commentator George Sokolsky, and filmed at Fox Movietone studios at a cost of $6336.99—which CBS paid. It was telecast April 6 at the regular *See It Now* time, and made a strong attack on Murrow.

MC CARTHY: Now, ordinarily I would not take time out from the important work at hand to answer Murrow. However, in this case I feel justified in doing so because Murrow is a symbol, the leader and the cleverest of the jackal pack which is always found at the throat of anyone who dares to expose individual communists and traitors.[7]

The sequence of *See It Now* McCarthy programs inevitably polarized opinion. Some felt that the Senator's reply had been totally ineffectual. Others felt quite otherwise.

Friendly was surprised when CBS president Frank Stanton, a few days afterwards, invited him into his office. He showed Friendly a survey which CBS had commissioned the Elmo Roper organization to make. It sampled opinion on the Friday and Saturday after the McCarthy rebuttal.

6. "Senator Joseph R. McCarthy," *See It Now,* CBS-TV, March 9, 1954; Friendly, *Due to Circumstances Beyond our Control* . . . , pp. 29–41.
7. CBS-TV, April 6, 1954.

Stanton seemed disturbed about the results, to which he evidently attached great importance. They showed that 59 per cent of those questioned had seen the program or heard about it; of these, 33 per cent believed that McCarthy had raised doubts about Murrow, or proved him pro-communist.

To Friendly, the findings had a different meaning than to Stanton. Friendly suggested that if the figures had been even more favorable to McCarthy, they would have demonstrated even more compellingly the need for the original program. It was not a reaction that Stanton was likely to appreciate.[8]

See It Now offered yet another program touching on McCarthyism—to many viewers, the greatest of the programs. In his reply to the Murrow broadcasts McCarthy had, as usual, shifted to new, sensational charges. He hinted that the hydrogen bomb, recently added to the United States arsenal of weapons, had been delayed eighteen months by "traitors in our government." Soon afterwards J. Robert Oppenheimer, who had been one of the creators of the atom bomb but had opposed a crash program to perfect the even more destructive hydrogen bomb, was suddenly stripped of his security clearance by the Atomic Energy Commission. He was, in a sense, forbidden access to scientific secrets he himself had unlocked and made meaningful. The board of inquiry went through the usual, perplexing ritual of stating that Oppenheimer's loyalty was beyond question; it was a matter of "security."

On January 4, 1955, *See It Now* offered "A Conversation With Dr. J. Robert Oppenheimer"—a long filmed interview made at the Institute for Advanced Study in Princeton. Oppenheimer did not refer to the security ruling but discussed on a philosophic plane the implications of increasing government control over research and its bearing on the freedom of the human mind and the future of man on earth.

Murrow took the unusual step of inviting board chairman William Paley to look at the film in workprint stage. Paley was moved, impressed, enthusiastic. Nevertheless the co-producers had to dig into their own pockets again to advertise the program. And once more *See It Now* precipitated furious attacks, including a denunciation from George Sokolsky in his Hearst newspaper column, and pressures on CBS and Alcoa. But to many it was one of the mightiest hours ever seen on television. There was no hint of the arrogance some scientists ascribed to Oppenheimer;

8. Friendly, *Due to Circumstances Beyond Our Control* . . . , pp. 59–60.

instead there was fragility, dedication, tension, and an unsparing urge to dig to the heart of issues. There were no easy slogans. He talked to Murrow and the audience as to equals. To *New Yorker* critic Philip Hamburger it seemed "a true study of genius." The reaction among educators was overwhelming; praises showered on the producers and on CBS. Among scientists, thought Friendly, "it was as though a stuck window had been opened." Prints of the film remained in demand year after year.[9]

The sequence of *See It Now* programs on McCarthyism—1953–55—had extraordinary impact. They placed Murrow in the forefront of the documentary film movement; he was hailed as its television pioneer. There were, of course, others: impressive documentaries came occasionally from the NBC *Project Twenty* unit that had produced *Victory at Sea;* from the *Omnibus* series; and at CBS from a documentary unit in the news division, led by Irving Gitlin. But their work was, for the moment, overshadowed by the triumphs of *See It Now.* Coming at the same time as the finest of the anthology programs, the Murrow documentaries helped to make television an indispensable medium. Few people now dared to be without a television set, and few major advertisers dared to be unrepresented on the home screen.

The McCarthy programs had many effects—the first, on McCarthy himself. They set the stage for televised hearings on his dispute with the Army. These began April 22, on the heels of the Murrow-McCarthy exchange, and proved the decisive blow to the Senator's career. A whole nation watched him in murderous close-up—and recoiled.

The hearings opened murky vistas of the great in action. David Schine, Roy Cohn's companion in harassments on behalf of McCarthy, and also heir to a hotel fortune, had been drafted, and Cohn had importuned the Secretary of the Army and his aides for special favors for Schine.

The Secretary was not entirely averse. He even, at David's invitation, paid a visit to the Schine suite at the Waldorf Towers in New York to discuss a possible assignment for the young man as "special assistant" to the Secretary. But the Army, while negotiating on these matters, was dilatory and coy, with the result that McCarthy began to berate it for using Schine as a "hostage" to obtain favors from McCarthy—such as calling off his inquiry into subversion at Fort Monmouth. The hearings became a Roman candle shooting out sparks of charge and innuendo; in the end,

9. *Ibid.*, p. 74.

they left a feeling of distaste for many of the participants, but especially for McCarthy.

Rod Steiger, star of the television production of *Marty,* was watching the hearings with his family when his mother turned to his grandmother and said, "Ooh, I think he's mean." Steiger knew then that the Senator was finished. The long close-up exposure had "cracked this man in half . . . it was a great moment in my life, because I was against this man. . . ."[10]

By the end of the year the Senate passed a vote condemning McCarthy, 67 to 22. He began a rapid decline.

The second effect was on the American Broadcasting Company. It had been the weakest of the three major networks since its separation from NBC in 1943. At the beginning of 1953 it had merged with the Paramount theater chain, which had been separated from Paramount Pictures by the anti-trust action *U. S.* v. *Paramount.* The merger had strengthened the network's position, but ABC-TV was still a very weak third, and was not even giving its affiliates a daytime schedule. It would have to do so to become competitive.

The Murrow-McCarthy conflict prepared the moment. In April 1954 it was the ABC-TV network that carried the hearings in full. CBS-TV and NBC-TV, already well provided with daytime programs and sponsors, carried only excerpts. ABC-TV, carrying the full hearings, riveted national attention. Winning impressive ratings, it began to make daytime sales and to challenge its rivals.[11]

Another effect was on News of the Day, the newsreel owned jointly by Hearst and MGM. Throughout the Murrow-McCarthy struggle the camera work, editing, and sound-recording for *See It Now* were done by Hearst-MGM personnel. Meanwhile Hearst newspapers, perhaps unaware of the irony, were especially virulent in their attacks on the series, as well as on other CBS news programs. Along with Murrow, a constant and favorite target was Don Hollenbeck, who broadcast a weekly analysis of the press, *CBS Views the Press,* which Murrow had helped launch and admired. Hollenbeck, ill and harried by the ceaseless attacks, committed suicide. Hearst columnist Jack O'Brian wrote that his suicide "does not remove from the record the peculiar history of leftist slanting of news . . . He was a special protégé of Edward R. Murrow, and as such, apparently beyond criticism or reasonable discipline. He drew as-

10. Steiger, *Reminiscences,* p. 26.
11. The small Dumont network also carried the hearings.

signments which paid him lush fees, pink-painting his news items and analysis and always with a steady left hand."[12] Murrow, a pallbearer at Hollenbeck's funeral, decided to end the Hearst relationship. All the technicians resigned from News of the Day to join *See It Now.*

A final effect was on Murrow himself and all he represented. The television excitements of 1953–55 and rising audience statistics were bringing many new sponsors into television. They wanted television time and programs, but generally not of the *See It Now* type. News and documentary were being pushed toward the edges of the schedule. Murrow himself, though at the height of his fame, and winner of more awards than any other broadcaster, felt a stab of ill omen on the night of June 7, 1955. Awaiting the start of a *See It Now* broadcast—on cigarettes and cancer—he watched the preceding program, the premiere of a new series. Horrified by what he saw, he predicted its overwhelming success. He said to Friendly: "Any bets on how long we'll keep this time period now?"[13] The program was *The $64,000 Question.*

PAY-OFF

For months Charles Revson, president of Revlon, had looked for a vehicle for a big, resounding television plunge, because Hazel Bishop lipsticks, on the strength of early television success, were "murdering" Revlon.

In 1955 Walter Craig, ex-vaudeville hoofer and writer turned television producer, became partner in a new advertising agency, Norman, Craig & Kummel. The moment he heard the idea for *The $64,000 Question,* brought to him by Steve Carlin of Louis G. Cowan, Inc.—an independent producer[1]—Craig could not contain his excitement. He managed to communicate some of his fever to Charles Revson, whose account he hoped to capture, and next morning got representatives of Revlon, of Louis G. Cowan, Inc., and of the CBS network into one office.

> I locked the door . . . and I said, very dramatically, "Nobody leaves this room till we have a signed contract." Well, I had the lawyers for everybody concerned in the room at that time—there must have been twenty of us. And about one o'clock a knock came on the door, and it was Norman, and he said, "How are you doing?" And I said, "You'd

12. New York *Journal-American,* June 23, 1954.
13. Friendly, *Due to Circumstances Beyond Our Control* . . . , p. 77.
1. See *The Golden Web,* pp. 161–2, 287, 290, 299.

> better have some lunch sent in, because we're going to stay till we get a contract." And we did have lunch sent in, and we signed the contract at four o'clock the next morning . . .[2]

The basic idea of the series was to hold contestants over several weeks to make possible an unprecedented cash award. For years quiz programs had given small cash prizes—such as $64—or merchandise prizes, donated by manufacturers in return for mention of the brand names. Now it was proposed to make a quantum jump in prize-giving. Before four o'clock in the morning, other details had been resolved. On *The $64,000 Question* a losing contestant, as "consolation prize," would get a Cadillac. Contestants would be entombed in a glass "isolation booth" as a security measure. A "trust officer" from a prominent bank would certify to the inviolability of the proceedings—in exchange for a program credit.

When *The $64,000 Question,* "biggest jackpot program in radio-TV history," opened, the results were sensational. The reviewer of the magazine *Broadcasting* described himself as in a dazed state. Ratings began high and climbed higher. On each program actress Wendy Barrie did stylish commercials for a new Revlon product, Living Lipstick, but in September the Living Lipstick message was suddenly omitted and a commercial for Touch and Glow Liquid Make-up Foundation substituted because, it was explained, Living Lipstick was sold out everywhere. Stores were phoning the factory with desperate pleas for additional shipments. Hal March, master of ceremonies, pleaded with the public to be patient. The program was drawing a 49.6 Trendex rating, with an 84.8 per cent share of audience. A Marine captain, Richard S. McCutchen, having survived several appearances as a contestant, seemed to have the whole nation rooting for him. Bookmakers were said to be quoting odds on whether he would answer the climactic $64,000 question. It was said that Las Vegas casinos emptied during the Tuesday evening programs. A convention of wholesale druggists in White Sulphur Springs, West Virginia, was halted for the announcement: "The Marine has answered the question!" The druggists cheered wildly before going on with their business. Louis Cowan, packager of the series, was asked to join CBS and became its vice president in charge of "creative services."

In January 1956 board chairman Raymond Spector of Hazel Bishop, Inc., explained ruefully to stockholders that the surprising 1955 loss was

2. Craig, *Interview,* pp. 17–18.

"due to circumstances beyond our control." He said that during the preceding six months "a new television program sponsored by your company's principal competitor captured the imagination of the public."[3]

Within months the series had imitators—*The Big Surprise* arrived late in 1955, followed soon by *The $64,000 Challenge*—under the same auspices as *The $64,000 Question*—and *High Finance, Treasure Hunt, Twenty-One, The Most Beautiful Girl in the World, Giant Step, Can Do, Nothing But the Truth.* At one point *The $64,000 Question* and *The $64,000 Challenge* held first and second places in rating lists. When Charles Van Doren, a Columbia University English instructor, began appearing on *Twenty-One,* the series climbed among the leaders. Winnings kept getting larger: Charles Van Doren's $129,000 on *Twenty-One* was quickly topped by Teddy Nadler's $152,000 on *The $64,000 Challenge* and Robert Strom's $160,000 on *The $64,000 Question.*

The atmosphere of television was changing. In 1955 Alcoa decided to drop *See It Now* for something different—perhaps fictional or "like the Ed Sullivan program." This was ascribed to an increasingly competitive consumer-goods market.

That summer William Paley had a suggestion for Murrow. Would it not be a fine idea, instead of having *See It Now* each week, to do it now and then? It might be a 60-minute program or even, occasionally, a 90-minute program. Wasn't thirty minutes, after all, too confining? Would it not be more satisfying to do fewer programs, in greater depth?

It was a shrewd approach. Murrow and Friendly were exhausted, and the notion of longer, fewer programs held attractions. Yet the move was the first step toward edging *See It Now* out of the picture.

The period long held by *See It Now* was sold to Liggett & Myers for a quiz program.

The Paley move had another element of shrewdness. The occasional or special program—the "spectacular"—was suddenly finding favor in television. In some quarters it was considered the wave of the future, thanks to the virtuoso salesmanship of an executive at another network —Pat Weaver of NBC.

PAT, BOB, AND BOB

Sylvester L. ("Pat") Weaver, Jr., had a brief, meteoric passage across the NBC skies. It had begun in 1949, when he had left a vice presidency

3. *Broadcasting,* September 19, 1955, January 23, 1956.

Arrival statement: Secretary of State John Foster Dulles.

U.S. Department of Defense

Wide World

Press conference: President Dwight D. Eisenhower.

United Press International

Marty: Rod Steiger and Nancy Marchant on Goodyear Television Playhouse.

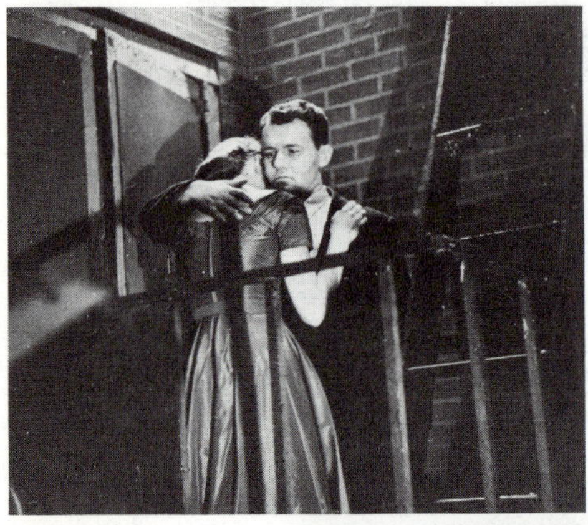

State Historical Society of Wisconsin

I Love Lucy: Lucille Ball and Desiderio Alberto Arnaz IV.

Of Human Bondage: Charlton Heston and Felicia Montealegre on Studio One.

State Historical Society of Wisconsin

Wide World

Senator Joseph R. McCarthy confers with Roy Cohen during Army-McCarthy hearings.

Edward R. Murrow in action for *See It Now* in CBS Grand Central studio.

Wide World

Crowd watches NBC's *Today* from 49th Street sidewalk and waits to see itself on screen.

NBC

NONCOMMERCIAL

Radio Liberty

Pals, Spain: Radio Liberation erects superpower transmitters on Mediterranean beach.

San Francisco, California: KQED survives by staging on-the-air auction of donated items.

KQED

at the Young & Rubicam advertising agency to become NBC vice president in charge of television.

Early in 1953 NBC found itself in extreme organizational difficulties, which led David Sarnoff, board chairman of RCA and its subsidiary NBC, to seat himself in the NBC presidential office to end the chaos. Looking desperately for signs of leadership among his executives, he found them in Pat Weaver. In December 1953 Weaver became NBC president.

General Sarnoff was frank about his plans. He had thoughts about his son, Robert W. Sarnoff, moving up to take over, but Robert first needed more grooming; he had, over a period of years, headed several divisions. "I knew," Weaver told associates later, "that I was just warming up the seat for Bobby."

Nepotism was a word often heard in NBC corridors, and often spoken with resentment—but not by Weaver. His attitude, always free-swinging and jovial, was that nepotism was as useful to him as to the General. When major plans needed approval, he took them first to Bobby, who became the spokesman when they went to his father.

There was protocol for such matters. Talking to Robert Sarnoff, no one at NBC ever said, "Your father. . . ." It was always, "The General . . . ," as though no connection existed.

Bobby Sarnoff, as everyone persisted in calling him—it was a mild kind of retaliation—was a pleasant fellow who quickly benefited from tutelage. Accustomed to dwelling at high levels, he was a relaxed executive. He was never pretentious. Appearing before government committees, he learned to handle himself well, even though his knowledge of the industry and its history was limited. Asked what had happened to the NBC Advisory Council, a highly touted body to which, according to early NBC statements to the Federal Communications Commission, a citizen could appeal "over the heads of the operating executives," Bobby Sarnoff looked nonplussed. He had never heard of it. He simply explained: "I have not boned up on that part of the history of NBC."[1]

Weaver made good his promise of leadership. In his days at Young & Rubicam and the American Tobacco Company, the advertising agencies and sponsors had controlled programming. Now Weaver was determined that control should shift to the networks. He pushed hard for the "magazine concept," under which advertisers bought insertions in programs

1. FCC Docket No. 12782. *Hearings*, p. 3386.

produced and controlled by the network. *Today* was set up on this basis, and so was *Tonight*, another Weaver creation, which began under Steve Allen and became a smash-hit under Jack Paar. The radio series *Monitor*, another Weaver innovation, applied the same principle in a weekend-long radio series.

Equally important was his espousal of the "spectacular." Contracts with sponsors for television time were revised by Weaver to allow the network to "withhold" occasional periods for special programs. Such pre-emptions had always been possible but had involved reimbursement of talent costs to the sponsor and of commissions to advertising agencies—all of which had discouraged special programs. Weaver institutionalized the special.

He devised names and banner phrases for every purpose. The radically revised relation to sponsors was called "the new orthodoxy." A special program—or "spectacular"—was designed to create "excitement and controversy and washday gossip," and to "challenge the robotry of habit-viewing." If it was of a documentary nature, like the series *Wide, Wide World*, it belonged to "operation frontal lobes." Weaver dictated fantastically long memoranda to NBC executives, which soon filled forty bound volumes in his office. He said, "Let us dare to think and let us think with daring." He radiated enthusiasm and communicated a lot of it to his subordinates. In a news release he was quoted as making a statement which seemed—even to Weaver—flamboyant, and he protested to his aide James Nelson. "But Pat," said Nelson, "those were your very words." Weaver regarded him reproachfully. "You have to protect me from myself."[2]

The magazine concept and the new sales orthodoxy were furiously resisted by some agencies. NBC's own sales people, who received the brunt of their protests, also gave him determined battle. Weaver welcomed this cheerfully. "There is some advantage," he said, "in having your enemy on your own payroll."[3]

Some of Weaver's specials were spectacular successes. Some were largely booking triumphs. *Peter Pan* with Mary Martin, telecast immediately after its Broadway run, was apparently watched by 65,000,000 viewers.[4] At first most specials were live, but film made an early appearance. *Richard III*, a Laurence Olivier film, became a television spectacular

2. Interview, James Nelson. The quoted phrases are from Weaver memoranda, 1953-55.
3. Interview, Sylvester L. Weaver.
4. *Broadcasting*, March 14, 1955.

by virtue of making its first United States appearance on television, rather than in theaters. The event suggested the precariousness of the live spectacular.

Taking the helm of a network that had long had morale problems, stemming to a large extent from ownership by RCA, Weaver effected an impressive change of atmosphere. He created talk and riveted press attention on NBC. He was one of the forces that made television indispensable.

When in 1955 David Sarnoff reverted to the subject of Robert Sarnoff becoming president, Weaver said, "Of course, but let's not do it yet." But the General felt the precise moment had come. In December it was arranged that Pat Weaver became chairman of the NBC board—he was already vice chairman—and Bobby moved into the presidency. The NBC news release quoted General David Sarnoff. Because things had been going so well the past two years, it "seemed to me a fitting time to recommend that Pat Weaver succeed me as chairman of the board of NBC. He, in turn, recommended that Bob Sarnoff succeed him as president of the company."

When Weaver came to his first meeting as chairman he is said to have noted—smiling—that General David Sarnoff had seated himself at the head of the table. "Why, General, that's my seat!" The General vacated the chair. A few months later Weaver, well provided, left the chairmanship, seeking other fields for spectacular achievement. Again Robert W. Sarnoff was promoted, becoming chairman of the board. Into the presidency of NBC moved Robert Kintner, who had been president of ABC. In place of "Pat and Bob" the trade press now referred to "Bob and Bob."

Behind the musical chairs were powerful economic changes. Weaver was a supershowman of the New York entertainment world, and had given valued leadership to a live-production era. If he suddenly seemed expendable, it was because the whole structure he represented showed signs of crumbling. Among the signs were events at ABC-TV, the upstart network—events that had been set in motion by Leonard Goldenson and Robert Kintner, and were centered in Hollywood.

WARNER BROTHERS PRESENTS

In 1954 the major film companies remained adamant. At Warner Brothers, Jack Warner still frowned on any appearance of a television set in a home scene in a Warner feature. The assumption seemed to be that if television could be banned from feature films, it could not survive.

But signs to the contrary were highly visible, even to Warner. His son-in-law, William Orr, back from an eastern trip, described miles of Chicago slums sprouting forests of antennas. History, thought Orr, might be passing them by.

Warner got similar persuasions from Leonard Goldenson, the ex-Paramount executive who, at the time of the split, had gone with the theater chain—and on into the ABC merger. As soon as the merger was complete, he and Kintner began wooing the film companies, where Goldenson as a film veteran had ready entree.

In April 1954 they won a foothold via a deal with Walt Disney for a *Disneyland* series, and possibly others. The terms looked so good to Jack Warner that they became the basis for a deal with Warner Brothers, under which Warner undertook to produce films for ABC-TV for the 1955-56 season. It was considered far more momentous than the Disney contract because Warner Brothers was one of the "majors"—the aristocracy.

The detail that clinched the deal for Jack Warner was that Warner could include in each one-hour film a 10-minute segment to be called *Behind the Cameras,* which would show Warner movie stars and crews at work on feature films soon to be shown in theaters. *Behind the Cameras* would be a glorified 10-minute commercial for Warner features. It was felt that this would ease theater exhibitors' anger over Warner dealing with the enemy.

Warner agreed to produce forty one-hour programs at $75,000 per program, all for use in the 1955-56 season. Twelve of the programs would be repeated during the summer, and Warner would get an additional $37,500 for each of these re-uses. Thus ABC-TV was assured of fifty-two programs, while Warner was assured of more than $86,000 for each film it made. They agreed on the over-all title *Warner Brothers Presents,* but this was really an umbrella for three series to be used in rotation, each based on a Warner "property"—a *Casablanca* series, a *King's Row* series, a *Cheyenne* series. The first two, based on outstanding Warner successes, were regarded as surefire prospects. But the *Cheyenne* series, derived from a comparatively unknown feature, was at once so successful that ABC-TV pressed Warner to increase the number of *Cheyenne* programs and reduce the others. *Casablanca* and *King's Row* were eventually dropped, and the series became *Cheyenne*—a network staple for seven years. It propelled into stardom Clint Walker, a spear-carrier before *Cheyenne.* It had seemed unthinkable to use a well-known actor.

The *Behind the Cameras* item was short-lived. The network showed Warner some survey statistics indicating that the audience disappeared during these sequences, which might therefore threaten the whole venture. The item was abbreviated and finally dropped.

Most *Cheyenne* films were shot in five days, with many economy measures.[1] For herds on the move, cattle stampedes, Indian battles, crowds, and even barroom scenes, the producers drew on leftover footage of old features. Hollywood quipped, "If you see more than two characters, it's stock footage."

Because the entrance of a major studio into television production was considered an historic event, sponsor and agency were at first very deferential, and were kept at arm's length. When an advertising agency expressed interest in being consulted as the work progressed, Jack Warner expressed incredulity. "They're going to tell *us* how to make pictures?" Protocol meetings were arranged, but there was no script review—not at first.

Another rebuff came. An agency wanted Clint Walker to do cigarette commercials but was told, "Cheyenne doesn't smoke." Warner would not let the actor be involved in commercials in any way, even to introduce them. This rule, too, held—for a time.

By 1956 *Cheyenne* was so successful that carbon copies became highly marketable, and Warner Brothers, with William Orr supervising television production—Warner, like Sarnoff, was a family man—began grinding out *Maverick,* followed by *Sugarfoot, Colt 45, Lawman*—all quickly acquiring sponsors.

Although Jack Warner, in negotiating, groaned over the financial arrangements and said that films could not be made on such a cut-rate basis, the early *Cheyenne* films were made within the allotted sums and even contributed to studio "overhead."[2] Receipts from residual uses were pure profit. Since residual profits were not yet shared with artists, the early films were a bonanza for Warner Brothers. The signs were noted elsewhere. Down the canyon, racing for buried gold, came others. The

1. A "low-budget" theatrical feature of the time generally cost between $300,000 and $600,000, so the television contract was felt to call for drastic economies.
2. Interview, William T. Orr.

years 1955–56 brought *Wyatt Earp, Gunsmoke, Tales of the Texas Rangers, Death Valley Days, Frontier, Broken Arrow, Adventures of Jim Bowie,* and more.

The stampede to westerns was also a stampede to the West. On the heels of *Warner Brothers Presents* came *MGM Parade* and *Twentieth Century-Fox Hour*—both exploiting studio properties, and both starting points for other series. It meant that the majors, though still with an air of condescension (television series could not have the best sound stages, nor the really big stars), were taking over. Paramount also announced television production plans.

Others were stepping up action. Columbia Pictures, through Screen Gems, was in high gear with *Ford Theater, Rin Tin Tin, Captain Midnight, Father Knows Best.* United Artists was negotiating for purchase of Ziv. MCA, acting as agent for many of the Hollywood great and less great while bursting at the seams with Revue Productions projects, was preparing to buy the Universal lot—eventually, Universal Pictures itself. MCA was growing into a Hollywood colossus.

While Warner and Disney were programming blocks of time on ABC-TV, a similar alliance was developing between MCA and NBC-TV. Reports told of an NBC meeting early in 1957 at which the following season was being charted. In the presence of president Robert Sarnoff an executive turned to MCA vice president David A. ("Sonny") Werblin: "Sonny . . . here are the empty spots, you fill them."[3] He filled them—with *Tales of Wells Fargo, Wagon Train, M Squad, and* others. CBS-TV was drawing on diverse sources, getting *I Love Lucy* and *December Bride* from Desilu, *Schlitz Playhouse* from MCA, and starting a *Perry Mason* series at Twentieth Century-Fox.

But film production was only one part of the stampede at the majors. In 1955 RKO—the only one of the big five not actively producing—decided to unload its backlog and studio. One $25,000,000 check from General Teleradio, offspring of the General Tire and Rubber Company, did it. By the end of the year, through various distributors, 740 RKO features were being offered to television stations, while the RKO studios were taken over by Desilu. Again the action broke a logjam. March 1956 brought announcement of a $21,000,000 deal covering distribution of Warner Brothers features. November brought word of a $30,000,000 deal covering Twentieth Century-Fox features. A few months later came a

3. *Fortune,* July 1960. The remark has been ascribed to Robert Kintner, but Kintner was not yet at NBC. Others attribute it to NBC vice president Emanuel Sacks.

$50,000,000 deal for Paramount features. Meanwhile Screen Gems began distributing Columbia Pictures features, and later a block of Universal features.[4] Most of these deals involved cash payments by distributors, toward guarantees. The distributors taking these gambles recouped their investments with astonishing speed as countless stations reduced staffs, closed expensive studios, and took up round-the-clock film projection, alternating with occasional sports events. WOR-TV, New York, which in 1954 had had live drama every night, had none two years later. In the fall of 1956 its schedule was 88 per cent film, and almost all of it consisted of feature films. The trend was followed at countless stations.

For many artists it was disaster; for others, upheaval. Some trekked to Hollywood, where all action now seemed centered. New York was for news and documentaries, a few variety shows, quizzes, daytime serials. As a television drama center it was dying. Hollywood was now the mecca.

For stations and networks, the road was clear. Film salesmen were lining up, and so were sponsors. Much had been settled, and the boom was on.

BOOM 1956–57

The boom that in 1956–57 was taking shape and direction touched every corner of American life. The 108 television stations of the freeze period had grown to over 500 stations, which forty million television homes—85 per cent of all homes—were watching some five hours a day. The programs were supported by tens of thousands of sponsors, to the tune of almost a billion dollars a year. These sponsors were also pouring half that sum into radio.[1]

For most television stations it had been a gold-rush period. "In the black in thirty days," was the report that came from WTVE, Elmira, New York, and such reports were not uncommon.[2] Many licensees became wealthy overnight. Prices paid for license transfers—in 1955 Dumont unloaded its WDTV, Pittsburgh, for $9,750,000, when the equipment was perhaps worth one million—were an indication of the increasing stakes.[3]

The range of sponsors plunging into the boom had exceeded all expec-

4. The films were pre-1948 films, owned outright by the studios, and involving no obligation for residual payments to artists. To release the later films the studios had to conclude negotiations with various guilds. The later films were also being held for possible network use.

1. *Television Factbook* (1960), p. 18.

2. *Broadcasting*, September 28, 1953.

3. Dumont gave up network operation at this time. The station became KDKA-TV, a Westinghouse station.

tations. Films of the pianist Liberace, who wore a velvet jacket studded with sequins, were at first sponsored on various stations by Breast o' Chicken Tuna, Maybelline, Serta Mattresses, Yes Tissues, and other consumer products. But when a Cleveland bank tried the program, offering a Liberace recording to new depositors, and gained $15 million in deposits in 1954, other bank sponsors flocked to him and apparently won a host of women depositors. Other phenomena included revivalist Oral Roberts, both sponsor and performer. He bought time on 125 stations and recouped the cost many times over via donations resulting from his on-camera faith-healing. There were complaints to the Federal Communications Commission about "undocumented" miracles, but Roberts declared that if the FCC took to evaluating miracles, it would violate the First Amendment.[4]

Products backed by the largest sums were headache tablets (Anacin *et al.*), stomach settlers (Alka-Seltzer *et al.*), cigarettes (Winston *et al.*). Among cigarettes there was a shift of budgets toward long, filter, and menthol cigarettes, in an effort to counteract an American Cancer Society report about effects of smoking. Interest in backing westerns, with their aura of fresh air, health, and vigor, received extra impetus from the cancer scare.

Radio was sharing in the boom. Regarded by many as doomed a few years earlier, it was experiencing an upswing on the basis of mobility and low costs. There were still losers, both in the AM and FM bands, but the advent of the transistor made for a rosier outlook. The 1955 sale of WNEW, New York, for $3,499,712 was symptomatic.

The new role of radio had come to hinge almost entirely on disk-jockeys. Help-wanted advertisements placed by radio stations in the trade press during this period reflected the transition:

> EARLY MORNING man. Minimum two years experience. Southerner preferred. No drunks or drifters. Send picture, audition tape and references. . . .
>
> EXPERIENCED RADIO pitchman who can pull mail for all-night trick by large midwest station. . . .
>
> ZANY DJ—glib. Ad lib. Humor. Actor background. Do character voices: "life of the party." . . .
>
> EXPERIENCED HILLBILLY and gospel DJ for live-wire 1000-watt daytimer. . . . No drifters. . . .

4. *Broadcasting*, February 7, 1955; October 14, 1957.

Top Hillbilly Disk Jockey . . . that certain touch which appeals to the common folk. Must be able to hit a commercial hard and sell it. . . .

Somewhere West of Erie and east of Laramie there is a man who is still old-fashioned and likes to work. He doesn't have long hair and he isn't theatrical. He is a sound program man who builds his programming to the needs of his audience. . . . Give full information in first letter.

Experienced Staff announcer for music, news station. Family man preferred. . . .

Colored Disk Jockey. Must be good southerner. Humorous. Good ad-lib. . . .

Another station wanted *several* Negro disk-jockeys, including a "Negro frantic type" and a "Negro spiritual and gospel smooth type." These stations reflected a growing tendency toward specialization.[5]

An advertisement in the classified section of the Birmingham *Times* was perhaps a reaction to all this:

Please—Pray that Christ will grant a new 5kw family radio station in Irondale, Ala.[6]

The boom was sprouting beguiling new advertising theories. Advertising agencies employed consultants like the high-priced Dr. Ernest Dichter, who fascinated them with analyses of latent psychosexual factors involved in a buyer's choice of a car, cigar, or brand of prunes. Advertising themes and program purchases were increasingly influenced by theories about subliminal associations. Dr. Dichter also gave advertising men a sense of destiny about their own role. In our culture, said *Motivations*, a Dichter periodical, "psychological demands are being made upon the family today which it cannot fulfill. There is a gap between human need and the capacity of the family institution to fill that need." This gap, according to *Motivations*, was being filled in part by the acquisition of consumer goods.[7]

The television boom inevitably entwined it with politics. The Lyndon Johnson family was a notable example. It was not clear whether the scores of sponsors who bought time on KTBC-TV, Austin, soon giving the family multimillionaire status, needed the advertising, or whether they liked to do business with a Senator who had become Minority Leader—

5. All the advertisements appeared in *Broadcasting* during 1954–55.
6. Quoted, *Broadcasting*, May 27, 1957.
7. *Motivations*, September 1957.

after 1955, Majority Leader—of the Senate. Such lines of interest inevitably converged.

The fight for channels—gold-mining claims—was bitter. Rumors of sharp practice and political pressure were rife. In FCC memoranda on a contested Wisconsin channel, Senator Joseph McCarthy's interest in the outcome was openly mentioned, although it should have been irrelevant. The channel finally went to a pro-McCarthy newspaper. There were even rumors about bribery. Such reports only seldom broke into print and were never heard on the air, but because of their persistence a New York University professor, Bernard Schwartz, was brought to Washington in 1957 by the House subcommittee on legislative oversight to conduct a probe of the regulatory agencies, including the FCC.

Among his first findings was that Representative Oren Harris, chairman of the commerce committee—which had jurisdiction over the subcommittee and therefore over the probe—had acquired a 25 per cent interest in television station KRBB, El Dorado, Arkansas, for $500 plus a $4500 promissory note, which was never paid. Shortly afterwards the station had applied for an increase in power—previously denied—and got it. Harris apparently saw no impropriety in this sequence of events. Professor Schwartz hardly knew how to proceed. He encouraged reporters to ask Representative Harris questions about his television coup; the press interest embarrassed Harris into selling his share, but he continued to have jurisdiction over the oversight probe. Schwartz wondered whether its task was to oversee or to overlook.[8]

The boom atmosphere also gripped the programming world and gave rise to varied corruptions that seemed to be taken for granted as suitable to an era of enterprise. But here and there concern was expressed. In 1956 Stockton Hellfrich, who headed NBC's censorship ("continuity acceptance") department, received a memorandum from a colleague.

> Memo from Carl Watson to Stockton Hellfrich:
>
> Is there a danger that an exposé of the growing payola enterprise in our industry in some reputable publication such as *Newsweek* will spark a government investigation of our industry? . . . In our day-to-day editing and monitoring, it is impossible not to notice the number of products getting a so-called free ride in the body of the established programs. . . .

8. Jaffe, "The Scandal of TV Licensing," *Harper's Magazine*, September 1957; Schwartz, *The Professor and the Commissions*, pp. 9–14, 96, 162.

> A completely separate operation is apparently in full swing whereby pay-offs to writers and producers are included for offering program plugs. This program is apparently handled by specialists in the field. . . .[9]

Writers knew well what Watson was talking about. Many had received word from a publicity agent that if they could arrange for potato chips to be eaten in a scene and notified him in advance—so he could tell the client—the writer would get a $100 check after broadcast. There would be no need to use a brand name. The client was intent on making potato chips a symbol of joyous partying and felt that any television scene of this sort would help. In 1956–57 writers and directors were receiving from agents whole lists of similar opportunities for pay-offs. Some required use or display of a brand name, others did not.

One such agency, Promotions Unlimited of Sunset Boulevard, Hollywood, was described by *Broadcasting* as soliciting clients at $250 "per insertion." Its promoters explained: "We work very closely with the writers, producers and stars of top-ranking coast-to-coast radio and television programs. . . .[10] The results kept network checkers busy with questions. Why did Bob Crosby on a CBS-TV variety show suddenly find it necessary to eat a Lifesaver? Did that dinner scene in the drama have to end with use of a Diners Club credit card? In that contract-signing scene, did the camera seem to linger on the Papermate Pen?

Sigurd Larmon, president of Young & Rubicam, was annoyed over the increasing presence of guns on dramatic series sponsored by his agency. He objected not because he disliked violence—although he apparently did—but because he felt the guns were, in effect, a huge free merchandising service for gun makers. He resented free services of this sort.[11]

Radio disk-jockey programs involved a special problem, generally called "payola." This type of programming had brought prosperity to phonograph-record companies, but the success of any new record depended on the favor of the disk-jockeys. Their friendship was cultivated with such gifts as hi-fi sets, but by 1956–57 these seemed to be inadequate and a system of cash donations had developed. The bigger the city, the higher the expected gifts. Some disk-jockeys accepted stock in record companies—an extra incentive to plug their records. As of 1956–57 these practices were considered normal fringe benefits.

9. *Hearings of Antitrust Subcommittee*, p. 6082.
10. *Broadcasting*, May 21, 1956.
11. Interview, David Levy.

At the FCC things were not so different. Professor Schwartz, browsing through files and expense vouchers, found that some commissioners made speeches for broadcasting groups, collected fees covering travel and other expenses from these groups, and then charged the same travel and expenses to the government. Commissioner Doerfer made a trip during which he addressed two groups and was reimbursed for travel three times —twice by the groups, once by the government. He also accepted transportation on company planes, and took a week-long yacht trip to Bimini at the expense of station owner George B. Storer at a time when Storer had at least one case before the commission. Doerfer later described all this as "the usual amenities."

Doerfer, who recommended that licenses be made permanent, was very popular with licensees. They constantly urged his promotion to the chairmanship, and in July 1957 President Eisenhower made him chairman.

The figure of Eisenhower, hovering remotely and benignly above this turmoil, seemed to assure that nothing much could be wrong. A television glimpse of Ike heading for the golf course was a comforting symbol of the time. He presided over a laissez-faire era that had let loose a flood of enterprise. Television had helped set it in motion, and was also its most spectacular expression. It was an upbeat era, and television was its upbeat voice.

It was not boom time for all of television. Theoretically the United States now had a dual television system; since 1952 the FCC had reserved noncommercial channels for most cities. But a license for such a channel was like a license to beg. Nothing much could be done without substantial funds. The fact that the system had survived to 1956–57 was mainly due to the Ford Foundation.

In 1952 the Ford Foundation had launched two major investments in television—one in commercial, the other in noncommercial television.

It set up a Ford Television Workshop under Robert Saudek—former ABC vice president for public affairs—to produce a network series to be available for commercial sponsorship. Titled *Omnibus,* it was designed to show that challenging material could be sponsored, and it more than proved its point. In five seasons of Sunday afternoon programs it often touched brilliance. One high point was a sequence of programs on the Constitution and its origins. The series always had sponsors, and by 1957 it had won countless honors. The Ford Foundation had spent on the project $8,512,109, of which $5,498,869 had been offset by sponsor payments. The series had a devoted following, but had scarcely influenced the habits of other producers or sponsors. The Foundation at this point decided to end its support, and turned *Omnibus* over to Saudek. The series continued and eventually—like *See It Now*—became an occasional special.

The Ford Foundation had also, since 1952, supported noncommercial television. Helping early stations with construction grants, it simultaneously made a grant to establish the Educational Television and Radio Center—later known as NET, National Educational Television.[12] Its job was to provide the stations with a program service. The Foundation hoped it was starting something that would soon acquire momentum and become self-supporting. But without repeated Ford Foundation transfusions, NET and the whole system might soon have collapsed.

There were several reasons for this. The system was invisible to most Americans. In such major cities as New York, Washington, Los Angeles, the channels in the standard VHF waveband (channels 2 through 13) had already been assigned. The available channels in the UHF band could not be seen on sets already sold in these markets (except by adding a converter), so the chance of developing an audience was minimal.

The New York State Regents nevertheless proposed state-supported stations in New York City and other locations, but Governor Thomas Dewey, who was frequently at odds with the Regents, sidetracked and buried the proposal.[13] In Washington efforts to find support for a noncommercial station likewise failed.

In Los Angeles a start was made—disastrously. A member of the board of trustees of the University of Southern California, Captain Allan Hancock, provided a grant to build a station on the campus of the University—KTHE, Los Angeles, channel 28. Launched in 1953, it was shaky from the start; few sets were equipped to view it. Its almost complete dependence on Captain Hancock made it all the more precarious. One of the station's features was a Hancock string quartet, in which Captain Hancock played the violin. The station lasted only a few months. Captain Hancock disagreed with decisions of the University trustees on other mat-

12. The 1952 organizing grant, made through the Ford Foundation Fund for Adult Education, was for $1,350,000. This was followed by annual Ford Foundation subsidies, beginning at $3 million in 1953 and reaching approximately double that amount a decade later.

13. The Governor set up a study committee, which recommended that the project proceed with "private finance."

ters and decided to discontinue his support. Station and string quartet vanished into thin air and noncommercial television had had a serious setback.

In San Francisco a noncommercial station was started in June 1954 on channel 9, in the standard waveband—a circumstance that offered more hope. Yet even here, in spite of a limited schedule—two days a week, KQED only broadcast one hour per day—the financial pressure seemed lethal. Early in 1955 the KQED board of trustees decided to dissolve the station. The action was stayed by pleas from the program staff, which asked for a chance to tackle the KQED financial crisis. In desperation it arranged an on-the-air auction in which celebrities turned with gusto to the business of auctioning donated items. As hundreds phoned in bids, noncommercial television turned the corner in San Francisco. The receipts alone did not save the station, but the community involvement that had been set in motion began to bring in new support. The auction became an annual event and was emulated by other noncommercial stations—with particular success in Boston and Chicago.[14]

Other stations survived; but some of the most stable were among the least promising. The University of Houston, a young institution, had expensive building plans on the drawing boards when the noncommercial reservations were proclaimed. The University promptly cut its building plans and, instead, built KUHT, launched in 1953. *Broadcasting* reported: "HOUSTON U SEES TV EDUCATIONAL STATION SAVING $10,000,000 IN BUILDINGS."[15] The rationale was that large lecture halls were now obsolete. A student could watch two lectures a week via television, then attend a seminar to discuss the implications. Television viewing could be done at home, in a dormitory room, or in special viewing rooms.

Lectures-by-television were probably no worse than lectures in large halls—in some cases, undoubtedly better. But KUHT's succession of lectures was scarcely a beacon light for noncommercial television.

In 1956 only two dozen noncommercial stations maintained a struggling existence. Some showed promise and vigor—WGBH-TV, Boston; KQED, San Francisco; WQED, Pittsburgh. At most stations, survival depended on arrangements with schools or boards of education, whereby selected courses were taught by television.

NET, kept alive by its Ford Foundation grants to provide stations with

14. Interview, James Day.
15. *Broadcasting*, April 15, 1953.

a skeletal program service, was also hard-pressed. Its average budget of $4500 per half-hour program forced it to rely to some extent on kinescope films of local productions. The stations, having no cable connections, had to be served through a cumbersome procedure of shipping films from station to station.

Many reserved channels were still unused. There was constant demand that the FCC release them for commercial use. Typical was the pressure from *Broadcasting*, which editorialized: "One day the FCC must take another look at the Communications Act in relation to these socialistic reservations. . . ."[16]

Some commercial broadcasters, while holding a similarly low opinion of noncommercial television, favored the reservations on the ground that they kept channels out of the hands of possible competitors. They also saw the stations serving a useful function comparable to that of London's Hyde Park. A lot of talk could go on there without doing much harm. To some extent fringe periods on commercial stations served a similar purpose.

By 1956 a hierarchy of restraints had evolved. Peak network hours, being virtually sold out, were most hostile to material dealing specifically with any current issue. If such material did enter these hours, it was most likely to be shown at 10:30–11:00, opposite a leading success on another network—that is, in a temporarily unsalable period.

More often, "controversial" material went into fringe periods like Sunday afternoon. Here *See It Now* came to rest, offering some of its most telling work in semi-banishment. This included "Clinton and the Law," which provided a brilliant vignette of a racist provocateur at work in a Tennessee town.[17] But most programs in the Sunday "cultural ghetto"—as it came to be called—took the form of round-table, panel, or interview. These had the virtue of economy. They also automatically eliminated many viewers, especially the young, who tended to shift at once to drama, which invited emotional identification. The limited audience of the talk program gave it a special permissiveness.

This too had limits. In 1957 Tex McCrary, leading a discussion program over NBC's channel 4 in New York, invited Dr. David M. Spain, who had done research indicating a link between cigarettes and lung cancer, to appear on the series. Soon afterwards the invitation was canceled with

16. *Broadcasting*, April 19, 1954.
17. "Clinton and the Law," *See It Now*, CBS-TV, January 6, 1957.

the explanation that no one could be found to present "the other side." As critic Jack Gould pointed out, the tobacco industry had apparently succeeded in vetoing the discussion, simply by declining to appear.[18]

A remoter fringe area, with even greater permissiveness, was radio, particularly in very late hours. Thus Tex McCrary in his radio series could talk at length with Helen Gurley Brown about sex life on campuses, and her observation that the diaphragm had become the new status symbol among co-eds. Such discussion would have been unthinkable on radio when its audiences were larger, but now caused little tremor.

The hierarchy of restraints made it always easy for industry leaders to cite their liberality, while at the same time keeping the peak hours as a world of refuge. Those who dwelt in that world, either as programmers or audience, could be—and apparently were—almost oblivious to problems of the fringe worlds. Accepting the magic hours as *the* world, they could scarcely believe in the reality of problems rumbling in the distance—poverty, race unrest. They could even be unaware of their unawareness.

In 1954 a black student at the University of California, Estelle Edmerson, completed a graduate study of the Negro in broadcasting. A lady in the CBS personnel department in Los Angeles, freely answering her questions, told her there was no racial discrimination at CBS; all jobs were open to qualified workers. Of course, she said, there were special circumstances to be considered. "There are certain positions where you feel it might not be advisable to use Negroes: one, receptionists; two, script girls who sit in on shows with the client. . . ." However, she concluded, "except where a company must be diplomatic in hiring, all jobs are open to Negroes." This diplomatic lady was certain she was racially enlightened.[19]

That same year the U. S. Supreme Court declared separate education to be "inherently unequal," and the following year—May 21, 1955—it called for integration of public schools "with all deliberate speed." The decisions set the stage for unrelenting pressure to end racial inequalities. That December a twenty-seven-year-old black minister, Martin Luther King, began to rally Negroes in massive nonviolent struggle against the might of the South. Beginning with a battle over bus seating in Montgomery, the struggle shifted to drugstore counters, restaurants, and other

18. New York *Times*, March 31, 1957.
19. Edmerson, *A Descriptive Study of the American Negro in United States Professional Radio*, p. 103.

fields. By the end of 1956, when Montgomery blacks began riding unsegregated buses, the Reverend Martin Luther King was world famous—less for his fantastic first successes than for the style of his leadership. He was arrested, spat on, imprisoned, fined, and reviled; but he told his followers:

> We must have compassion and understanding for those who hate us. We must realize so many people are taught to hate us that they are not totally responsible for their hate. But we stand in life at midnight, we are always on the threshold of a new dawn.[20]

At first television and radio paid little attention to King. But soon his Gandhi-inspired crusade, which always ran the risk of bloodshed, began to draw cameramen and tape recorders, sometimes resulting in 2-minute items on newscasts. The issues were also discussed on Sunday-ghetto talk programs, but seldom penetrated to the citadel of the peak hours. The commercials remained purest white, and the surrounding dramas were kept in harmony.

The roster of the great and famous visited in 1956 on *Person to Person*, even by so enlightened a man as Edward R. Murrow, gives some indication of prime-time criteria. In his alter ego as establishment figure he found occasion for visits to Liberace, Pat Weaver, Eddie Fisher and Debbie Reynolds, Jane Russell, Billy Graham, Hal March, Dr. George Gallup, Jayne Mansfield, Rocky Marciano, the Duke and Duchess of Windsor, Admiral Richard Byrd, Lawrence Welk, Anita Ekberg, and others—but not Martin Luther King.[21]

The obliviousness of the peak hours applied especially to issues affecting broadcasting itself, including blacklists. The subject was almost never mentioned on the air and therefore, for the larger public, scarcely existed. The death of Senator Joseph R. McCarthy in May 1957 led even people in the industry to think that McCarthyism might be dead. That it was not became clear from the Faulk case.

In January 1956 CBS newsman Charles Collingwood and WCBS disk-jockey and panelist John Henry Faulk took office as president and vice president of the American Federation of Television and Radio Artists, New York chapter. They had been elected as a "middle of the road" slate, declaring themselves non-communist but also repudiating the tactics of

20. New York *Times*, February 23, 1956.
21. The 93 guests during 1956 included two show-business Negroes—"Dizzy" Gillespie and Cab Calloway.

Aware, Inc. Their election aroused the anger of Aware. Although Faulk had never appeared on any blacklists, not even those of Aware, it suddenly issued a bulletin denouncing Faulk with "citations" of various "communist" activities. The bulletin had the usual result. Prodded by Syracuse supermarket executive Laurence Johnson, sponsors quickly deserted Faulk. In June 1956 Faulk brought suit against Aware, Inc., its leader Vincent Hartnett, and its patron Laurence Johnson. Faulk still had his WCBS disk-jockey stint. But CBS, perhaps under sponsor and supermarket pressure, fired him.[22]

Edward R. Murrow was outraged. Along with Collingwood, he had sought desperately to stay the action. He had argued that CBS should finance the Faulk suit; having lost this argument, he had himself sent Faulk a check for $7500 so that he could retain the famed attorney Louis Nizer. To Murrow, the case seemed a chance to open and expose a festering sore that had long infected the industry.

Some of the "citations" against Faulk were false and easily disproved. Others, in their use of half-truths, showed even more tellingly the towering malice of the "anti-communist" crusade as conducted by Aware—and abetted by networks, agencies, sponsors. One item charged:

> A program dated April 25, 1946, named "John Faulk" as a scheduled entertainer (with identified communist Earl Robinson and two non-communists) under the auspices of the Independent Citizens Committee of the Arts, Sciences and Professions (officially designated a communist front, and predecessor of the Progressive Citizens of America).

Aware did not mention the following facts, which it knew but omitted. The event was a first-anniversary salute to the United Nations, and was sponsored by numerous organizations, including the American Association for the United Nations, the American Bar Association, the American Association of University Women, the American Jewish Committee, the Young Men's Christian Association, and others. Speakers included U. S. Secretary of State Edward R. Stettinius. Presiding over a portion of the program was United Nations Secretary General Trygve Lie. Ambassadors of many countries were present. CBS broadcast the event, and asked Faulk to take part.[23]

22. For earlier—and effective—pressures on CBS, see *The Golden Web*, pp. 273–83.
23. Nizer, *The Jury Returns*, pp. 298–302. Robinson composed "Ballad for Americans," featured at the 1940 Republican National Convention, and "The Lonesome Train." See *The Golden Web*, pp. 120–21, 210, 266n.

Louis Nizer took the Faulk case, but Faulk meanwhile became unemployable. Most listeners knew only that he had vanished from his CBS spot, and had been replaced by someone less interesting. Boom-time euphoria was preserved. The lawsuit, with little public attention, dragged on for years.

The year 1956 brought a presidential election. Re-election of Dwight D. Eisenhower, on a wave of prosperity, seemed certain. The campaign as conducted on television and radio was an almost perfect reflection of the boom environment.

The Republican National Committee, as in 1952, enlisted the Batten, Barton, Durstine & Osborn advertising agency, which also handled United States Steel, Du Pont, General Electric, American Tobacco, Armstrong.

They developed a new strategy. In previous campaigns, commercially sponsored programs had been canceled for political speeches, but this had sometimes caused resentment. In 1952 Adlai Stevenson, having pre-empted an *I Love Lucy* period, got letters saying: "I love Lucy, I like Ike, drop dead." The Republicans felt the tides were running with them, and that all would be well if they could avoid stirring up trouble. Leading sponsors were persuaded to surrender the last five minutes of their programs for political appeals. The sponsor paid for 25 minutes, the party for 5 minutes. The most popular programs became lead-ins for political appeals, although technically there was no relationship. For the viewer, the temptation to switch elsewhere was minimal. For the party, costs were reduced. For the network, disruption and loss were eliminated. Thanks to the use of these 5-minute "hitch-hike" programs, supplemented by station-break spots, the 1956 campaign hardly disturbed the television boom.

The mood of the day also affected Democrats. The Democratic National Committee could hardly find an advertising agency willing to take its account. A number of major agencies, apparently feeling their clients would look with suspicion on them if they dealt with Democrats, rejected overtures.[24]

The eventual solution carried irony. Norman, Craig & Kummel was the young agency that had secured for Revlon the spectacularly successful *The $64,000 Question*. In spite of this, Revlon decided shortly afterwards to switch its business to Batten, Barton, Durstine & Osborn. Norman,

24. *The Reporter*, September 6, 1956.

Craig & Kummel, stung by the loss and facing uncertainties, decided to risk the Democratic Party account. Besides, it turned out that Walter Craig was a Democrat—a rare bird among advertising men. The agency also handled Schenley, Consolidated Cigar, Chanel Perfumes, Cook's Imperial Champagne, Bon Ami, and Maidenform Brassieres.

Although Adlai Stevenson, the Democratic nominee, considered the merchandising of candidates "the ultimate indignity to the democratic process"—as he said in his televised acceptance speech—he was persuaded to emulate the Republicans with some 5-minute spots titled *The Man From Libertyville.*

The Democratic National Convention produced a television dispute. The Democrats decided that the keynote address as an institution could be modernized by presenting a portion of it on film. The networks were told of this, and, according to party chairman Paul Butler, raised no objection; all were planning convention coverage. Young Massachusetts Senator John F. Kennedy—rising rapidly in party favor—narrated the film, which was produced by Dore Schary. But when it came time for the keynote presentation, CBS-TV decided not to take the film portion, and cut away to a round-up of news analysts. Paul Butler denounced the action as "sabotage." NBC-TV and ABC-TV, however, carried the film, and its impact may have helped spur a sudden surge of support for Kennedy for the vice presidential nomination, for which he was narrowly edged out by Senator Estes Kefauver.[25]

Both conventions were tangled in masses of television cables. In the hotels they slithered into the rooms of the great; at the convention hall they writhed down the aisles, which were also crawling with walkie-talkies and creepie-peepies. To reporter Marya Mannes the newsmen with battery backloads and weird antennae looked like "displaced frogmen." Equipment statistics were dizzying. CBS had a hundred television cameras; NBC had ten thousand pounds of equipment.[26]

One result of the convention was that NBC-TV, for the first time in years, seemed to get the better of CBS-TV, largely because of the work of NBC newsmen Chet Huntley and David Brinkley. This later led to their installation as a team of anchormen on the NBC-TV early-evening

25. It was Edward R. Murrow who suggested Kennedy as narrator; as a possible alternative he suggested Senator Edmund S. Muskie. Arthur Schlesinger, Jr., and Norman Corwin contributed to the script. Correspondence, Norman Corwin.
26. *The Reporter*, September 6, 1956.

news program, replacing the breezy John Cameron Swayze. This did not at once affect the flimsiness of the 15-minute newscasts, but it did set the stage for changes.

The Republicans were apparently worried that Vice President Nixon was not as popular as they felt he should be. This led to a curious White House maneuver. Each cabinet member received a note signed by Jack Martin, presidential assistant:

> The President has asked that each Cabinet Secretary wherever possible give Dick Nixon a boost in his speeches.
>
> I enclose a suggested approach.

The enclosure had blanks to be filled in by the individual cabinet member.

> *Richard Nixon—suggestions as a basis for remarks by cabinet members*
>
> During the last   years as Secretary of   , I have come to have a deep personal respect and warm affection for Dick Nixon.
>
> Sitting in meetings of the highest councils of the land, the Cabinet and the National Security Council over which he has often presided, I have been impressed with his wisdom, his self restraint. . . .

Secretary of State John Foster Dulles penciled his copy, "file and forget."[27]

The Republicans spent $2,739,105 on the presidential race; the Democrats $1,949,865.[28] The Eisenhower-Nixon team won a resounding victory —35,581,003 to 26,031,322 in popular vote, 457 to 73 in electoral vote.

The inauguration, for the first time, was recorded by a new process—*videotape,* which could record both picture and sound magnetically. At once superior to kinescope film, it doomed the kinescope. Ampex videotape recorders were expected on the market late in 1957 at a price of about $45,000.

The second-term inauguration of Dwight D. Eisenhower was probably the high water mark of Republican confidence. Not since Coolidge had business and government been so closely meshed.

In the Coolidge days the press, overwhelmingly Republican, had been almost an arm of government. During the Eisenhower period the broadcasting industry was edging into a similar position.

Prime-time programming, in particular, reflected the alliance—not only

27. Dulles, *Papers.*
28. *Broadcasting,* February 4, 1957.

in its restraints and taboos, but also in ideas it furthered. And "entertainment," rather than news programs, seemed to play the dominant role in this respect.

Accepted doctrine had it otherwise. The word "entertainment" was used to imply relaxation for an idle hour, apart from the world's business. And of course, entertainment had been that.

For the young, once upon a time, movies were a weekly gap in a learning schedule. But telefilms had become the learning schedule. Hours each day, they told of a larger world, and defined the good and great.

Networks played down the influence. They made a point of proclaiming that news programs were done under their "supervision and control," suggesting that only those were crucial. The others were something else—"entertainment."

But if television was playing a formative role, it was scarcely through news, which operated on the fringes—seldom watched by the young—but through a rival form of journalism—telefilms.

Their role might be suggested by a paraphrase of Jefferson. "Let him who will write the nation's laws, so long as I can produce its telefilms."

TELEFILMS

By the end of 1957 more than a hundred series of television films—*telefilms*—were on the air or in production. Almost all were Hollywood products, and most were of the *episodic series* type. They came from majors and independents alike. The films processed by film laboratories were now mainly for television.[1]

In 1957 the various family-comedy series that had followed *I Love Lucy* were being submerged by tidal waves of action films. These came in several surges but were essentially the same phenomenon, in varying guise. Their business was victory over evil people.

A crime-mystery surge, on the *Dragnet* model, already included *Big Town, The Falcon, Highway Patrol, The Lineup, Official Detective, Racket Squad, The Vise,* which were joined in 1957 by *M Squad, Meet McGraw, Perry Mason, Richard Diamond, Suspicion,* and others.

An international-intrigue surge, exploiting unusual backgrounds and shadowing the success of *Foreign Intrigue,* included *Biff Baker USA, Cap-*

1. Signs appeared in laboratories: "Unless otherwise specified, all film will be processed for TV." This meant that contrasts between light and shade were to be reduced, to compensate for the fact that television accentuated the contrasts.

tain Gallant, Captain Midnight, Dangerous Assignment, The Files of Jeffrey Jones, I Led Three Lives, A Man Called X, as well as the more fanciful *Superman* and *Sheena, Queen of the Jungle;* which were followed in 1957 by *Assignment Foreign Legion, Border Patrol, OSS, Harbor Command, Harbor Master, Passport to Danger, The Silent Service,* and others.

And a mighty western surge, on a trail blazed by Hopalong Cassidy, Lash Larue, Gene Autry, Roy Rogers, and Tex Ritter, included *Adventures of Jim Bowie, Annie Oakley, Brave Eagle, Broken Arrow, Cheyenne, Cisco Kid, Davy Crockett, Death Valley Days, Frontier, Gunsmoke, The Lone Ranger, Tales of the Texas Rangers, Wild Bill Hickok, Wyatt Earp, Zane Grey Theater*—joined in 1957 by *The Californians, Colt 45, Have Gun—Will Travel, Restless Gun, Sugarfoot, Tales of Wells Fargo, Tombstone Territory, Trackdown,* and others. Somewhat related were series in which animals were heroes but evil men supplied the occasion for drama —*Fury, Lassie, My Friend Flicka, Rin Tin Tin,* and others.

By 1958 thirty western series were in prime-time television, dominating every network. An administration that had begun with the Eisenhower advice, "Read your westerns more," had achieved a stampede of westerns.

Although many fine films throughout film history have dealt with internal character conflicts, such conflicts were seldom important in telefilms. Telefilms rarely invited the viewer to look for problems within himself. Problems came from the evil of other people, and were solved—the telefilm seemed to imply—by confining or killing them.

Simplistic drama was probably fostered by the shortness of playing time—usually 24 or 48 minutes—and by the function of providing a setting for a commercial. Dr. Ernest Dichter, who gave advertisers socio-psychological rationales to go with his recommendations, had additional observations. In the western series he saw a defense against frustrations of modern society. Most people felt a great hopelessness, he wrote, about the world's problems. But in westerns "the good people are rewarded and the bad people are punished. There are no loose ends left. . . . The orderly completion of a western gives the viewer a feeling of security that life itself cannot offer."[2] In Dichter's view, the western seemed to serve the same emotional needs as consumer goods, and their alliance was presumably logical.

But his explanation also indicated a political dimension. It seemed to say that the American people, exasperated with their multiplying, un-

2. *Broadcasting,* September 2, 1957.

solved problems, were looking for scapegoats, and that telefilms provided these in quantity; also, that frustrations were making Americans ready for hero solutions—a Hickok, an Eisenhower.

Telefilm writers thought little about such matters. The "market lists" issued regularly by WGA-w—Writers Guild of America, west[3]—tabulated the announced needs of producers and made it clear that hero-villain drama was about all that was wanted. Producers, in turn, saw little network or sponsor interest in other kinds of drama.

Along with hero-villain conflicts, another ingredient had become standard. At the end, the evil man was not merely arrested—as in radio or theater. He almost always resisted or "made a break," precipitating a final explosion of action in which he was subdued by fistfight, gunplay, knife battle, lariat strike, karate action, or secret weapon. A few series, like the Perry Mason courtroom dramas, did not go in for climactic combat; in most, it was a formula requirement.

An unspoken premise seemed to be that evil men must always, in the end, be forcefully subdued by a hero. In the telefilm the normal processes of justice were inadequate, needing supplementary individual heroism.

No such implication was, of course, intended. Films used the violent climax because they, and only they, could do so. Physical combat was always impractical in theater, seldom going beyond a ritual ballet. Radio could only offer a few seconds of sound-effects clatter and grunts. Only film could make use of what has been called the "pornography of violence." In Hollywood, where love scenes were severely restricted in the 1930's and 1940's, film seemed to gravitate toward violent climaxes as a substitute. A standard element in low-budget features, they passed naturally into telefilms and were welcomed by audiences. Inevitably, telefilm producers of 1956–57 were ransacking studio storehouses for interesting instruments and techniques for action. Thus they came up with series like *Yancy Derringer, Colt 45, Adventures of Jim Bowie.* The trade press took such interests for granted. *Broadcasting,* reviewing the ABC-TV *Jim Bowie* premiere in September 1956, quoted it without any hint of disapproval. Bowie, having been bested in a wrestling match with a she-bear, goes to a master cutler, telling him:

3. Writers Guild of America was formed in 1954 to represent film, television, and radio writers, who had previously been in various Authors League of America units. WGA was organized in two regional divisions, WGA-e and WGA-w. In separating from the Authors League, they moved in a labor-union direction, with emphasis on collective bargaining.

BOWIE: I want a knife, not just a skinning knife too small to pierce a bear's fat and so brittle it will snap off if it strikes a bone, but a real knife that a man can depend on—so long, so thick, and with a double-edged tip, curved so and balanced for throwing.

Bowie soon has a chance to test his knife in combat—not with a bear but with an evil man. *Broadcasting* considered the series off to a good start. If all episodes proved as action-packed, fans would be happy.[4]

Action telefilms were becoming the chief inspiration for toys. Toy manufacturers had franchise arrangements with scores of series, including *Dragnet, Wyatt Earp, Davy Crockett, Ramar of the Jungle.* The trend was only beginning.

A few voices of concern were heard. Senator Estes Kefauver, holding hearings on the rising juvenile crime rate, wondered whether television violence was contributing to it. *Reader's Digest,* quoting Kefauver, published an article titled "Let's Get Rid of Tele-Violence."[5] It was promptly answered by an officer of the National Association of Broadcasters, Thad H. Brown, who called the article "vicious." The Academy of Television Arts and Sciences was said to be drafting a voluntary code for telefilms, which it was widely assumed would take care of the whole problem—if there was one.[6]

Telefilm artists saw little cause for concern—for obvious reasons. The ingredients of the telefilm were not new. The only revolutionary element —one which did not involve them or their work—was its constant presence in the home, as the center of the home environment. They were conscious of this, but it seemed a vote of confidence. A writer was concerned with one script at a time—not with possible effects of a ceaseless barrage.

Telefilms were showing increasing interest in foreign settings, real or imaginary. One reason may have been that the evil of evil men could more readily be taken for granted in such settings. The trend also took advantage of increasing American interest in the world scene. The series *The Man Called X,* produced by Ziv, tried to give the impression that it was based on exploits of the Central Intelligence Agency. Its stories were "from the files of the man who penetrated the intelligence services of the world's Great Powers." Advertisements for the series included numerous references to the CIA. A CIA trainee, said one advertisement, must learn

4. *Broadcasting,* September 17, 1956.
5. *Reader's Digest,* April 1956.
6. *Broadcasting,* April 9, 23, 1956.

to kill silently when necessary "to protect a vital mission." *The Man Called X* was full of vital missions and silent killings. "Secret agents," said the same advertisement, "have molded our destiny." Ziv's *I Led Three Lives,* a widely distributed success, was derived from *I Was a Communist for the FBI,* which the company had also distributed in a radio version. Its local sponsor received promotion material proclaiming him a member of "the businessman's crusade" against the communist conspiracy. An early MCA series, *Biff Baker USA,* combined similar material with family drama. It offered "an American husband and wife behind and in front of the iron curtain." Described as "full of overseas intrigue and color," it was also "safe and satisfying for the kids."[7]

If this was livelier journalism than the *Camel News Caravan,* it must also have had more political impact, especially "for the kids." By 1956-57 the significance of this was growing. For the telefilms that came like a cataract were not only submerging United States prime-time television. They were flowing in a steadily increasing stream onto television screens of the world.

7. *Ibid.,* November 9, 1953; January 9, 1956.

2 / EMPIRE

Power always thinks it has a great soul. . . .

JOHN ADAMS

The people should have as little to do as may be about the government.

ROGER SHERMAN

If the United States was a felt presence throughout the world, the rise of broadcasting had something to do with it. Almost everywhere American voices could be heard—by people of all nationalities, including Russians and Chinese.

The American broadcasting activity took many forms, some a heritage of World War II.

There was, for example, the Voice of America. An international broadcasting system launched in 1942 for war purposes, it was later retained to fight the "war of ideas"—*for* what or *against* what, was not clear. Divisions in the country were reflected within the Voice of America.

Some Voice leaders saw their country as a natural champion of those looking for independence from colonial rule. The inspiration drawn by freedom movements from American experience was a theme of interest to a commentator like Raymond Swing, who saw it as an asset to American prestige.

But others preferred to see America as a champion of private enterprise in a world crusade against communism—even if it meant friendliness with military dictators—and were more interested in stability and commerce than in independence movements.

The McCarthy purge put the anti-communist crusaders firmly in the saddle and stepped up the cold war. On the air the war of words with the Soviet Union grew harsher, as both sides contributed to the increasing bitterness.

In 1953 the Voice of America and other international information activities were placed in a new agency, the United States Information Agency, but policy control remained in the State Department. USIA set out to increase the power and number of American transmitters.

Meanwhile transmissions met increasing interference in communist countries from local jammers, which in all major cities set up squeaks and howls, particularly against broadcasts in their own languages. (English-language broadcasts were not jammed.) The Soviet Union was thought to have over a thousand local jammers,[1] and they were a subject of constant protest by the United States, which did no jamming. American protests to Moscow and at the United Nations cited the Universal Declaration of Human Rights, to the effect that freedom of opinion included the right "to seek, receive and impart information and ideas through any media regardless of frontiers."[2]

But the Soviet Union also protested. It cited a 1936 Geneva resolution, reiterated in 1954 by the General Assembly of the United Nations, condemning broadcasts calculated "to incite the population of any territory to acts incompatible with internal order. . . ."[3] American broadcasts seemed to be moving decisively in that direction.

Although few Americans realized it, the American role in the war of words was in many ways more aggressive than that of the Soviet Union. Soviet broadcasts to the United States came from Soviet soil and were limited to short-wave transmissions. The average American remained unaware of them unless he sought them out among the short waves which, because of range and by agreement, were an international broadcasting arena.

American policy was different. The Voice of America had become a system ringing the communist countries—it had a "Ring Plan"—and addressing their people from all directions via various broadcasting bands. In 1953 Russian-language programs used not only short waves but also long and medium waves—the bands allocated by international agreement to domestic broadcasting. A Russian listener thus came across American programs in his standard wave band, often "cuddling up" to local stations. He could hardly help being conscious of American broadcasts. Transmissions from the United States were multiplied via relay trans-

1. Barrett, *Truth Is Our Weapon,* p. 121.
2. *Universal Declaration of Human Rights,* Article 19.
3. General Assembly Resolution 841, December 17, 1954.

mitters in the United Kingdom, West Germany, Italy, Greece, Tangier, the Vatican, and on a ship in the Mediterranean equipped with three transmitters—including a 150,000-watt standard-wave transmitter, three times as powerful as any station in the United States, where the limit was 50,000 watts.[4]

In August 1953 American policy went still further. *Broadcasting* reported:

> Few laymen realize it but Voice of America . . . has made its strongest and most dynamic play in cold war propaganda field. Its new (and only) long-wave transmitter in Munich actually has rolled back Radio Moscow's 173 kc. operation. U. S. operation uses same frequency and what is said to be most powerful long-wave radio transmitter in the world.[5]

What *Broadcasting* did not mention and perhaps did not realize—few Americans were made aware of it—was that 173 kc. was a wave length which the Copenhagen Convention of 1948 had assigned exclusively to Moscow with maximum power of 500,000 watts. In usurping this frequency—with 1,000,000 watts of power—the United States was engaged in international piracy. A McCarthy-dominated leadership easily rationalized the action as part of a holy war against an unprincipled enemy.

Along with radio bombardments at Russians and Chinese, the USIA was stepping up its attentions to friendly and non-aligned countries. Here it girded for a huge expansion in television. By 1954 it was supplying films free of charge to television stations in nineteen countries in Europe, Asia, Africa, and Latin America. Some films were "cultural," but the emphasis was doctrinal. The USIA, according to its director Theodore C. Streibert—formerly of Mutual Broadcasting System—was using every means at hand "to fight international communism."[6]

Another inheritance from World War II was the Armed Forces Radio Service, likewise launched in 1942 and retained after the war.

Its character had changed substantially. Its early career had featured 50-watt transmitters, often housed in Quonset huts, along with mobile units serving advance lines and wired systems on ships and at air fields. Many stations were audible only to troops they served. But on entering Germany American forces took possession of a 100,000-watt transmitter

4. Barrett, *Truth Is Our Weapon*, p. 128. The ship was the U.S.S. *Courier;* in 1953 it was anchored off the Greek coast.
5. *Broadcasting*, October 12, 1953.
6. *Ibid.*, August 30, 1954.

in Munich, a 100,000-watt transmitter in Stuttgart, and a 150,000-watt transmitter in Frankfurt am Main—all exceeding in power any United States domestic station. These became units in an interconnected chain of AFRS outlets reaching most of Europe and North Africa. AFRS meanwhile took over as its European headquarters the Von Bruening castle near Frankfurt am Main. In the fourteenth century the owners had collected tolls from boat traffic on the Main River. More recently it had belonged to the founders of the I. G. Farben chemical trust, which had played an important role in the rise of Hitler. AFRS, surrounded by such symbols of power, acquired a different aura than it had had in its Quonset days. Although its mission remained theoretically unchanged—that of informing, educating, and entertaining troops—it inevitably became an enterprise conceived in larger terms. It was the soft-sell arm of United States propaganda, radiating friendliness and popularizing swing and baseball. On Sundays there were hymns and religious programs. Because many servicemen abroad had wives and children with them, AFRS took up children's programs and personal-advice broadcasts like *Dear Abby*. Stations received news material from AFRS headquarters and could use anything from the wire services, but were forbidden to develop their own programs of "comment, criticism, analysis or interpretation." They were forbidden to carry political speeches except those furnished through official channels. The State Department often provided policy statements by Secretary of State John Foster Dulles and other official spokesmen. Thus the AFRS stations, while maintaining a nonpolitical stance, served to some extent as an administration sounding board.[7]

As a result of successful tests at Limestone Air Force Base in Maine, the AFRS in 1954 became the AFRTS—Armed Forces Radio and Television Service. The experimental station at the base had apparently caused a sharp drop in the AWOL rate.[8] By 1956 twenty AFRTS television stations were in operation in various parts of the world, joining the more than two hundred radio outlets. Each television station received a weekly allotment of seventy hours of filmed programming, including *Dragnet, Gunsmoke, Omnibus,* and the children's series, *Howdie Doodie*—all by special authorization of artists, production companies, guilds, networks, advertisers. In 1956 AFRTS even announced with pride that commercials were

7. *The American Forces Network*, pp. 3–6; *Policy Memorandum*, U. S. Department of Defense, November 4, 1949.

8. *Report on the Morale Effectiveness of AFL-TV*, pp. 1–2. The Limestone base was renamed Loring.

being run intact, giving service men a taste of home, but protests restored the practice of eliminating commercials.

Virtually every army, navy, and air force installation outside the United States had an AFRTS outlet—radio or television or both. In far corners of the earth, voices in the ether were a constant reminder of American power.[9]

A similar reminder, though in a separate category, was RIAS—Radio in the American Sector—in Berlin. Its origin was a Nazi-built wired-radio system which American occupation forces had taken over and subsequently linked with a number of transmitters—short, long, and medium wave. A well-financed German-language broadcasting organization, with continuing financial support from the United States government, it was able for some years to maintain its own symphony orchestra. It also served as a training ground for a reviving broadcasting system in West Germany, and meanwhile became a symbol of the United States presence in Berlin—and as such was assailed and jammed in East Germany. It relayed selected Voice of America broadcasts.

Still another group of stations involved a World War II heritage. Wartime listeners had heard many "black stations"—stations which were not what they pretended to be. Thus, a station giving the impression that it represented a German underground group behind German lines was really a Psychological Warfare project of the United States forces. Many nations engaged in such projects; the FCC monitored sixty of them and may not have been sure which were American.[10] The Office of War Information (OWI) under Elmer Davis was said to have unfavorable feelings about this technique, but the Office of Strategic Services (OSS), headed by William J. Donovan, was an enthusiastic supporter.[11]

When the technique reappeared in the cold war and became an instrument of United States policy, it was under the auspices of the Central Intelligence Agency (CIA), successor to the OSS.

The CIA, originally a fact-gathering organization, was able to enter international broadcasting on a world-wide scale because of an extraordinary law passed in 1949—the Central Intelligence Agency Act. It

9. In 1956 the United States had military installations in Japan, Marshall Islands, Okinawa, Philippines, Saudi-Arabia, South Korea, Taiwan, Turkey, Eritrea, Libya, Morocco, Azores, France, Greece, Greenland, Iceland, Italy, Spain, United Kingdom, West Germany, Bermuda, Canada, Cuba, and the Panama Canal Zone.
10. See *The Golden Web*, pp. 158, 201–2.
11. Sorensen, *The Word War*, pp. 11–12.

marked a sharp departure from democratic principle. The Act empowered the CIA to undertake unspecified activities abroad, and, although the Constitution requires public reports on federal disbursements, the new Act—with dubious constitutionality—exempted the CIA from "publication or disclosure of the organization, functions, names, official titles, salaries, or number of personnel employed by the agency." It could even spend money "without regard to the provisions of law and regulations relating to the expenditure of Government funds." Its budget was mysteriously buried in the budgets of other agencies, and its personnel could infiltrate other agencies and organizations, governmental or private. Most of the congressmen who voted for the law did not know what the agency would be doing, or how much it would spend. A cold-war product, the law soon led to escalation of the cold war.[12]

In broadcasting, the first of several projects was Radio Free Europe.[13] Government agencies had for some time considered the problem of refugees from communist countries. Radio Free Europe was one of several projects designed to weld them into a political force. Anti-communist intellectuals from Poland, Czechoslovakia, Hungary, Bulgaria, and Romania were assembled as a staff of broadcasters in Munich, to maintain a barrage of critical analysis of the communist regimes in their homelands. The organization acquired medium-wave transmitters in Germany and, later, powerful short-wave transmitters in Portugal—a location considered suitable for distant short-wave transmissions and for "security" purposes.

The operation was announced as a "private" one, financed by American donors. Fund appeals, including spots on American television and radio, seemed to confirm the "private" nature of the venture. Actually they brought in only a fraction of costs and served mainly as a "cover story." To assist the camouflage, donations were treated as tax-deductible —although, according to tax rules, gifts to propaganda organizations were not deductible. Government funds from the CIA were channeled through a Free Europe Committee, which also sheltered other activities. It got policy direction from the Department of State.

Radio Free Europe was constantly described as the autonomous voice of exiles, but the exiles themselves were aware of policy controls by the

12. Central Intelligence Agency Act, Public Law 110, June 20, 1949. See Appendix B.
13. The CIA role in Radio Free Europe became a subject of wide newspaper discussion in 1966, and was not denied. See especially New York *Times*, April 27, 1966.

United States government. An internal "policy handbook" rationalized this situation as follows:

> As a non-governmental station responsible to the millions of Americans who support it, RFE cannot take a line contrary to United States Government policy or to the beliefs of the people of the United States and American institutions.[14]

By 1955 Radio Free Europe had twenty-nine transmitters addressing the target countries, using a number of transmitters simultaneously for every broadcast. The transmitters in Portugal used up to 100,000 watts.

Arguments made for a "private" propaganda system were of several sorts. General Lucius Clay, a director of the Free Europe Committee, explained it in these terms: unlike the Voice of America, Radio Free Europe could be "less tempered perhaps by the very dignity of government; a tough, slugging voice, if you please."[15] Others explained that a private set-up avoided problems with Congress.

The slugging voice was, of course, periodically protested by the target countries as an incitement to disorder, and they jammed it—with only partial success. The United States meanwhile denounced the jamming. The target countries also proclaimed that Radio Free Europe was a creation of the Central Intelligence Agency, but this information reached few American readers, radio listeners, or television viewers. Some, hearing of the charge, assumed it was a lie.

Half a world away, and paralleling Radio Free Europe, a Radio Free Asia was launched in 1952 with short-wave transmitters in Taiwan and the Philippines. Set up in similar fashion, it was shielded by a Committee for Free Asia, which also carried on propaganda activities in other media, not only in Asia but also in the United States.

Radio Free Asia faced severe problems. Because of vast distances to be covered in Asia, only short-wave transmission was feasible, but the target areas of the Asian mainland had few short-wave radios. Also, Radio Free Asia failed to win support of the sort that Radio Free Europe received from refugee intellectuals.

Robert Goralski, traveling for the Committee for Free Asia to various Asian countries in 1953 to produce a series of radio programs titled *Voices*

14. *Radio Free Europe Policy Handbook*. The "millions of Americans who support it" were doing so involuntarily as taxpayers, rather than as donors as the words were apparently meant to suggest.
15. Quoted, Wise and Ross, *The Invisible Government*, p. 321.

of Asia, for use on educational stations in the United States, wrote to headquarters to explain some of his problems. One was the committee name. To the Vietnamese people, he pointed out, "Free Asia" could only mean freedom from the French. That a committee with such a name could have something quite different in mind and could even cooperate with French colonial rulers baffled many Vietnamese.[16]

The Committee for Free Asia was later replaced by an Asia Foundation, similarly financed. Radio Free Asia continued for about two years.

An air already crowded with American transmissions received another —of the same sort—in 1953. Radio Liberation, addressing the people of the Soviet Union from Germany—and later from other bases—began broadcasting two months after President Eisenhower took office. Although plans for it had been made earlier, Radio Liberation became the epitome of the foreign policy of the following years, a policy dominated by John Foster Dulles of the Department of State and Allen W. Dulles of the Central Intelligence Agency—two remarkable and complex men, differing yet working in harmony. They made a fateful impress not only on American diplomacy but also on its broadcasting—at home and abroad.

DULLES DUO

To their boyhood in a Watertown, New York, parsonage, where the children took notebooks and pencils to church to take notes on father's sermon so they could discuss it later at home, John Foster Dulles and Allen Welsh Dulles seem to have reacted differently. Allen Dulles, the younger brother, became genial and socially adaptable. In later years he conveyed a country-squire impression. He smoked a pipe and loved to read spy stories. Taking up espionage, he seemed to make it a gentleman's occupation.

John Foster Dulles, as a boy, was earnest and self-absorbed. He went through Princeton but later his classmates could scarcely summon up a memory of his presence there. To Foster, as his family called him, the Christian context of his childhood remained ever-present and often entered his vocabulary. When he decided to become a lawyer rather than a minister, he explained that it should be possible to be a "Christian" law-

16. Letter, October 15, 1953. Goralski, *Papers*. The series *Voices of Asia* was distributed in the United States through the National Association of Educational Broadcasters, which—like Goralski—was apparently unaware of the source of the Committee for Free Asia funds.

yer and he would try. He remained active as a layman in Presbyterian affairs, quoted the Bible, and liked to refer to communism as "atheistic communism." A righteous tone came easily to him. Some newsmen felt he used his church connections for political ends. But President Eisenhower praised his moral fervor and compared him to an Old Testament prophet.[1]

The President hardly knew Dulles when he appointed him Secretary of State. Dulles had been foreign-affairs adviser to Governor Thomas E. Dewey of New York, Republican candidate for the presidency in 1948, and had been expected to become Dewey's Secretary of State. Eisenhower inherited him from Dewey.

Both as adviser and later as Secretary of State, Dulles devised apocalyptic phrases that seized headlines and signaled a sharp change in foreign policy. The keynote was *liberation.* The Truman policy of merely "containing" communism was swept aside for something far more militant. The Republicans would "roll back" the iron curtain. People of "the captive nations" would no longer be abandoned to "godless terrorism." In the Far East Dulles hoped to "unleash" Chiang Kai-shek. Aggressive moves by communists could expect "massive retaliation." Unrelenting pressure would make communist rulers "impotent to continue in their monstrous ways." It would bring "hope for all mankind."[2]

How the rollback would be accomplished was not clear. But because Dulles spoke often of the war of ideas, observers were puzzled that he did not show more intense interest in the Voice of America. A reason may have been his plans for Radio Liberation.

One of his first official moves was to ask that General Walter Bedell Smith, CIA chief, be made Under Secretary of State. This was done, creating a strong CIA-State Department link—all the stronger because the same move opened the top CIA job to the man who had held its No. 2 position—Allen Dulles, veteran of OSS. From then on foreign affairs were firmly in the hands of a curious duo: an elder brother known for fervent moral rhetoric, aided by a younger brother in charge of clandestine operations—the ramifications of which were still unguessed by most Americans.

The two were in constant telephone communication, and spent many

1. See Challener and Fenton, "Recent Past Comes Alive in Dulles 'Oral History,'" *Princeton Alumni Weekly,* March 14, 1967. The following draws on various interviews in the Dulles Oral History Project.
2. Drummond and Coblentz, *Duel at the Brink,* pp. 43–80.

evenings conferring. In the setting of goals, Allen punctiliously deferred to Foster. Means might be left to Allen.

In the field of broadcasting alone, the arrangement put Secretary of State Dulles in an extraordinarily powerful position for what he considered his climactic struggle with the Soviet Union. By presidential directive he had complete policy control over the Voice of America. His policy statements dominated the armed forces stations. He also controlled Radio Free Europe, Radio Free Asia, Radio Liberation, and any other voices the CIA might create—all "privately" operated and beyond the scrutiny of Congress and its committees, with their budgets unknown to Congress and taxpayers alike.

Howland Sargeant, a former State Department official who had also been an intelligence aide to the Joint Chiefs of Staff, was put in charge of Radio Liberation. Like Radio Free Europe, it enlisted refugees—in this case, from the Soviet Union—and operated under a buffer committee.[3] Its program staff in Munich received direction via committee headquarters in New York. The organization began with transmitters in Germany but later added powerful transmitters on a stretch of Spanish Mediterranean beach, and others in Taiwan. Acquisition of the Spanish broadcasting site, considered ideal for short-wave transmission, involved diplomacy at highest levels, as indicated by a note from Allen Dulles to John Foster Dulles thanking him for help given by Henry Cabot Lodge to Howland Sargeant to secure the Spanish base.[4]

While acquiring control over a varied battery of international voices —of increasing aggressiveness—the Dulles team also began to have a strong influence on broadcasting media in the United States. The Secretary of State became a skillful manipulator of news and a constant broadcaster.

As broadcaster he acquired an effective, devoted assistant. Appearing on a Chicago television program in 1954, Dulles was impressed by the businesslike young man running the program. Before the broadcast David Waters suggested that Dulles, instead of remaining at a desk, walk to a large globe and point to world trouble spots as he discussed them. To prepare for this suggestion, Waters had had a magnificent globe brought

3. Originally American Committee for Liberation of the Peoples of Russia, it became American Committee for Liberation from Bolshevism, then American Committee for Liberation, and eventually Radio Liberty Committee.

4. Letter, September 9, 1957. Dulles, *Papers*. Lodge was U. S. Ambassador to the United Nations at the time.

from the prop room. Somehow this impressed Dulles, who said, "You know your business." It was suggested that the young man visit Dulles in Washington, if he cared for a job in the State Department. Waters for his part was awed by the formidable, crusty man in the black Homburg hat, and a few days later turned up in Washington and was made television and radio aide to the Secretary of State. Overnight he became a leading arranger of news events. In the next few years John Foster Dulles traveled over half a million miles and every trip became an occasion for "departure statements" and "arrival statements" and, in between, press conferences and speeches. Waters was constantly on the phone with NBC and CBS and ABC to make sure there would be film cameras at the airstrip. Waters regularly predicted policy statements on which the free world might depend. Waters went along on trips, carrying a collapsible lectern that could be set up instantly anywhere. En route, in plane or hotel room, he got Dulles to rehearse statements. The Secretary made extraneous lip movements—perhaps remnants of a childhood stammer—that sometimes made him seem "out of sync" on film, and gave a fuzzy impression. Dulles worked hard, under the young man's tutelage, to sharpen the delivery, and acquired a very decisive manner. Waters often urged cameramen to shoot from a low angle, which he felt gave Dulles an "American eagle look." Although an upstate New Yorker, Dulles pronounced *communism* as a southerner would—*commonism.* Waters did not try to change this.

In 1955 Eisenhower began admitting newsfilm cameramen to his regular press conferences, and Dulles followed suit. Sensitive film that reduced the need for oppressive light made all this possible—to the annoyance of pad-and-pencil reporters, who tended to become extras in a television show. Dulles developed into a bravura camera performer. No film was permitted to be used until he had had a chance to examine transcripts and order cuts. This gave him firm control, and he came to relish the sessions. Beforehand he would say to Waters, "Well, let's go down and see the lions." En route Waters would find the elder man's steps quickening. "He'd get a pace and excitement about it . . . he loved the exchange." NBC, CBS, ABC, Fox Movietone, MGM-Hearst News of the Day, and Voice of America covered the sessions.[5]

Some correspondents, including David Schoenbrun of CBS, distrusted Dulles. Dulles, aware of their skepticism, seemed to enjoy baiting and

5. Waters, *Interview*, pp. 3–26.

exasperating them. To a reporter who intervened with a new question, Dulles said: "Schoenbrun hasn't finished working me over. You've interrupted his train of thought."[6]

Dulles also made skillful use of off-the-record "background" sessions for favored reporters. There were several groups of these; one met with Dulles periodically at the home of Richard Harkness of NBC. Harkness would be the host; there would be drinks. Dulles fascinated reporters with his habit of stirring his highball with a forefinger. With apparent frankness he would discuss faraway crises—not for quotation. Foreign newscast items attributed to a "high government source" often stemmed from Dulles himself. The sessions seemed to give reporters a pipeline to inside information but also made them to some extent tools of the Secretary. The special role of Harkness, reflecting an honored status with the Secretary, no doubt strengthened his standing at NBC, but it also tied him closely to the Secretary.

For Dulles and the public, the importance of all this was that world events were seen to a large extent through the eyes of Dulles. Crises erupted during these years over Guatemala, Vietnam, mainland China and other places without network bureaus. The troubles were, in any case, not of a sort that could yield their essence to newsreel cameras, even if available. This meant that a filmed press conference excerpt, or a newsman's report "from a reliable source," or a filmed statement by Dulles from a lectern at the edge of an airstrip, *became* the news. For networks he often seemed a welcome *deus ex machina.* In a 15-minute newscast, a 90-second report on Southeast Asia by the Secretary of State himself seemed grand and took care of Southeast Asia nicely. That television was beginning to pay a high price for its dependence on pseudo-events was guessed by few.

The façade held firm. Events all fitted into a world drama of good against evil—or, as Schoenbrun put it, of "Christ against anti-Christ."

What went on behind the façade was still largely a mystery. A handful of congressmen—apparently selected for their acceptable attitude toward military ventures—received fragmentary briefings on CIA activities. Among them was Senator Richard Russell of Georgia, who in 1956 learned things—as he reported with admiration—"which it almost chills the marrow of a man to hear about."[7] There were hints of this sort but—as yet—few facts.

6. Schoenbrun, *Interview*, p. 53.
7. Dulles (Allen W.), *The Craft of Intelligence*, pp. 260–61.

When information later began to emerge, it was partly because of leaks and partly because Allen Dulles decided that it should. Several CIA interventions abroad had been so masterfully conducted that he felt that they should become known. The effectiveness of the CIA would be further heightened, he felt, by making it an "advertised fact." The information thus revealed—mostly during the 1960's—gave hindsight glimpses of how press-conference façades differed from hidden events.

Allen Dulles began to take bows for the overthrow of Premier Mohammed Mossadegh of Iran in 1953 and of President Jacobo Arbenz Guzman of Guatemala in 1954. Both had come to power, Allen Dulles conceded, "through the usual processes of government and not by any Communist coup." But did that absolve the United States, he asked, of the responsibility "to right the situation?" He was sure it did not.[8]

The Guatemala intervention proved to be especially important. As the Eisenhower administration took office, President Arbenz of Guatemala, who had been elected on a reform platform, was preparing to expropriate United Fruit Company lands—not lands planted in bananas, but reserve areas. In a country where 90 per cent of the land was held by a few dozen owners and most people were indescribably poor, the move was expected and widely supported. The right of expropriation was recognized by United States policy, provided payment was prompt and adequate. The Arbenz government offered prompt payment—in 3 per cent government bonds—but the United States said it was not adequate, and negotiations began.

Then Secretary of State Dulles began telling reporters that a conspiracy directed from Moscow was taking over Guatemala and that a "reign of terror" was in progress. At a press conference he mentioned intelligence reports that a "beheading of all anti-communist elements in Guatemala" was imminent.[9] As Dulles reported subsequent events, this grisly development was averted by a spontaneous patriotic uprising and the overthrow of Arbenz. Dulles hailed it as a victory of free men over "red colonialism." He ordered the theme trumpeted throughout the world.[10]

Later, in retrospect, events acquired a different look.

While diplomatic negotiations on the expropriation were in progress,

8. *Ibid.*, pp. 51, 232–5.
9. Press conference, June 8, 1954. Dulles, *Papers*.
10. Memorandum by Theodore Streibert to USIA posts; quoted, *Broadcasting*, July 19, 1954.

the CIA assembled, armed, and supported a military force in neighboring Honduras and Nicaragua, both ruled by dictators amenable to such activities. The CIA also arranged an "air cover" of United States planes, to be flown by United States pilots. To head the ground forces the CIA selected a right-wing Guatemalan colonel, Carlos Castillo-Armas, trained at the Fort Leavenworth army command school. President Arbenz, unable to buy arms from his usual source, the United States—which had halted all trade—ordered arms from Czechoslovakia; this was later cited as proof he was under Moscow orders. In mid-June 1954 the CIA-Armas force entered Guatemala. It did not have to fight, however. The unmarked American planes, flown by Americans, attacked Guatemala City to prepare the way; some planes were lost but they were promptly replaced by order of President Eisenhower. President Arbenz, startled at the power arrayed against him, fled, and Armas was installed. He was promptly promised $6,500,000 in United States aid, and land reform was halted.[11]

At the United Nations, Ambassador Henry Cabot Lodge, in answer to protests, categorically denied any United States connection with the invasion.

USIA executive Thomas C. Sorensen later confirmed that the Voice of America "made much" of the Arbenz downfall and "of course" did not mention "the behind-the-scenes role of the CIA."[12]

The operation had been so smooth that it inevitably led to other such adventures. They seemed to promise cheap and easy victories for the "free world."

Vietnam offered such a possibility—though the details were perhaps more marrow-chilling. When in 1954 the French at Dienbienphu faced defeat by Vietnamese forces under Ho Chi Minh fighting for independence, Dulles told the U. S. Joint Chiefs of Staff that the French position must be saved at any cost. He was advised that three "small atom bombs" would do the job and let the French march out, as one chief put it, with the band playing "The Marseillaise." According to French Foreign Minister Georges Bidault, Dulles twice offered him atom bombs, but he demurred.[13] However, when the war was halted by a Geneva peace agreement, calling for internationally supervised elections, with interim administration in two regions, the Dulles team embarked

11. Wise and Ross, *The Invisible Government,* pp. 165–83; Bernays, *Biography of an Idea,* pp. 744–73. Bernays was a public relations counselor to United Fruit.
12. Sorensen, *The Word War,* p. 53.
13. Drummond and Coblentz, *Duel at the Brink,* pp. 121–2.

on a program of its own. With the CIA going into action, the southern region was to be turned into a bastion of the free world under Ngo Dinh Diem, who had recently been living at the Maryknoll Seminary in Lakewood, New Jersey, and whose views had won the ardent support of Cardinal Spellman. The United States began a determined effort to bolster Diem as a rival to Ho Chi Minh; Diem was offered millions—it became billions—in United States aid. Meanwhile the proposed elections were blocked by Diem, with American approval. Later, in his memoirs, Eisenhower told why this was considered essential: his advisers said that "possibly 80 per cent of the population" would have voted for Ho Chi Minh.[14]

Again all this was cloaked in free-world rhetoric. In May 1954 Dulles told a news conference that "the United States should not stand passively by and see the extension of communism by any means into Southeast Asia."[15] The words *by any means* sounded sinister and suggested that Dulles would not permit a communist coup, but they really meant that he would not permit a Ho Chi Minh victory at the polls, even though 80 per cent of the population might want him.

If Secretary Dulles dominated reporting on many remote events, it was partly because of a thinness— sometimes a vacuum—in available information. Dulles often seemed determined to keep it that way—especially with respect to mainland China.

The Chinese civil war had halted organized reporting from China. But when, in August 1956, the communist regime of Mao Tse-tung announced a willingness to admit American newsmen, in return for United States willingness to admit an equal number of Chinese newsmen, Secretary Dulles flatly refused. Severely criticized, he shifted his position: he would permit the arrangement, but the United States would have to approve each Chinese reporter individually. He intimated, at the same time, that it could not admit communist reporters. This ended the negotiation. In the United States, news and opinion about China continued to rest heavily on items fed by the State Department plus reports from other countries, including those of Hong Kong "China watchers"—an absurdity comparable to reporting United States affairs from an observation post in Tijuana, Mexico.

Three reporters, in defiance of the State Department ban, nevertheless

14. Eisenhower, *Mandate for Change,* p. 449.
15. Press and radio news conference, May 11, 1954. Dulles, *Papers.*

went to China. One was William Worthy of the Baltimore *Afro-American*, and CBS took advantage of his presence in China by broadcasting a short-wave report by Worthy from Peking. According to Worthy, Under Secretary of State Robert Murphy then telephoned William Paley of CBS and persuaded him to carry no further broadcasts of this sort. CBS not only complied; it also killed a 5-minute discussion on the subject by Eric Sevareid, criticizing the State Department policy—the first time in seventeen years that a Sevareid program had been vetoed in entirety. The episode angered Edward R. Murrow, who spoke on the subject on his own radio news series—and was rebuked by CBS. The industry, except for newsmen, seemed to take the issue calmly. *Broadcasting* considered it purely a CBS matter, to be settled "within the family."[16]

Murrow thought otherwise. When Worthy's passport was revoked on his return and he took the matter to court, Murrow drafted a brief in support of Worthy, stressing his rights as a citizen, but also stressing the public's right to be informed. Again he felt that CBS, which had carried the Peking broadcast, should itself be fighting the case. Murrow wrote in his brief:

> Democratic governments cannot survive unless the electorate has at its disposal all available information on which to reach intelligent conclusions about the future of our country, and indeed our world. A leadership responsible only to an uninformed, or partially informed electorate, can bring nothing but disaster to our world.[17]

Such ideas could be argued in court but, for the moment, prime time was guarded against them. It was the Dulles era.

Only in later years would the extraordinary reach of the Dulles team into channels of communication begin to be revealed. When the two surviving newsreels, Fox Movietone and the Hearst-MGM reel, faced extinction, a secret government subsidy was arranged—under the code name "Kingfish"—which provided that propaganda items prepared under USIA direction would be inserted in foreign editions of the reels.[18] The secret subsidy helped the newsreels remain alive. Secret CIA subsidies were also used to infiltrate labor unions, youth groups, foundations, and to influence the content of books and magazines. Apparently no medium was immune.

16. *Broadcasting*, February 11, 18, 25, 1957.
17. Kendrick, *Prime Time*, pp. 417–18.
18. Sorensen, *The Word War*, p. 65; *Variety*, May 7, 1969.

In 1955 the New York writer Philip Reisman, Jr., who had been script editor of the television series *Man Against Crime, I Spy,* and *The Hunter,*[19] and had written the feature film *Special Delivery*—dealing with international intrigue and shot in Europe—received a phone call asking if he would like to work on a film in Mexico. He said he was busy but the caller said it was urgent and asked him to come to a room at the Hotel Roosevelt to discuss the matter. In the room he found several people seated in a semicircle facing an empty chair—for him.

One man said, "I suppose you know who we are." Reisman did not. The man, before starting his explanation, requested Reisman not to reveal for at least ten years what would be said. He agreed.

The men explained they were from the Central Intelligence Agency, and were tackling a problem relating to Latin America. They said the Soviet Union, through financial help to producers, was influencing the content of films produced in Mexico; a number of anti-American films had appeared. The CIA proposed to meet this threat in kind. It was arranging through an intermediary to finance an independent Spanish-language feature. The independent producer did not know the source of the funds, but his associate producer—a CIA agent—did. The CIA now wanted to select the writer; he and his associate producer would have to carry out the plan.

After further meetings, investigations, briefings—in New York and Washington—the arrangements were concluded.

> So the next thing I knew I was living at the home of the CIA agent, in a beautiful residential area of Mexico City. I had a guest house all to myself, and I was a little bit startled to discover, when I went to bed, there was a large lump under the mattress, which I learned was a loaded Luger pistol.

In the main house were other concealed weapons. The house was surrounded by a stone wall topped with broken glass. Plans were discussed in the middle of the garden or in a moving car. Reisman was given a cover name. Money conferences were held with a man "from the embassy," who was picked up on street corners and taken for car rides. They completed the film, a comedy in which Americans—it was ingenious strategy—were treated with satiric humor but the ultimate villainy came from Russians. Titled *Asesinos, S. A. de V. P.* (Assassins, Inc.), the film

19. *I Spy* was a series featuring Raymond Massey; it was unrelated to a later series of the same title. *The Hunter* was also a spy series, in which each program depicted a United States agent at work in an iron-curtain country.

was finished before the expiration of Reisman's contract, so the CIA agent suggested that Reisman accompany him to Guatemala—supposedly to scout forest locations for another film. They did so, but the scouting appeared to be a cover story: the agent really wanted a chance to disinter money buried for the overthrow of Arbenz but not needed for that purpose.[20]

The Dulles team felt no noticeable qualms over activities of this sort. The United States was not "really" at peace, wrote Allen Dulles later, since "Communism declared its own war on our system of government and life." Thus the rules of war applied and it was necessary to "mobilize our assets and apply them vigorously." He spoke of a Moscow "orchestra of subversion" which included secret subsidies to communication media. It seemed to him axiomatic that this and other threats must be met by an American orchestra of subversion.[21]

John Foster Dulles added the sanctifying note. It was a "basic fact," he told one of his first press conferences as Secretary of State, that Soviet leaders "do not admit the existence of such a thing as moral law."[22] Most Americans—especially those relying on television and radio for news—had little reason to doubt that American foreign policy was in moral hands. Even liberals troubled about McCarthyism at home were satisfied to leave foreign affairs to Dulles. Only later did it appear that Dulles policy was—as Archibald MacLeish came to see it—McCarthyism on a global scale.[23] To Dulles, as to McCarthy, anti-communism was in itself a policy. Dulles, like McCarthy, demanded that everyone—every nation—stand up and be counted. It was a holy crusade.

The Dulles moral and religious rhetoric won ready response. "I am grateful," said a 1954 fan letter from Vermont, "that our country has a true Christian in a top government post." The writer applauded him for using "Christ's way of settling disputes and dealing with other countries." In 1955 an Indiana writer told him, "God is trusting you to help redeem the world." An Idaho mother wrote:

> My small son sits now with a globe in his hands and says, "Mama, what is an iron curtain?" May God give men such as you the strength and insight to push back that curtain before my son is old enough to understand what it is.[24]

20. Reisman, *Interview*, pp. 48–57.
21. Dulles, *The Craft of Intelligence*, pp. 54, 230–31.
22. Press statement, April 3, 1953. Dulles, *Papers*.
23. New York *Times*, January 21, 1967.
24. Correspondence files. Dulles, *Papers*.

Dulles even precipitated a song sung by Carol Burnett on a Jack Paar program over NBC-TV, "I Made a Fool of Myself Over John Foster Dulles." NBC officials were afraid it might be considered in bad taste but were relieved to learn that the Secretary of State had found it "marvelous" and hoped for a repeat.[25]

The religious note seems to have been an element in his closeness to other world leaders. Speaking of Dulles, Chancellor Konrad Adenauer of West Germany said: "Our joint goal was the fight for freedom, and defense against atheist communism. That was the basis of our entire policy."[26] And when diplomat George V. Allen at a small gathering made a deprecating remark about Chiang Kai-shek of Taiwan and Syngman Rhee of South Korea, suggesting they were less than democratic, Dulles turned on him sharply. Those men, said Dulles, were Christian gentlemen who deserved to rank among early leaders of the Church.[27]

But where was the crusade heading? What did Dulles hope to achieve? His television and radio aide David Waters, a total believer, was convinced that Dulles foresaw an early collapse of "communism," to be achieved mainly through propaganda pressure.[28] But in Cabinet meetings Dulles insisted on large military appropriations to add to the "stresses and strains." When Secretary of the Treasury George Humphrey called for cuts in military budgets, Dulles would protest—as he did during reports of turmoil in Moscow after the death of Stalin—that "we ought to be doubling our bets, not reducing them. . . . This is the time to *crowd* the enemy—and maybe *finish* him, once and for all."[29]

Some critics said that the talk of liberating the Russian people or satellite nations meant the United States was ready to go to war with their governments—or else it meant nothing. This became the issue in a 1956 crisis involving Hungary—and Radio Free Europe.

THROUGH THE CURTAIN

By 1956 Radio Free Europe employed about 2000 people, including some 600 German technicians. In Munich 150 Americans were in key positions, but the all-important programming staff consisted of about 450 Poles, Czechs, Slovaks, Hungarians, Romanians, and Bulgarians.[1]

25. Continuity Acceptance memorandum, NBC, August 27, 1957.
26. Drummond and Coblentz, *Duel at the Brink,* p. 43.
27. Allen, *Interview,* p. 23.
28. Waters, *Interview,* p. 21.
29. Hughes, *The Ordeal of Power,* p. 137.
1. Broadcasting, April 9, 1956.

They were united in their dislike of communism, but otherwise diverse. They included liberals, royalists, disaffected socialists. Some were suspected of former nazi or fascist affiliations. There were noblemen who had lost lands and privileges; they provided a courtly hand-kissing atmosphere. Political differences were accentuated by certain ethnic rivalries—as between rural Slovaks and the more cosmopolitan Czechs. Such differences caused administrative difficulties but on the whole were submerged in the over-all purpose—undermining Soviet power.

The American policy of hiring disaffected exiles involved risks. *Exile* became a profession rather than a status. The disaffected acquired a vested interest in disaffection. Their presence on the air, on powerful transmissions, encouraged other discontented to seek security in anti-communist careers. The migrations were accepted as signs of success and a vindication of purpose.

The exiles brought memories that inevitably receded. Their understanding of home conflicts became less certain and, in some cases, more doctrinaire. Their basic views—now a source of livelihood—sought support from fragments of news. When reinforcement was provided by new refugees, they were inevitably like-minded. The staff suffered from isolation.

Funds for the Munich operation—about $3,000,000 a year— came from New York. According to the American administrator in Munich, he and others assumed this was mainly government money, "but we were not supposed to talk about that." He was at times directed to add to the Munich payroll a person elsewhere in Europe. When asked what function should be indicated on the payroll records, he was usually told "writer" or "research." Some sent research reports, while some sent nothing.[2]

In 1956 morale in the organization was high. Unrest in the communist countries had been rising since the death of Stalin. In February Nikita Khrushchev, one of the group that had taken control of the Soviet government, made his world-shaking three-hour attack on the dead dictator. Radio Free Europe and Radio Liberation—as well as the Voice of America—had the satisfaction of broadcasting to the communist countries the text of his denunciation. A process of de-Stalinization was under way. To many at Radio Free Europe it augured the freedom of their homelands.

2. Interview, Sumner Glimcher.

The Republican Party platform that year proclaimed: "We are going to continue our efforts to liberate the captive countries." The more cautious Democrats said: "We look forward to the day when the liberties of all captive nations will be restored to them. . . ." Radio Free Europe quoted both statements in numerous transmissions.

Its writers were sometimes cautioned not to create "exaggerated hopes of Western intervention." On the other hand, the policy handbook instructed them to foster "hope of eventual liberation through a convincing display of the skill, resources, and military strength of the West."[3]

The exiles watched with growing anticipation as a process of liberalization moved through the communist countries. Perhaps their own work had had something to do with it. Yugoslavia, long at odds with Stalin, won major concessions from his successors. Poland won concessions, and these in turn encouraged Hungarian agitation for liberal reforms.

At this point the Soviet leaders appear to have become alarmed that de-Stalinization was getting out of hand, and Soviet policy became more repressive. In response a spirit of rebellion erupted in Hungary. In October 1956 rebel groups seized radio stations and began to broadcast demands. Street fighting broke out. The defiance brought rumors of more drastic Russian intervention.

On October 27 a Hungarian commentator on Radio Free Europe, speaking under the name "Colonel Bell," implied that if resistance fighters succeeded in establishing a "central military command," there would be foreign assistance. The following day he told stories of how Yugoslav partisans, during World War II, had held off larger forces of nazis. On October 30 another Hungarian commentator gave instructions in anti-tank warfare, again with examples from World War II.

On November 4 another read quotations from the British newspaper *The Observer*—such quotations were a favorite technique—and added comments of his own. He read this quotation:[4]

> "If the Soviet troops really attack Hungary, if our expectations should hold true and Hungarians hold out for three or four days, then the pressure upon the government of the United States to send military help to the freedom fighters will become irresistible. . . ."

3. *Radio Free Europe Policy Handbook* (unpaginated).
4. This and following script excerpts are from Michie, *Voices Through the Iron Curtain*, pp. 233–66. Michie was a Radio Free Europe staff member in Munich. The book appeared in 1963—seven years after the events.

At this point he spoke these words of his own:

> This is what *The Observer* writes in today's number. The paper observes that the American Congress cannot vote for war as long as the presidential elections have not been held.

He then resumed reading, quoting *The Observer:*

> "If the Hungarians can continue to fight until Wednesday, we shall be closer to a world war than at any time since 1938. . . ."

He concluded, again with words of his own:

> The reports from London, Paris, the United States and other Western reports show that the world's reaction to Hungarian events surpasses every imagination. In the Western capitals a practical manifestation of Western sympathy is expected at any hour.

During these days "freedom stations" erupted in various parts of Hungary. They included amateur stations, some broadcasting in Morse code. At 12:34 p.m. on November 4, one station broadcast:

> Special appeal to Radio Free Europe. Early this morning Soviet troops launched a general attack on Hungary. We are requesting you to send immediate military aid in the form of parachute troops over the Trans-Danubian provinces. S O S. Save our souls.

At 1:32 p.m. a "Radio Free Vac" broadcast:

> Attention, Radio Free Europe, attention! We request immediate information. Is help coming from the West?

At 2:35 p.m. "Radio Free Rakoczi" was saying:

> Urgent! Urgent! . . . We are breaking off, for we are in immediate danger! We ask urgently for immediate help. Free Europe!

At 3:30 p.m. "Radio Free Csokonay" asked Radio Free Europe to relay an appeal to the United Nations. During the next few days some freedom stations shifted wave length repeatedly, while others vanished. On November 7 "Radio Free Rakoczi" was still on the air, but its tone was increasingly desperate. At 9:35 a.m.:

> In the name of all honest Hungarians we appeal to all honest men in the world. Must we appeal once again? Do you love liberty? . . .

On November 8 a last, almost unintelligible voice was heard: "Do not give up your arms." It was the end of freedom stations and the revolt. It came as Republicans celebrated the sweeping re-election of President Eisenhower.

The Hungarian tragedy brought anguish at Radio Free Europe. On November 9 *Freies Wort*, a West German publication, denounced the organization for its "opportunistic agitation" and blamed it to a large extent for the "bloodbath." This view was echoed in other German newspapers and a few American ones—mainly the German-language press. Chancellor Kurt Adenauer, whose consent had permitted Radio Free Europe and Radio Liberation to come into existence, appointed a review committee, which quickly concluded that only a few broadcasts had exceeded proper bounds.

Alan Michie of Radio Free Europe later published the information that of 1007 Hungarian refugees who were asked whether they had expected aid, 96 per cent answered yes. Their answers showed, according to Michie, that broadcasts had played a large part in creating and building the expectation.[5]

Ironically, communist statements denouncing Radio Free Europe as a CIA creation, though discounted elsewhere, had clearly found credence in Hungary. Accepting Radio Free Europe as an official voice of the United States government, Hungarians were all the more certain that help would come.

Radio Free Europe executives were quick to note that, judging by events, Radio Free Europe reached and influenced an audience. It became an argument for continuing, but both at Radio Free Europe and Radio Liberation, policy controls were tightened. In 1957 weekly "directives" from New York replaced the previous occasional "guidances."[6] The name Radio Liberation was later changed to Radio Liberty, which sounded less activist. Talk of liberation was avoided as the stations assumed the role of liberalizing influence. They stressed stories of bureaucratic failures and of news suppressions in the communist countries.

It was a game that others could also play. An East German station, Radio Berlin International, directed some of its programs to American Negro soldiers stationed in Germany. A girl—apparently a black American—broadcast reports of United States racial conflicts, scarcely mentioned, or ignored, on American government stations, including AFRTS stations.[7] Material for such broadcasts was in ample supply during the late 1950's: Southern governors defying the U. S. Supreme Court, black children turned away from school doors amid threatening screams from

5. Michie, *Voices Through the Iron Curtain*, p. 266.
6. Holt, *Radio Free Europe*, p. xi.
7. Mahoney, *Broadcasting Communism*, p. 15.

white mothers, Negro churches burned, Negro rights demonstrators driven back with fire hoses and dogs. Such broadcasts could likewise claim to be beaming truths through the iron curtain, and could assume the role of liberalizing influence, putting American stations under pressure to take cognizance of painful events they preferred to ignore.

This game required each side to concentrate on sins and failures of the other and to remain reasonably complacent concerning its own. Each managed to do this.

The American public was scarcely aware of the distant war of words. It had no inkling of what was being said on either side. They knew Radio Free Europe mainly through advertisements and spot announcements of the Advertising Council, appealing for "truth dollars."

But the international role of American government stations, whether acknowledged or unacknowledged, was beginning to be overshadowed by something else—of larger propaganda significance, although not so planned. On several continents the export of American telefilms was reaching a breakthrough stage.

BONANZA WORLD

The start of commercial television in Britain late in 1955 was the turning point, opening a crucial market for American advertisers and their agents. That they were ready to leap in was not surprising, for they had worked hard to bring this transition about.

Before television many American advertisers had promoted British sales via powerful Radio Luxembourg, which provided efficient coverage of the United Kingdom. The shrinking of radio audiences threatened these advertisers with shrinking British sales—unless alternative television facilities became available. Television coverage from locations like Luxembourg was not feasible, because television, unlike radio, was limited to a line-of-sight range. The strategy for achieving commercialization of British television is said to have been master-minded by the London branch of the J. Walter Thompson advertising agency, working with a group in Parliament. Their well-financed campaign did not emphasize commercial advantages but shrewdly attacked the British Broadcasting Corporation at its most objectionable point—its monopoly status.[1]

Commercial television brought to Britain the same sort of explosion

1. The campaign is described in Wilson, *Pressure Group.*

that had taken place in the United States. A television license, said the enthusiastic Roy Thomson, the Canadian who in 1956 won the television franchise for Scotland, is "like having a license to print your own money!" His franchise soon won him a fortune which he parlayed into worldwide holdings in communication media, including television stations in such places as Aden, Australia, Ethiopia, Mauritius, Sierra Leone, and Trinidad.[2]

British commercial television, shaking up economics and customs at home, also had reverberating effects elsewhere, as other nations followed the British example. All instantly became purchasers of American telefilms. They also became outlets for advertising of many American companies and their subsidiaries.

The advertising agency Foote, Cone & Belding, studying foreign markets for its clients early in 1958, reported that commercial television was already in operation in twenty-six countries and in the planning stage in others.[3] Television systems accepting advertising included both government and private systems. At all these systems, salesmen of telefilms were converging. They were reporting sales in clusters—seven series here, five series there. Some countries with non-commercial systems, such as Denmark, the Netherlands, and Sweden, also began buying American telefilms.

To translate telefilms, dubbing operations were springing up far and wide. Practices evolved through trial and error. Screen Gems found that films dubbed into Spanish in Madrid were considered unusable in Latin America, the chief Spanish-language market. Thus Mexico and Cuba became the main centers for dubbing into Spanish. Similarly, Brazil became headquarters for dubbing into Portuguese. On the other hand, dubbing into French was done in France; powerful French unions made this necessary, and the films so dubbed proved usable in French Canada and other French-speaking areas. Egypt and Lebanon became preferred sites for Arab dubbing, Hong Kong for Chinese dubbing. Japanese versions were prepared in Japan, Italian versions in Italy—where dubbing

2. Braddon, *Roy Thomson of Fleet Street*, pp. 200–214, 326–33. Not being a United States citizen, Thomson could not own stations in the United States, but he did purchase newspapers in many American cities, including Austin, Texas.

3. In operation: Australia, Bermuda, Brazil, Canada, Colombia, Cuba, Dominican Republic, El Salvador, Finland, France, Guatemala, Hong Kong, Italy, Japan, Mexico, Nicaragua, Panama, Philippines, Portugal, South Korea, Thailand, United Kingdom, United States, Uruguay, Venezuela, West Germany. For the Foote, Cone & Belding report, see *Sponsor*, April 5, 1958.

had long been a skilled specialty in the feature-film field. A number of smaller countries used English-language versions, in some cases with subtitles.

For the dubbing procedure a film was cut into innumerable short sequences. A 10-second sequence, spliced to form a loop, would be projected so that an actor would see it over and over and over without pause. After memorizing the rhythm of the lip movements he would record the translation, written to fit the original. In the late 1950's this time-consuming operation—which employed many former radio actors—cost $1200 to $1400 per half-hour program in most parts of Europe, much less in Asian or Latin American countries. But costs soon began to rise.[4]

While some comedy series, like *I Love Lucy* and *Father Knows Best,* were international successes, series emphasizing action rather than dialogue were far more translatable. Climactic gun battles or fist fights, following a chase up a fire escape or down a canyon, needed no "looping" and conveyed their meaning readily in any city or hamlet on the globe. The networks and major film companies, who soon dominated the foreign distribution, acquired an extra incentive toward production of action series—which in any case stood high in United States ratings. By 1958–59 a television writer could scarcely find a market for any other kind of material. That winter's *Television Market List,* issued for its members by WGA-w, listed 103 series, of which 69 were in the action-crime-mystery category.[5] They were peopled by cowboys, policemen, and detectives whose terse words would issue in many languages from the screens of the world.

4. *Broadcasting,* May 6, 1957.
5. *Adventures in Paradise, Alfred Hitchcock Presents, Bat Masterson, Black Saddle, Bold Venture, Border Patrol, Bourbon Street Beat, Buckskin, Cannonball, Cheyenne, Cimarron City, Colt 45, Flight, Fury, Gunsmoke, Have Gun—Will Travel, Hawaiian Eye, Heave Ho Harrigan, Highway Patrol, Jefferson Drum, Laramie, Lassie, Lawman, Lineup, M Squad, Mackenzie's Raiders, Man Without a Gun, Markham, Maverick, Mike Hammer, One Step Beyond, Perry Mason, Peter Gunn, Philip Marlow, Rawhide, Rescue 8, Restless Gun, Richard Diamond—Private Detective, Riverboat, Rough Riders, Sea Hunt, 77 Sunset Strip, State Trooper, Steve Canyon, Sugarfoot, Tales of Wells Fargo, The Alaskans, The Blue Men, The D. A.'s Man, The Enforcer, The Rebel, The Rifleman, The Texan, The Thin Man, The Troubleshooters, The Web, Tombstone Territory, Trackdown, 26 Men, Twilight Zone, U. S. Marshal, Wagon Train, Wanted—Dead or Alive, Whirlybirds, World of Giants, Wyatt Earp, Yancy Derringer, Zane Grey Theater, Zorro.* Other series included surviving anthology series using similar material, and children's series like *Mickey Mouse Club,* similar in many ways. Contrast was provided by two religious series—*This Is the Answer* and *This Is the Life,* both seen in fringe periods—and a number of family comedy series, such as *Dennis the Menace, Father Knows Best, Leave It To Beaver,* etc.

The worldwide explosion seemed to Hollywood a show-business phenomenon, but it was more. To every nation, along with telefilms and their salesmen, came advertising agencies—often branches of American agencies. (By 1958 J. Walter Thompson had 34 branches abroad, including 8 in South America, 8 in Asia, 5 in Africa.) And along with them came new advertisers—often companies affiliated with or owned by American companies. Television was merely the highly visible crest of a huge economic surge.

There were reasons why American companies, at this moment of history, turned by thousands to investments abroad. In the period after World War II they had made substantial foreign sales—by-products, to some extent, of the Marshall Plan and other aid programs. Some of the earnings were in "blocked" funds and could not be transferred to the United States. Even when they could, there were tax advantages in investing them abroad. Labor costs, low compared with pay scales in the highly unionized United States, were an additional incentive. Another was the attraction of markets with huge growth possibilities; the United States was, in comparison, product-saturated.

Beyond these incentives was another factor. The foreign policy of Secretary of State Dulles seemed to promise the American investor protection abroad, by force if necessary. The risks of foreign investment seemed to decline.

The investment trend, once launched, created earnings which led to further investments, in a growing tangle of relationships. In the television and film fields, the ramifications were endless. Some of the companies marketing telefilms also sold receivers and transmitters; some sold consultant services; some invested in foreign stations, production companies, dubbing services, animation studios, theaters.

To lobby for their interests throughout the world, the major motion picture companies in 1959 formed a television division in the Motion Picture Export Association. The following year the networks formed the Television Program Export Association to serve similar purposes. The moves reflected rising expectations.

These moves by media men were possible only because manufacturers in many fields were making similar moves. The launching of commercial television in a new market was often done by a consortium of interests—a group of set manufacturers, advertisers, and program distributors could virtually guarantee success.

If newly developing nations yielded generous franchises to such groups, they were moved by enticing vistas. Entrance into the television age had symbolic values. In addition, studies from UNESCO and others offered heady visions of what television might do for a developing nation. From one television studio, it was said, classrooms of a whole nation could learn physics or chemistry. No longer would a school have to rely on its own pitiful supply of laboratory gadgets; such inadequate methods could be left behind in one leap into the television age. It was also said that adults could, in similar fashion, have instruction in scientific farming, soil conservation, family planning. There was also talk about cultural exchange. Their television stations would show American films, and by and by their own films would be shown in America. And all of this would cost almost nothing; advertising would pay for it, as in the United States.

Using such arguments as these Robert E. Button, deputy director of the Voice of America, made a television-promotion tour in 1956 that took him to Indonesia, Iran, Lebanon, Pakistan, Taiwan, and other countries. He came back in a state of heady optimism. "If I ever saw anything that would lick the communists on their own front, this is it," he said. "Talk about jumping from camel to jet plane, this is jumping from papyrus scroll to television." Aside from the explosion of private enterprise involved in the process, he saw the stations providing added exposure for USIA films.[6]

American aid programs gave impetus to television in a number of developing countries. Under the Marshall Plan as launched by Truman, aid was without strings. Under Eisenhower all aid came to have strings. Some aid was in the form of loans; in other cases, aid funds had to be used for equipment or materials or services purchased in the United States. In many cases military aid had to be accepted along with other aid. This tended to fortify the existing regimes in aided countries.

The various forms of "tied aid" made for continuing American involvement. Harvesters, tractors, transmitters, office equipment, cars, planes, machine guns all required spare parts—purchasable from the United States. In addition, transmitters needed telefilms; office machines needed paper; machine guns needed ammunition. Much "aid" was thus, in effect, a subsidy of American exports, and tended to create a continuing dependence on the part of the aided country.

6. New York *Times*, February 26, 1956.

In the mid-1950's, television, like missionary expeditions of another era, seemed to serve as an advance herald of empire. Implicit in its arrival was a web of relationships involving cultural, economic, and military aspects, and forming the basis for a new kind of empire.

All this was not entirely unplanned.

The United States had never fully emerged from the Depression of the 1930's. Although a variety of legislation had stemmed the business collapse, the basic economic problems had not been solved; instead, they had been masked by a war economy. Afterwards the cold war, by maintaining a wartime rate of military expenditure, had helped prevent a major recession. With large numbers of men in uniform, unemployment was kept at levels considered manageable, although still serious. But underlying problems remained unsolved.

One proposed solution, favored among others by Clark Clifford—adviser to several Presidents and at other times to RCA, Du Pont, and other large corporations—was an expansion of the American economy via worldwide investments and markets, safeguarded by ample military expenditures. Only in this way, it was argued, could the slack in the American economy be taken up.

Such a blueprint for empire was becoming a reality in the 1950's, with television spearheading the process.

Inevitably the television invasion brought strains of many sorts. Reports from telefilm salesmen usually reflected an enthusiastic reception of all that was offered. Sales figures suggested complete acceptance of the avalanche of action films. But this picture was incomplete.

The telefilms that distributors—CBS, NBC, MCA, Screen Gems, and others—sold to Australia in 1956 had already earned back their production cost—in some cases, several times the production cost. Any price paid by the Australians would be profit. And it would further profit American companies to get the Australians started. Prices were therefore set at a most attractive level. "We gave them some series," said John McCarthy of the Television Program Export Association at a UNESCO conference in New York, "for as little as a thousand dollars for a one-hour program, for all of Australia."[7]

The price scale was unquestionably helpful to Australians; television station managers and advertisers were grateful. After a few years the

7. Conference on "The Free Flow of Ideas," Carnegie International Center, October 6, 1966.

price edged up to $3000 per one-hour episode, $1500 per half-hour episode—still reasonable.

The same sequence of events was experienced quite differently by writers, directors, and producers in Australia's small film industry. Inspired by the advent of television, they began—American-fashion—to make proposals to Australian sponsors. A series of half-hour films could be produced for $20,000 per episode, they explained—if every possible economy was used. Magnificent themes from Australian history and present-day life were available. But the sponsor—perhaps a subsidiary of an American company—received such offers with comments like: "Look—I can get *Restless Gun* for $1400 an episode. It had a Videodex rating of 31.1 in the United States. According to *Sponsor* magazine, it had a CPMHPCM rating of $2.34. That's some CPMHPCM rating, isn't it? Now what can you offer?"[8] The Australian film maker could offer ideas, perhaps talent, but certainly not a bargain certified by Videodex and CPMHPCM ratings.

In Canada, with its brilliant record of film production, the situation was similar and perhaps even more tragic. In the late 1950's American telefilm series were being sold in Canada for $2000 per half-hour episode—for Canada-wide rights. No Canadian film maker could quote a price lower than ten times that sum. In Canada, as in Australia, sponsors moved into prime-time periods with telefilms of the American West, the highway police, and struggles against gangsters and spies. The writer-director Henry A. Comor, president of the Association of Canadian Television and Radio Artists, told a New York forum of the Academy of Television Arts and Sciences: "You've made it impossible for us to earn a living." He added with a studied casualness: "By the way, my young son thinks he lives in the United States."[9]

Both Australia and Canada introduced quota systems to limit foreign programming and protect home programming. But native-content rules could be taken care of with football matches, round-table discussions, cooking lessons—rather than by expensive drama. Film drama tended to become largely a United States preserve, taking the best hours. Before long the situation was taken for granted.

8. *Sponsor* magazine listed CPMHPCM (cost-per-thousand-homes-per-commercial-minute) ratings as a basis for comparing program costs. Videodex ratings were based on viewing diaries maintained by a fixed sample of homes, via arrangements made by the rating organization.
9. Forum on "American TV: What Have You Done to Us?," October 6, 1966.

The prices that were winning for American telefilms a dominant position in half the world were profitable prices because of the large economic base from which they came. But to foreign producers they seemed "cut-throat" prices, stifling possible competition. Britain, with a very stringent quota system—only 14 per cent of British schedules could originate outside the British Commonwealth—had some success in holding back the television invasion and even in finding foreign markets for its telefilms. Japan later achieved a similar position.

In Toronto in the early 1950's a brilliant group of television artists was at work—Lorne Greene, Norman Jewison, Harry Rasky, Reuven Frank, Christopher Plummer, Art Hiller, and others. Like similar groups elsewhere, they knew the exhilaration of discovery, and saw talents recognized. But the waves of telefilms swept around them. To escape being stranded, they migrated to Hollywood and New York. Soon Art Hiller learned to direct programs like *Gunsmoke,* Reuven Frank and Harry Rasky were creating New York documentaries, and Shakespearian actor Lorne Greene became the cowboy-patriarch of *Bonanza,* soon seen in prime time around the world along with *Cheyenne, Wyatt Earp, Hawaiian Eye, 77 Sunset Strip.*

If artists in many lands grew restive, so did educators. The vision of a country united by a television schoolroom was replaced by a different reality. Instead of scientific farming, a mythology of violent struggle riveted attention, followed by cola drinks, cigarettes, headache tablets, soaps, laxatives, hair tonics, deodorants.

In every country American television tended to create a division. It won enthusiastic adherents among station entrepreneurs, advertisers, advertising agencies, distributors, merchandisers, retail outlets—the whole complex of modern distribution—and captured large segments of audience. On the other side were artists, teachers, social workers. They tended to feel a culture was being wrenched loose from its moorings. There were also isolated local businessmen, facing a razzle-dazzle competition linked with overseas enterprise of unimaginable resources.

But their fears and objections were not very audible, even at home. For one thing, they were not on television. And their objections, which seemed petulant and small-spirited, were waved aside by others. These films, were they not merely entertainment?

The divisions abroad were counterparts of those at home. The networks, surveying television interests that were becoming worldwide, were increasingly confident. As their telefilms flowed into ever-expanding markets, the rising prosperity was shared by others—stockholders, advertisers, advertising agencies, distributors, producers—who firmly supported the prevailing order.

It was in mid-1958 that William Paley felt he could get along without *See It Now,* and dropped it. Among series that served a safety-valve purpose, *See It Now* with its $90,000-per-program budget was by far the most expensive. In contrast, such a series as *Face the Nation*—a press-conference type program—was cheap, and often made news. In the confident atmosphere of 1958, *See It Now* seemed expendable.

John Crosby, television critic for the New York *Herald-Tribune,* commented:

> *See It Now* . . . is by every criterion television's most brilliant, most decorated, most imaginative, most courageous and most important program. The fact that CBS cannot afford it but can afford *Beat the Clock* is shocking.[10]

Edward R. Murrow, though still prosperously busy with *Person to Person,* was deeply troubled over the trend. He compared prime-time schedules with Nero and his fiddle, Chamberlain and his umbrella. He told a group of television and radio news directors: "If we go on as we are, history will take its revenge."[11]

But empire television continued to spread with irresistible momentum. In thatched huts and villas men watched cattle stampedes and gunfights, amid the clatter of hoofs and the ricochet of bullets. Precisely what it all meant to them, no one could be sure. Perhaps they had a sense of sharing a destiny with a breed of men who could make decisions and make them stick. Perhaps they felt the world was in good hands, that could not lose.

USIA, in charge of American propaganda abroad, was aware that it faced new problems. A 1958 memorandum projecting its future in the television era envisioned continued expansion in USIA's film output. But it was finding—as were educators in many countries—that "sponsored programs are beginning to edge cultural and informational programs aside." This trend, said the forecast, would challenge USIA's ingenuity.[12]

10. Crosby, *With Love and Loathing,* p. 184.
11. Quoted, *The Reporter,* November 13, 1958.
12. *A Forecast of USIA Television,* pp. 4–5.

There was also some concern about the "image" of violence dominating the commercial output.

American telefilms, radio transmitters, consumer goods, military bases —much of the world was living in an atmosphere shaped by them. The orientation of all this toward peace was assumed by most Americans and many allies.

To Russians and Chinese, hearing and seeing, the purpose often seemed different. Nikita Khrushchev, gradually emerging as the leading figure in the Soviet government, was telling Jawaharlal Nehru of India: "We feel like a besieged people."[13]

ENTER MR. K.

In 1957 Nikita Khrushchev, to the amazement of practically everyone, accepted an invitation to appear on the CBS-TV series *Face The Nation.* Allowing his Moscow office to be turned into a film studio tangled in cables and ablaze with lights, he answered questions from Daniel Schorr and Stuart Novins of CBS and B. J. Cutler of the New York *Herald-Tribune,* while CBS cameramen shot 5400 feet of film, and a Soviet crew made its own footage.

The rotund Khrushchev chose the role of genial persuader and, according to Roscoe Drummond, "played it like a Barrymore." His theme was "peaceful competition." America preferred capitalism and the Soviets preferred communism—very well, let them enter into peaceful competition. He and other Soviet leaders were confident of the result. The program went on the CBS-TV network on June 2 at 3:30 p.m. It was not prime time but a fringe period in the slack time of the year, but it was a first step, and caused a sensation. *Time* called it "the season's most extraordinary hour of broadcasting." President Eisenhower did not quite know what to make of it. He called it a "unique performance" but intimated that CBS was merely trying to improve its commercial standing. Murrow considered the President's remarks "ill-chosen, uninformed." CBS repeated the interview at night on its radio network.

Some congressmen were outraged that CBS had given a propaganda platform to the Soviet leader, but the majority—along with most of the press—acclaimed the network for its journalistic initiative. Senator Lyndon B. Johnson, Majority Leader of the Senate, said there should be reg-

13. Mehta, *Interview,* p. 7.

ular exchange programs of this sort, an "open curtain" in place of an iron curtain. Secretary John Foster Dulles, apparently taken aback by the applause greeting this suggestion, said he had been trying for eighteen months to promote something of this sort. Senator J. William Fulbright of Arkansas, a member of the Senate foreign relations committee, scoffed at this remark: if the State Department had been pressing Russia with an "open curtain" proposal for eighteen months, he said, "it is the best kept secret since the first atomic bomb was made." CBS, flushed with praise, arranged further interviews with unorthodox leaders, including Nehru of India. It was a fine idea, many people agreed, to be familiar with the attitudes of foreign leaders. President Frank Stanton was asked if CBS had plans to interview a leader of communist China. "We have no such plans," he said firmly.[1]

Khrushchev was meanwhile traveling. He turned up in various capitals, always available to reporters and cameramen. He became one of the most photographed of leaders, a familiar television figure around the world. From some angles he could look like Eisenhower, but in profile his sagging chin and figure gave a different impression. With somewhat baggy pants, outstretched hand, jovial manner, he looked more like a traveling salesman than a mysterious dictator. He was ready to put on funny hats for cameramen. He had a large supply of aphorisms. He displayed a ready familiarity with American affairs. Telling an American reporter that their conversation would probably lead to a subpoena from the House committee on un-American activities, he promised smilingly to vouch for the reporter's orthodox character. Ready for questions, Khrushchev slipped in his own messages. In good humor, he reminded questioners that the United States had sent troops to Russia in 1917 in an effort to stifle the Soviet Union at the hour of its birth, and that it currently had bases all around the Soviet Union without any similar provocation from her. But that, he seemed to say, was foolishness. War would solve nothing. Both nations were advanced and mighty and could learn from each other. They should compete and trade.

Competition—trade—they were honored words in the capitalist lexicon. Khrushchev was making them his own, along with *peace* and *negotiation.*

With neat timing, and underlining his apparent confidence, the Soviet Union on October 4 put the first man-made satellite in orbit around the earth. The world marveled and the United States was stunned.

In a matter of weeks Khrushchev seemed to have made a mockery of

1. *Broadcasting*, June 3, 10, July 8, 1957.

the Dulles depiction of the Soviet world. It did not seem on the point of collapse. It was not rigid. It exhibited a wish for peace and friendly competition.

In 1958 Khrushchev became Premier of the Soviet Union. Late that year John Foster Dulles became ill and began a series of checkups and treatments. The job of Secretary of State fell to Christian Herter, who was himself not in good health; foreign-policy problems fell increasingly on President Eisenhower and Vice President Nixon. Eisenhower was troubled; he wanted to think of himself as a man of peace. In January 1958 he approved a cultural exchange program, and the United States decided to stage an exhibit in Moscow in the summer of 1959. It would include color television, videotape, and other marvels of modern America. Vice President Richard Nixon would open the exhibit and would meet with Khrushchev. Nixon would be accompanied by George V. Allen, career diplomat who had become director of USIA.

Diplomacy, Allen told an audience before leaving for Moscow, used to be conducted by diplomats.

> They were sent abroad to various foreign countries, sometimes supplied with a pair of striped pants and a top hat. They dwelt in foreign capitals and dealt with a small group of people in the foreign office of that country; and that was the link between nations.

All that had changed, said Allen. A great nation now addressed people of other nations over the heads of their rulers, appealing to public opinion.[2]

The new diplomacy had for years used the radio medium but now, as Nixon and Allen left for Moscow, they were thinking especially of television—as was Khrushchev. Nixon and Allen hoped to create an event that would find a place on Soviet television. Khrushchev had his eye on American television.

The world was entering a period when the planning of television spectaculars, designed to seize the attention of nations, was becoming a central activity of rulers.

KITCHEN TALK

When Vice President Richard Nixon arrived in Moscow in July 1959 he was received by Khrushchev at his office and shown a cantaloupe-sized

2. Allen, Address, American Municipal Association Convention, December 1, 1958.

ball—a duplicate of the first man-made satellite, Sputnik I. Nixon expressed interest. When Khrushchev later visited the American exhibit, Nixon was the host.

To lead a cautious venture into intercultural relations, Nixon no doubt seemed a fine choice, politically speaking. A one-time member of the House committee on un-American activities, he had played a leading role in probes of Hollywood and Alger Hiss. As early as 1954, when the French position in Vietnam was imperiled, Nixon was already suggesting the possibility of "putting our boys in" to prevent communist rule.[1] Nixon could not be considered soft on communism.

At the exhibit, Nixon guided Khrushchev first to a building housing an RCA closed-circuit color television set-up and Ampex videotape equipment.[2]

They entered a room with an overhead gallery on four sides, crowded with Russian spectators and a few visiting Americans.

Below, as Nixon and Khrushchev entered, the color cameras were on them, and they could see themselves on a monitor. They were also being videotaped, so that they would immediately be able to see a replay—also in color—of their live performance. Nixon, well briefed, began an explanation of the marvels.

Khrushchev, seeing the cameras, the tape, the rapt attention of Russians in the gallery, clearly did not relish the role in which he was being cast—a mute country bumpkin listening to explanations from the centers of progress and civilization.

He interrupted. The Soviet Union had such marvels too, he said, and then began to discuss something he felt was of greater interest: America's foreign bases. Russians did not maintain bases around the world, so why should Americans? Nixon tried to steer the discussion back to the marvels of RCA color television, but Khrushchev plowed ahead, getting occasional applause from the gallery. He asked why America held back on trade with the Soviet Union. Why did it permit only limited cultural exchange? By way of emphasis, he wagged a finger at Nixon.

But now Nixon, too—with a sense of being trapped on videotape—saw the need for resolute action, and began jabbing a finger toward Khrushchev. Meaningful competition, said Nixon, required a free ex-

1. Schlesinger, *The Bitter Heritage,* p. 25.
2. The following is based on Hearst, Considine, and Conniff, *Khrushchev and the Russian Challenge,* pp. 167–82; other sources as noted.

change of ideas. "You must not be afraid of ideas. . . . You don't know everything." Nixon again brought the discussion back to color television and videotape, and recommended them as ideal instruments for an exchange of ideas. Khrushchev was dubious, saying Americans habitually suppressed things. Nixon rose to the challenge. He suggested that both countries televise the taped discussion in full, without cuts. Nixon was scarcely in a position to make such a commitment, but apparently felt sure it would be honored by American broadcasters.

Khrushchev grinned. "Shake!" Their right hands met in a resounding clasp.

So they were off on a running debate. The 16-minute taped harangue[3] in the television exhibit hall was followed by further tangles elsewhere, starting with the kitchen exhibit. They seem to have paused here because Tex McCrary, ex-broadcaster turned public relations specialist, had gathered a cluster of people to halt their progress. As a result, photography of the continuing debate showed products of his clients and also showed his wife Jinx Falkenburg, former Rheingold beauty contest winner and television personality. A public relations man had his moment on the world's stage.

The debate rose and fell. Each debater grew indignant and just as abruptly calmed down and smiled. When Khrushchev held the floor too long, Nixon called him a "filibusterer" and then had to explain the term. When Nixon seemed to launch on a major speech, Khrushchev turned to the crowd: "Who is filibustering now?" He appeared to have huge reserves of good humor—and information. Introduced to William Randolph Hearst, Jr., he said: "Ah, my capitalist friend. The American newspaper monopolist." Coming to a crowd of Russians, he asked Nixon if he would care to meet some "slaves." He asked them: "Are you captive people?" On such an occasion Nixon said: "You never miss a chance to make propaganda, do you?" Khrushchev answered: "I don't make propaganda. Just truth."

Later in the visit Nixon was given a chance to address people of the Soviet Union more formally on television and radio. Much of his talk was indistinguishable from a Khrushchev speech.

> NIXON: The American people and the Russian people are as one in their desire for peace . . . each of us is strong and respects the strength of the other. . . .

3. This material was donated by Ampex to the Library of Congress.

But that also meant, said Nixon, that neither could tolerate being "pushed around." He then attacked the Russian jamming of American broadcasts.

> NIXON: Let us put a stop to the jamming . . . so that the Soviet people may hear broadcasts from our country just as the American people can hear forty hours of broadcasts a day from the Soviet Union.

Reviewing the kitchen debate, USIA executives were not sure who had come out ahead, but they distributed its text to all USIA posts.[4] The videotaped exchange in the RCA-Ampex exhibit was telecast in the United States by all networks, and became the basis for Nixon's claim of "standing up" to Russian leaders. But the big show was not over. At the Moscow airport, inspecting an American plane, Khrushchev was asked by a reporter: "Would you like to fly to the United States in a plane like this?"

"This one or some other one."

"When?"

"When the time is ripe."

Within weeks the time was considered ripe, and American television prepared for its greatest spectacular to date: the American tour of Nikita Khrushchev.

But before it began, the television industry faced a painful crisis.

"SAY IT AIN'T SO"

During the rise of Khrushchev as television personality, the smoldering rumors of "fixed" quiz programs and other broadcasting irregularities blazed into scandal. The revelations were dismaying to the industry and the nation, and the timing was embarrassing.

There had long been hints and allegations of malpractice. Articles on the subject had appeared in *Look* and *Time*. In August 1958 Herbert Stempel, an early winner on *Twenty-One*, said the program was "fixed." At this time Charles Van Doren, most famous of the *Twenty-One* winners, was presiding over the *Today* series on NBC-TV as summer replacement for David Garroway. On the air, Van Doren said he knew of no irregularitities. But a New York grand jury began to look into the

4. Sorensen, *The Word War*, p. 110.

Denies . . .
October 14, 1959, quiz winner Charles Van Doren meets the press.

Wide World

Admits . . .
November 2, he reads statement to House subcommittee.

Wide World

'I would give almost anything I have . . ."

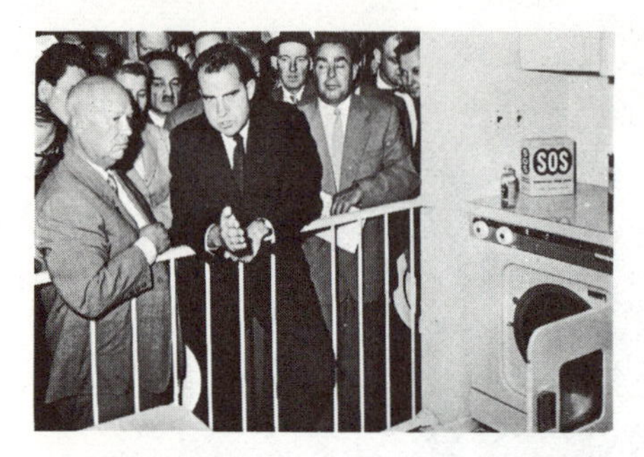

1959: Moscow, kitchen debate. Front row: Nikita Khrushchev, Richard M. Nixon, Leonid Brezhnev.

1959: Iowa, farm tour.

1960: *Open End* telecast.
David Susskind, Nikita Khrushchev, Victor Sukhodrev.

Wide World photos.

Willis Conover and Louis Armstrong on *Music USA*.

1959: Willis Conover visits jazz festivals in eastern Europe.

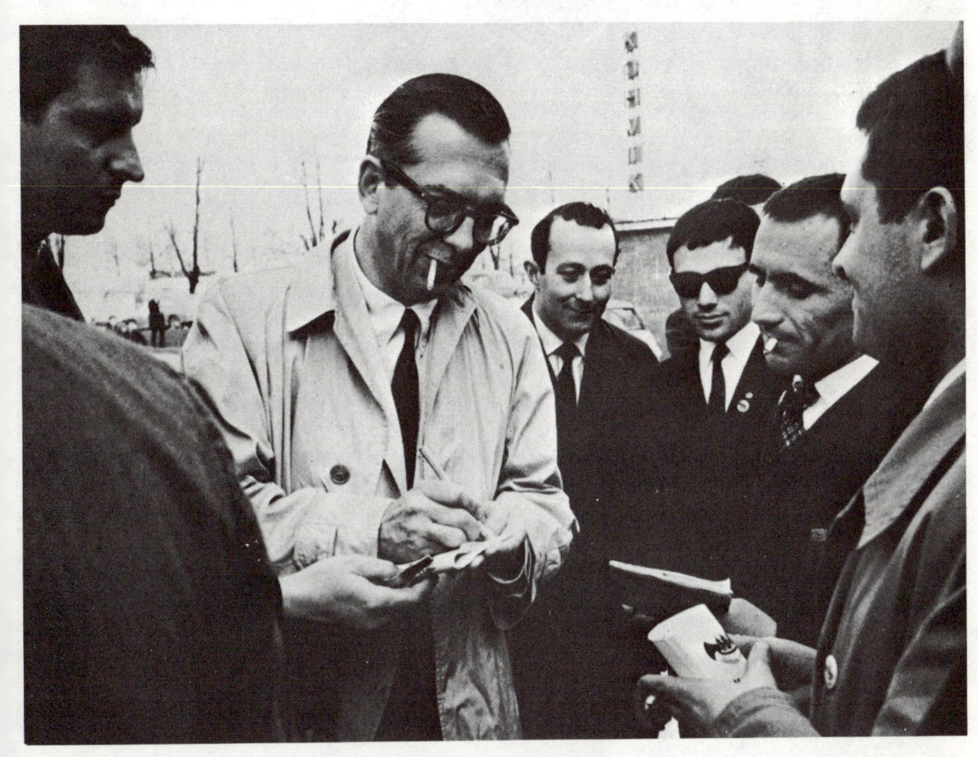

Voice of America booth at National Stadium, Tokyo. Left to right, Yoshihisa Maeda, Frank S. Baba, Haruyasu Nakayama.

U.S. Information Agency photos

William H. Birch

"Let's get America moving again . . ."

Left to right: CBS cameraman Don Norling; wire service still photographer; NBC cameraman William H. Birch — covering John F. Kennedy in 1960 Minnesota primary.

"I assume you two gentlemen know each other . . . ?"

First debate: Don Hewitt of CBS flanked by candidates John F. Kennedy and Richard M. Nixon — both without makeup.

CBS

Wide World

"A golden age of poetry and power . . ."

Inauguration, 1961: Robert Frost is troubled by the glare. At extreme right, Lyndon B. Johnson.

matter. In January 1959 Van Doren, along with many others, again gave assurance that all was on the level, but the grand jury found discrepancies in their statements. Meanwhile the House of Representatives special subcommittee on legislative oversight—under Representative Oren Harris—became interested and called witnesses.

Early that fall, amid charges and denials, Charles Van Doren dropped from sight, and then suddenly reappeared. Before the oversight subcommittee, in deep anguish, he read a long statement.

> I would give almost anything I have to reverse the course of my life in the last three years. . . . I have deceived my friends, and I had millions of them.

He recounted how he, being good at games, had been urged by friends to appear on a quiz program and had gone to the office of Barry & Enright, producers of *Tic Tac Dough* and other programs, to apply. He had passed a written examination and then another, much harder. He apparently did well, for he was told he would appear on the company's new night-time series, *Twenty-One*. This was regarded as a high honor. He began earnestly to memorize miscellaneous facts and attended several broadcasts as "stand-by contestant." But before his first actual appearance on the air he was asked by Albert Freedman, producer of the series, to come to his apartment.

> He took me into his bedroom where we could talk alone. He told me that Herbert Stempel, the current champion, was an "unbeatable" contestant because he knew too much. He said that Stempel was unpopular, and was defeating opponents right and left to the detriment of the program. He asked me if, as a favor to him, I would agree to make an arrangement whereby I would tie Stempel and thus increase the entertainment value of the program.
>
> I asked him to let me go on the program honestly, without receiving help. He said that was impossible. He told me that I would not have a chance to defeat Stempel because he was too knowledgeable. He also told me that the show was merely entertainment and that giving help to quiz contestants was a common practice and merely a part of show business. This of course was not true, but perhaps I wanted to believe him. He also stressed the fact that by appearing on a nationally televised program I would be doing a great service to the intellectual life, to teachers and to education in general, by increasing public respect for the work of the mind through my performances. In fact, I think I have done a disservice to all of them.

It became the practice for Freedman, before each program, to go over the scheduled questions with Van Doren and give him the answers. Van Doren found he knew the answers to some questions, not to others.

> A foolish sort of pride made me want to look up the answers when I could, and learn as much about the subject as possible.

Freedman was concerned about the manner of answering. He suggested —said the Van Doren statement—that he pause before some answers, skip parts and return to them, hesitate, build suspense. After the first program, on which he tied Stempel three times, Van Doren was told he would win the next week, and be the new champion. At this point he had won $20,000, but the earnings mounted rapidly. So did his fame. Thousands of letters poured in: invitations to lecture, write articles, appear in films. He almost persuaded himself he was helping "the national attitude to teachers, education, and the intellectual life." But at the same time the tensions within him mounted. He testified that after several months, which earned him $129,000, he pleaded for release. When finally informed that he would lose to Mrs. Vivienne Nearing—after a few ties, to build suspense—he described his own reaction as: "Thank God!"

The celebrity aura clung to him. He read poetry on the *Today* series, discussing it as he might in a freshman English class. At Columbia University he finished his Ph.D. and became assistant professor of English.

When Stempel's accusations were reported in the press, Van Doren was "horror-struck." Thousands of letters, including many from school children, were expressing confidence in him. "I could not bear to betray that faith and hope." While asserting his innocence on television and before the grand jury, he hired a lawyer—but did not tell him the truth. When the House oversight subcommittee began its probe, NBC demanded that Van Doren offer to testify, but he delayed. He was suddenly surrounded by harassments. Newsmen waited outside his door.

> Everywhere I went, there were people trying to interview me and flashing bulbs in my face. . . . My wife and I drove up to New England. I drove aimlessly from one town to another, trying to come to some conclusion.

When he finally decided to tell the truth, he began with his lawyer, who said, "God bless you."[1]

1. The full statement was in the New York *Times*, November 3, 1959.

Many quiz figures had apparently committed perjury before the grand jury. New York District Attorney Frank S. Hogan said later that of the 150 grand jury witnesses "maybe fifty" had told the truth.[2] Their statements had led to continued investigations.

The revelations ripped into other major quiz series. Production personnel of *The $64,000 Question* and *The $64,000 Challenge* told the oversight subcommittee that the sponsor, Revlon, had often instructed them which contestants should be disposed of, and which allowed to continue. If they were dull, they had to be eliminated. Details were left to the producers, but they could expect angry reprimand if they failed to follow instructions. Some contestants had frustrated executive decisions. The psychologist Joyce Brothers was considered dull and marked for extinction, but survived the questions meant to do the job.[3] Most contestants said they had accepted help gladly. Among these was the band-leader Xavier Cugat, who said, "I didn't want to make a fool of myself."

While Revlon suffered discomfiture, its rival, Coty, decided to run a series of advertisements about honesty in business, condemning the "shameful TV quiz hoax."[4]

Albert Freedman, who after early evasion had been frank with the grand jury, defended quiz programs as a breath of fresh air in schedules "saturated with murder and violence," and pointed to other prevailing deceptions:

> Is it any great shock to learn that important national figures generally hire "ghost writers" to write their speeches, and in many intances to write their books? [5]

The probes prompted lawsuits by losing contestants, which remained in litigation for years.

The year 1959 also brought revelations of "payola"—bribes to disk-jockeys. One was found to have received $36,050 from eight record companies, from June 1958 to October 1959, to favor their records. Disk-jockey Stan Richards of WILD, Boston, admitted accepting money and other gifts but, like Freedman, pointed to precedent. It was like a political contribution, he said. The giver pays "in the hope that something good will happen." He said he had not been influenced.[6]

2. *Ibid.*, November 9, 1959.
3. *Broadcasting*, November 9, 1959.
4. New York *Times*, October 27, 1959.
5. *Ibid.*, November 9, 1959.
6. *Broadcasting*, February 15, 1960.

To those who heard this testimony, the members of the House oversight subcommittee, it must have had a familiar ring. Its own probe conducted by the energetic Professor Bernard Schwartz had—in spite of obstructions by the subcommittee itself—recently led to the resignation of FCC Chairman John C. Doerfer, whose constant acceptance of "amenities" was found imprudent; also of FCC Commissioner Richard Mack, who had accepted a bribe for his vote on a disputed Florida channel; and finally, of Sherman Adams, assistant to the President, whose pressure on the Federal Trade Commission on behalf of a man from whom he had accepted gifts embarrassed the White House.[7]

These men were all Eisenhower appointees. But the President, who expressed deep shock over the quiz and payola revelations, did not link them with the political scandals. Instead he thought of the White Sox baseball scandals of 1919, when the World Series was fixed, and of the fan who said to one of the players, Shoeless Joe Jackson, "Say it ain't so, Joe." The quiz deceptions, said the President, were a "terrible thing to do to the American people."[8]

Another 1959 broadcasting scandal, involving bribery in the newscast field, disclosed bizarre international intrigue on the outer margins of the cold war. During 1958 Hal Roach Studios, flush with telefilm pioneering successes and with several series in international distribution—*Code 3, Racket Squad, Passport to Danger, Oh Susanna, My Little Margie*—had merged with the F. L. Jacobs Company, described as a "Detroit-based industrial complex" headed by Alexander Guterma. Its involvements ranged from automobile parts to lace.

Soon afterwards the complex acquired, for $2,000,000, the Mutual Broadcasting System. Hal Roach, Jr., became MBS board chairman, and Alexander Guterma became MBS president.

7. Schwartz, *The Professor and the Commissions*, pp. 195–201. The subcommittee often seemed determined to scuttle its probe. When Schwartz first outlined facts he had turned up, including evidence of bribery, Representative Hale of Maine said: "Mr. Chairman. . . . Dr. Schwartz's memorandum imputes misconduct on the part of members of the FCC which might result in criminal charges or, in the least, in the removal of the officials concerned. I am shocked. I had no idea when I voted to set up this subcommittee and agreed to serve on it that we would go into this sort of thing. It is none of our business." Representative Harris seemed to agree, saying they should stick to a "general survey" of the FCC and other agencies. Schwartz leaked his findings to the New York *Times* and was fired by the subcommittee, but his action forced continuation of the probe. See New York *Times,* January 22, 1958.

8. *Broadcasting,* October 26, November 9, 1959.

MBS had been in financial trouble, but Guterma expressed unbounded optimism. He considered broadcasting a "depression-proof" business, and added: "Mutual has a newscasting staff that can't be improved on. What it needs is strong guidance." He proceeded to provide it.

In January 1959, accompanied by Hal Roach, Jr., and other associates —Roach later said he slept most of the time—Guterma flew to Ciudad Trujillo, capital of the Dominican Republic, and negotiated an agreement with representatives of the dictator Rafael Trujillo. For $750,000, paid in advance in a sackful of United States currency, Guterma agreed that for eighteen months MBS would broadcast a "monthly minimum of 425 minutes of news and commentary regarding the Dominican Republic." There would be nothing contrary to Dominican interests.

A figure who helped arrange all this was Porfirio Rubirosa, Dominican playboy-diplomat who had been bridegroom to Woolworth heiress Barbara Hutton and escort to other famous ladies. Rubirosa said that Trujillo was not getting the right kind of publicity in the United States, and that something had to be done about it.

During negotiations Guterma displayed fancy salesmanship. He asked the Dominican representatives to suggest an item of news they would like to have broadcast, so that he could show how the system would work. They could think of none, so Guterma made a suggestion: "Since we have Mr. Rubirosa here and Mr. Roach here, why not say Mr. Rubirosa is going to make a picture for Hal Roach in the Dominican Republic and they are negotiating that." Guterma got in touch with MBS, and next day they heard Walter Winchell proclaim the invented news item over the network.

During February 1959 Guterma lost control of the Jacobs company and began to run into a succession of legal troubles—stock fraud indictment, failure to register as a foreign agent and, eventually, a suit by the Dominicans to get their money back. In the course of these difficulties, which forced his resignation as MBS president, the details of the negotiations became a matter of record.

It was never clear how far Guterma could have carried his corruption of MBS news. But that a nationwide network could so casually be purchased was thought-provoking.

Participants in the Dominican agreement pointed out, in its defense, that it included the clause: "We will not carry any news extolling the communist cause but agree that the primary purpose is to exemplify the stability and tranquillity of the Dominican Republic and its unequivocal position and stand against communism."[9]

Could anyone object, except perhaps a communist?

It was a bad year for broadcasting. The eruption of scandals raised questions as to who controlled what, and how. Various congressional committees, and the FCC through its network study committee, showed increasing interest in such questions. Network schedules were harshly criticized by many witnesses, and the situation called for desperate measures. CBS president Frank Stanton announced that his network, in the fall of 1959, would start a new series of documentaries on important subjects. The network that only a few months earlier had felt confident enough to wash out *See It Now* hurried to prepare a similar series—*CBS Reports*. NBC likewise moved to organize a "creative projects" unit in its news division.

Among other reform moves, Stanton also announced that thenceforth everything on CBS would be "what it purports to be." For example, canned laughter and applause would have to be identified as such. However, this radical idea was abandoned a few weeks later.

Meanwhile, statesmanlike pronouncements of a familiar sort were heard in hearing rooms. Legislation was not the answer. Self-regulation in the American tradition was called for. There was some reorganizing and renaming of network surveillance units. At NBC the "continuity acceptance" unit became the "standards and practices" unit.

Possibly the most telling result of the scandals was that the networks scrapped big-prize quiz programs and filled the gaps mainly with telefilms. Thus one of the principal remnants of New York television disappeared, yielding further hours to Hollywood. New York still had assorted variety programs, some drama, and news programming. But the 1959 fall schedules brought more than thirty new Hollywood telefilm series, including *Hennesey* with Jackie Cooper, *Johnny Staccato* with John Cassavetes, *Riverboat* with Darren McGavin, *The Alaskans* with Roger Moore, *The Untouchables* with Robert Stack, *The Detectives* with Robert Taylor, *Bonanza* with Lorne Greene, *The Deputy* with Henry Fonda, *Wichita Town* with Joel McCrae, *Shotgun Slade* with Scott Brady, *Twilight Zone* with Rod Serling—and some twenty others.

9. Cater and Pincus, "The Foreign Legion of U. S. Public Relations," *The Reporter*, December 22, 1960. *Broadcasting*, June 2, September 15, 22, 1958; September 7, October 5, 1959.

Prime-time television settled to a steady diet of telefilms—with spectacular interruptions.

BUT NOT DISNEYLAND

When Nikita Khrushchev on September 15, 1959, at Andrews Air Force Base in Maryland, stepped from a Soviet plane into two weeks of prime-time television, he faced enormous risks, as great as those involved in his denunciation of Joseph Stalin.

Emerging from an era in which the Soviet ruler had been a Kremlin recluse, the new leader was exposing himself in a land regarded by many in Russia as its mortal enemy. He was facing questions he could not censor, risking answers he could neither edit nor call back. He was doing so over the objections of some leaders in the Soviet Union and of the leadership of its most important ally, China.[1]

If Khrushchev commanded American prime-time television, it was precisely because of the risks involved in the drama—and because he was prepared to give every act a bravura performance. Seemingly heedless of pitfalls, he joked, argued, clowned, denounced. Irrepressible good humor at times gave way to bursts of anger; but, as suddenly, good humor rebounded. His fund of maxims seemed inexhaustible. "Two mountains never meet, but two people can."

The Khrushchev gamble was that he could puncture stereotypes of American political thinking, ease the cold war and the arms race, and free Russian resources for consumer goods. This was a desperate Russian need, and the goal to which he had committed himself. His political future depended on the degree of his success.

Once more, in a triumph of timing, scientific achievement set the stage. Three days before his arrival a Soviet missile hit the moon, depositing on its surface a Soviet flag. It enabled Khrushchev during the arrival ceremonies, in the presence of President Eisenhower and other American dignitaries—political and military—to strike a magnanimous and at the same time condescending note:

> KHRUSHCHEV: We have come to you with an open heart and good intentions. . . . We entertain no doubt that the splendid scientists, engineers, and workers of the United States of America who are engaged in the field of conquering Cosmos will also

1. Crankshaw, *Khrushchev: A Career*, pp. 272–82.

> carry their flag to the moon. The Soviet flag, as an old resident of the moon, will welcome your pennant and they will live there together in peace and friendship as we both should live together on the earth in peace and friendship, as should live in peace and friendship all peoples who inhabit our common Mother Earth—who so generously gives us her gifts.[2]

Scores of cameras turned and clicked, along with videotape machines, which were becoming a prominent tool in television reporting. A floating army of some 375 reporters began to follow Khrushchev and remained with him almost continuously during the following two weeks. The result was scores of short sequences on regularly scheduled news programs, plus dozens of special programs under such titles as *Mr. Khrushchev Abroad* (ABC-TV), *Eyewitness to History* (CBS-TV), and *Khrushchev in America* (NBC-TV). NBC chartered a plane so that its fifteen-man crew of reporters and technicians could fly ahead and be ready for Khrushchev at each new stop.[3]

Khrushchev was sometimes met by pickets with signs like "FREEDOM FOR HUNGARY" and "KHRUSHCHEV, THE BUTCHER OF THE UKRAINE." Crowds were curious but wary. They were seldom demonstrative. Undoubtedly they were puzzled.

There were occasions away from cameras, like a White House state dinner, at which Khrushchev is said to have buttonholed Allen Dulles and asked him questions about CIA operations abroad—which Dulles did not care to answer. But Khrushchev seemed to know quite a bit about them, so Dulles asked: "Perhaps you have seen some of our reports?"

Khrushchev grinned. "We get much of our own information from the same people as you do. Maybe we should share the expenses."[4]

Meetings with Eisenhower at Camp David were likewise off-the-record. The public was told about an encouraging "spirit of Camp David" and the possibility of a summit meeting of world leaders.

The drifting hordes of reporters became ludicrous—and a security nightmare. Reporters on the edges of the swarm scarcely knew what was happening. Some carried transistor radios to keep up with the news. When Khrushchev visited an Iowa farm, host Roswell Garst became exasperated with the trampling army and pelted reporters with silage.

2. Hearst, Considine, and Conniff, *Khrushchev and the Russian Challenge*, p. 202.
3. *Broadcasting*, September 21, 1959.
4. This and the following are based on Hearst, Considine, and Conniff, *Khrushchev and the Russian Challenge*, pp. 204–23; other sources as noted.

When Khrushchev plunged into a city supermarket, photographers climbed over counters and onto piles of cans for good angles. Customers were crushed and bruised and a woman fainted across a meat counter. "Is she unconscious?" a policeman asked. A photographer answered: "Either that or on sale."

The comedy climax was the Hollywood visit. There had been uncertainty as to how its elite would treat the event. Hearst columnist Louella Parsons, once the law of the land, favored boycotting the visit, but Hollywood took its cue from the State Department, which asked for a hundred seats at the big banquet. Stars and executives began to fight for invitations. The struggle over seating arrangements became the battle of the century, with claws bared. The final seating—as at appearances of the presidium in Moscow—reflected current standing. Seats of special prominence went to Marilyn Monroe, Frank Sinatra, Marlon Brando, Elizabeth Taylor, Dean Martin, Gary Cooper, Audrey Hepburn, Gregory Peck, Bob Hope, Debbie Reynolds. Among the few who showed no interest in attending was Hungarian Eva Gabor.[5]

The attentions showered on Khrushchev, by an industry that still blacklisted many artists—while it employed them *sub rosa*—made the Hollywood visit an extravaganza of hypocrisy. A recent Oscar for "best screenplay" had gone to one "Robert Rich," who never came forward to receive his Oscar because he was really Dalton Trumbo, writer of such screenplays as *A Guy Named Joe* and *Thirty Seconds Over Tokyo.* He had been considered publicly untouchable since the Hollywood hearings of the House committee on un-American activities but survived by writing behind "fronts"—a practice condoned by the hierarchy so long as no "trouble" erupted, such as protests or boycotts. Dalton Trumbo, not on the guest list, was expected to keep well out of sight while Spyros Skouras took charge of honoring the communist leader. Skouras, presiding at the white-tie banquet, decided to insert a bit of message and explained to Khrushchev about democracy. He said he came from an obscure family of Greek immigrants.

> SKOURAS: When we came to this country, my two brothers and I worked as humble busboys. But because of the American system of opportunity for all, now I am president of Twentieth Century-Fox, an international organization employing 10,000 people.

5. Schumach, *The Face on the Cutting Room Floor,* pp. 149–55.

Khrushchev absorbed this and finally said he was not greatly surprised. He himself had worked from the time he could walk, had been a sheep herder till the age of twelve, worked in mines and factories, and now, thanks to the Soviet system, he was the Prime Minister of a nation of 200,000,000 people.

Skouras felt challenged and stood up, holding the microphone. "How many Prime Ministers in Russia?"

Khrushchev looked puzzled, and a trifle irked. "How many Presidents in the United States?" This brought some laughter and applause.

Skouras shouted: "We have two million presidents—presidents of American corporations . . . their stockholders!"

It threatened for a time to escalate into hot debate. Skouras spoke of American relief efforts in Russia after the Russian revolution. Khrushchev said America should by all means continue to live under capitalism. "But don't inflict it on us."

At a Twentieth Century-Fox studio Khrushchev watched the filming of a can-can number; later he burlesqued the dance and said he did not think much of it.

Then he had a good time about Disneyland. On the spur of the moment he suggested a visit there, but the State Department, with no time for security arrangements, vetoed it. Khrushchev asked the reporters, armed with their notebooks and cameras and tape recorders: "What do they have there—rocket launching pads? . . . Have gangsters taken control of the place?"

Wherever he went, Khrushchev could expect hostile questions. Labor leaders gave him a hard time about the right to strike. Questions about the jamming of American broadcasts brought varied answers. If stones are thrown in a window, he said on one occasion, isn't it logical to protect oneself? He also suggested that if the Voice of America were the "real" voice of America, he would welcome it. He also said: "You also jam American voices sometimes." He reminded them that the great American singer Paul Robeson had been denied a passport to go abroad, where engagements were awaiting him. "Why was his voice jammed?"

Questions on Hungary nettled him more, and brought angry retorts. "We also have some dead cats we can throw at you."

All in all, Khrushchev had won what the business-oriented *Broadcasting* magazine called "million-dollar coverage." Much footage of the visit was also being seen abroad, profitably syndicated by various companies.

In the United States he had at least had a jolting effect. He seemed to have set changes in motion both in the United States and the Soviet Union. Voice of America transmissions were not jammed during the Khrushchev visit, and were then resumed selectively, on a lesser scale. Films of the Khrushchev visit were shown widely in Russia, with the clear purpose of building support for Khrushchev and his policy.

A few days after his departure, as a punctuation to the television special, the Russian Lunik III photographed the other side of the moon.

Some weeks later the White House announced that American, Russian, British, and French leaders would join in a summit conference in Paris in May 1960, and that President Eisenhower would afterwards tour the Soviet Union. The floating spectacular would go on. Preparations began in a fever that seemed to infect the world.

The White House chose the Christmas season of 1959 to make the announcement, and it was received as a message of peace and hope.

VOICE OF JAZZ

It was in 1959 that an executive of the Voice of America suggested to young Willis Conover that he take a trip abroad.

Since 1955 Conover had been broadcasting a nightly program called *Music USA*. Top officials at the agency generally seemed to ignore it. It was a "cultural" program, considered an advisable ingredient in the schedule but probably not important. Political commentaries got the executive attention.

The Conover series had a connoisseur quality. In the jazz portion of the program he played examples from various periods, and discussed them informatively, pacing his speech slowly and easily. He often interviewed famous jazz musicians, such as Louis Armstrong.

He had received fan letters from various parts of the world but it was impossible to tell how large a following these represented. It was suggested to Conover that he visit a number of countries and try to make contact with listeners, to gauge reactions. He started with Tunisia and moved north through Europe to Finland. Then he heard about a jazz festival in Poland and decided to apply for a visa. A trip into communist countries had not been contemplated, but a visa was promptly granted.

Soon he was sitting in Warsaw in the midst of an audience of jazz enthusiasts, listening to a succession of young jazz musicians and combinations. He talked to members of the audience, introducing himself. His name appeared to bring a click of recognition, and word of his presence spread through the hall. Then, to his amazement, he heard himself spoken of from the stage. He heard the speaker, in a stream of enthusiastic Polish, mention "Music USA" and "Willis Conover" (but apparently not "Voice of America"). His name brought a deafening roar and he was waved to the stage. When the ovation subsided he spoke—overcome—of how impressed he was with their music. "It is *I* who should applaud *you.*" He seemed to be in the presence of hundreds who had studied jazz for several years through his programs. That week they arranged two concerts "to show what we have learned from you." Posters announced the event: "MEETING WILLIS CONOVER, NATIONAL PHILHARMONIC HALL." The printed programs mentioned *Music USA.* They featured a photo of Conover at the microphone—its initials VOA were clearly visible. Later they made LP records of the concerts and sent him copies. Crowds saw him off at the airport, with hugs, kisses, flowers, cheers, weeping. The extraordinary event led in later years to Conover appearances at jazz festivals in Belgrade, Tallinn, Leningrad—where he was met at the station by the Leningrad Dixieland Band, marching in splendor—and to a proliferation of *Music USA* fan clubs, served by jazz news bulletins sent out by Conover. The clubs sprang up by dozens in Poland, Czechoslovakia, Yugoslavia, Hungary, the Soviet Union—as well as in Spain, Finland, Turkey, India, Indonesia. The groups began to send recordings of their performances to Conover in Washington, and he used them on the air over the Voice of America.

What did it mean? Authorities in most countries called it nonpolitical. But there seemed to be in these handclasps something more—affirmation of a bond, perhaps repudiation of something else that rattled around them on all sides, often in the voices of commentators.

Conover's *Music USA,* being in English, had never been jammed in the communist countries.

A group of Americans in Moscow on a 1959 intercultural visit was asked about music in the United States. One American replied that Shostakovich, Khatchaturian and Prokofiev were all very popular. A Russian asked: "And is Willis Conover highly regarded in the United States?" The American looked puzzled; he was a Harvard research librarian but had never come across the name. Yet Conover was already an institution in much of the world—outside the United States.[1]

1. New York *Times,* September 13, 1959; June 26, 1960. Interview, Willis Conover.

FLOATING SPECTACULAR

The floating spectacular was due in Paris and then in the Soviet Union. Officials and television technicians were planning a fantastic coverage. James Hagerty, presidential press secretary, flew ahead to Moscow.

Eisenhower was taking charge of American plans. During the early months of 1960 he journeyed to several South American countries to confer with their leaders. From each of these visits came footage of ceremonial receptions and pronouncements on the search for peace. Kurt Adenauer, Harold Macmillan, Charles de Gaulle flew to Washington for discussions—and more ceremonial footage. The peace quest seemed to dominate all comings and goings.

Some concern was expressed over "summit" meetings. Open meetings inevitably encouraged speeches for home consumption and made compromise difficult. Could an event telecast to all continents produce results? President Eisenhower, in his centrifugal press-conference style, commented on this.

> EISENHOWER: There is one more thing I would like to say about these summit meetings. If the summit meetings are all plenary meetings—sessions—with the room full—as a matter of fact, you have a room full of people about like this, you have a big square table and you have around it as many people as you can crowd, and behind this you have two or three rows of so-called advisers (*Laughter*)—everybody is talking *at* everybody else instead of talking *with* them. . . .

Only part of it, said the President, should be like that, "for show." There should also be occasion for four world leaders, with no one present but their interpreters, to sit together and explore each other's thinking. "What do you *really* want to do? What can *we* do?"[1]

Eisenhower depended heavily on staff work. Plans were presented to him and approved or disapproved. In the army, where there was usually agreement on the objective—defeat of the enemy—this worked well: it was a choice between alternative methods. Now he was surrounded by assistants—able men—who seemed to work in totally different directions. In one cabinet meeting after another, speech-writer Emmet John Hughes had seen a complete disparity of objectives, and all left dangling—unfaced, unresolved.

1. Press conference, March 30, 1960.

In the early months of 1960, things were no different. Platoons of aides were preparing possible peace moves. Other platoons were working in totally opposite directions.

One group was pressing him about Fidel Castro. In 1959 Castro had overthrown the gangster-ridden regime of Fulgencio Batista and proclaimed an era of reform. His revolution, beginning as a guerilla operation in the Sierra Maestra mountains, had seized the imagination of large sections of the American public—particularly the young—and his regime had won prompt diplomatic recognition from Washington. But the official honeymoon ended quickly. Washington offered friendship and aid, but Castro felt that the techniques of American aid assured a continuance of the vassalage under which Cuba had lived. His Cuba was to be built for the exploited Cuban—including the long-suffering black Cuban—rather than for the foreign investor and those of the Cuban middle class who fed at his table. Castro planned expropriation of various United States holdings and of Cuban interests linked to them—including television and film. Castro's policy toward these media affected their subsequent history in Latin America.

Cuba already had a large television system—the largest of any Latin American country—firmly launched on United States lines. Its leading tycoon, Goar Mestre, happened to be entertaining a visiting Screen Gems executive, William Fineshriber, when news came of the fall of Batista. Mestre said he would certainly have to leave the country, and he soon took off for Argentina.[2] With cooperation from United States interests, he launched commercial television operations in Argentina and Venezuela. Cuban television meanwhile became an instrument of the Castro revolution. So did its film studios, which had been a major center for pornographic production and now turned to documentaries in support of his social reforms.

Washington, protesting Castro expropriations, backed the protests with economic pressure: it cut sugar import quotas. Castro apparently faced ruin or retreat, but he showed no sign of accepting either. A visit from Anastas Mikoyan suggested to United States observers that Castro was depending on Soviet aid to underwrite his declaration of independence from United States hegemony.

But President Eisenhower was already—only few knew of this—embarking on steps to retain that hegemony by military means. In March 1960, two months before the Paris summit meeting, he approved plans

2. Interview, William Fineshriber.

drawn up by the CIA for invading Cuba and overthrowing Castro.[3] No hint of this disturbed the search-for-peace talk on television.

In April, along back streets of Miami, the CIA began quietly recruiting an army among Cuban refugees. It included ex-Batista adherents, disenchanted ex-Castro followers and others. The CIA approach was based on plausible lies. The recruiters said they had nothing to do with the United States government but were employed by a group of wealthy capitalists who were fighting communism and had influential friends in Washington. Recruits were offered $175 a month; if married, they got $50 more, plus $25 for each dependent. They were assured of "all the weapons you need." Tanks and airplanes were mentioned. Some recruits felt the United States government must be behind it and this reassured them. There were jokes about the Cuban billionaire, perhaps named "Uncle Sam." Recruits were taken in closed vans in the dark of night to a disused airport—Opa Locka—and hurried into a plane with windows blacked out. The planes took off quickly, and when they landed they were at a training camp high in rugged mountains. They learned later it was in Guatemala. A jet airstrip was being built, although Guatemala had no jets. They were on the Pacific slope of a vast coffee plantation owned by Roberto Alejos, whose brother, Carlos Alejos, was Guatemalan Ambassador to Washington. The CIA found compact regimes of this sort convenient to deal with and had made the arrangement promptly after President Eisenhower's approval. Quick approval came also from President Miguel Ydigoras Fuentes of Guatemala—successor to Armas, who had been assassinated. Ydigoras had collaborated with the 1954 coup, which was the model for the new enterprise. As in the Guatemala intervention, there would be air strikes from fields in Nicaragua, where the Somoza dictatorship was again ready to cooperate with the CIA. But there was an added feature: a CIA radio station was being built on a small Caribbean island, Swan Island. This would be a "liberation" station to foment rebellion in Cuba. It went on the air with 50,000 watts in May 1960—about the time the dignitaries of the great powers converged on Paris to defuse the cold war and usher in peaceful competition.[4]

3. Months after his retirement, at a June 13, 1961, press conference, Eisenhower himself mentioned the date when he had approved the CIA plan. See also Schlesinger, *A Thousand Days*, p. 164.

4. Details of the recruitment are based on Johnson, *The Bay of Pigs*. Subtitled "the leaders' story of brigade 2506," the work includes accounts by Manuel Artime, José Peréz San Román, Erneido Oliva, Enrique Ruiz-Williams. The story of Radio Swan is told in Wise and Ross, *The Invisible Government*, pp. 328–37.

Another platoon of Eisenhower aides was busy with Vietnam. By the spring of 1960 it was clear that Ngo Dinh Diem was widely and ardently hated. Far from introducing reforms—the supposed condition for United States aid—he had abolished elected village government, centered power in his hands and those of his brothers, and was operating a despotism. In five years the United States had provided $2.3 billion in aid funds to bolster his regime; this had enriched a Saigon elite and built an army, but it had also brought an inflation which greatly aggravated the problems of the peasant. Under cover of a Michigan State University advisory group on political science, the CIA had equipped Diem with an internal-security police force which was exterminating political opposition. In spite of the absence of reform, the Eisenhower administration felt it had no choice but to go along with Diem, who in State Department television and press statements remained the Vietnam leader of the free world.[5]

As the leaders assembled in Paris, another platoon of Eisenhower aides made an unscheduled appearance in newscasts. On May 1, 1960, a CIA spy plane, on a flight over the Soviet Union from a United States base in Pakistan, failed to return. The CIA had a cover story—fabrication, lie—ready for this. An "official" government bulletin—issued, to its discredit, by the National Aeronautics and Space Administration—announced that a weather observation plane flying over Turkey had apparently had "oxygen difficulties" and might have made a forced landing. Three days later Khrushchev announced that an American spy plane had been shot down over the Soviet Union.

Most Americans at this time knew nothing about U-2 spy planes. These planes had been used for several years for reconnaissance over the Soviet Union, flying at 80,000 feet, but were still a hush-hush subject. Eisenhower had been repeatedly assured by John Foster Dulles that the flights involved no diplomatic risks because the Russians, humiliated by inability to shoot down planes at this height, would never make public protest. Top officials—the few who knew about the U-2 flights—had clung to this confidence even after a Soviet missile had hit the moon. Besides, the U-2 planes had a self-destruct mechanism. If anything went wrong, not a fragment of evidence was supposed to remain.

This confidence led the State Department, even after the Khrushchev announcement, to stick to the CIA fabrication. In a crucial moment in

5. Schlesinger, *The Bitter Heritage*, pp. 32–7; Scheer, *How the United States Got Involved in Vietnam*, pp. 33–8.

American diplomatic history, millions of American television viewers saw Lincoln White, State Department press chief, appear before a battery of cameras and declare that there had been *"absolutely no—N-O—deliberate attempt to violate Soviet air space . . . never has been!"* The President had told an earlier press conference that he personally had forbidden provocative flights over the Soviet Union. Viewers were left with the impression that the Soviet talk about a spy plane had been another outrageous Russian falsehood.

The United States had walked into a trap. The pilot, Francis Gary Powers, was not dead. His plane had not been obliterated. He had been captured and had confessed the nature of his mission.

Presently television viewers in many lands were treated to a filmed tour of the U-2 wreckage, guided by Khrushchev himself.

The unfolding drama included poignant details. The captured pilot, according to word from Moscow, had with him his social security card, a half-used pack of filter cigarettes, an unused suicide needle, currency of several countries, a girlie magazine, two watches, and seven gold rings. Khrushchev, whose spirits were at their highest, quipped: "Maybe the pilot was to have flown still higher to Mars and was going to lead astray Martian ladies."

The administration, having been exposed in a lie, compounded it with confusion and further lies. The State Department now admitted the U-2 was a spy plane but said the flight had not been authorized. The President contradicted this and said he had authorized all such flights. He defended the flights as a duty and said they would continue. When Khrushchev demanded an end to the flights, Vice President Nixon, interviewed by David Susskind on the television series *Open End,* scoffed at the demand and said the United States would certainly not discontinue them. Then the President in a press-conference statement said that he had already discontinued them.

To some people the government's confusion was more dismaying than its falsehoods, but not to everyone. Most Americans had always assumed that a government bulletin, while perhaps not giving the whole truth, was true. For many, the U-2 incident was the first in their memory in which their government had clearly been shown lying.

The incident had a fantastic television climax. Audiences throughout the world saw Paris pageantry. There were red-carpet arrivals: the magisterial President Charles de Gaulle of France; the discreet, gentlemanly

Prime Minister Harold Macmillan of the United Kingdom; the waddly-gaited Premier Nikita Khrushchev of the Soviet Union; and then President Dwight D. Eisenhower of the United States, looking lined and sobered. With each came retinues of advisers, interpreters, security officers —followed by throngs of reporters and cameramen. It was difficult for a cameraman to take a shot that did not include other cameramen; and they were, in fact, a central ingredient in the scene.

For many viewers it was the most glittering array of world leaders yet assembled in one telecast. At times there were three in one picture: Eisenhower, Macmillan, de Gaulle.

Khrushchev, holding aloof, captured the main attention. He announced that the summit conference could not begin until Eisenhower had apologized and punished the guilty. When this brought no results, a Khrushchev press conference was announced. Before the world's cameras, one of the most extraordinary scenes among television spectaculars unfolded.

Khrushchev had the world's spotlight and made the most—or the worst—of it. Translated by the able young interpreter Victor Sukhodrev, who had an exceptional feeling for idioms, Khrushchev excoriated the United States as a war-mongering power, denounced Eisenhower, reviewed the history of US-USSR relations, fumed, raged. He withdrew the invitation to Eisenhower to visit the Soviet Union. The summit conference had apparently ground to a halt.

Stunned television audiences were given glimpses of other leaders, solemn-faced, conferring by twos and threes, entering or leaving stately buildings; then Eisenhower in a short airport statement, saddened, saying the quest for peace must necessarily go on. Its dignity won respect. For most people he had the best of it. Khrushchev, irrepressible, had overplayed his role—at least, for much of the world.

Eisenhower started for home. En route he paid state visits—complete with filmed ceremonial handshakes—to Franco of Spain and Salazar of Portugal, erstwhile admirers of fascism and nazism and still unpalatable to many Americans. Why he chose to do this went unexplained. If it was related to American air and naval bases on the Iberian peninsula and the sites for those powerful transmitters that beamed America's cold-war propaganda into communist countries, most Americans could not have guessed it. Most were scarcely aware of the tangle of relationships with Spain and Portugal—matters seldom mentioned on American television and radio—and knew nothing about the auspices and organization of Radio Free Europe and Radio Liberty.

The floating spectacular had one more sequence. In the summer of 1960 Castro began to claim that the United States was training an invasion army in Central America to overthrow his regime, and asked that the matter be put on the United Nations agenda for September. He himself would speak for Cuba. At the meeting a number of new African nations were to be admitted and the admission of communist China was to be debated. The combination of items brought an extraordinary representation. Khrushchev announced he would head the Soviet delegation. The United States announced an address by Eisenhower. Other world leaders decided to attend: Tito, Nehru, Nasser—the array was unprecedented, unbelievable. In mid-September they began to arrive and to drift back and forth across the television screen. Khrushchev and Castro were notified by the State Department that they would be confined to Manhattan. Castro had disputes with a midtown hotel; he said his rooms were bugged. Others—including Hearst writers—said he was turning his suite into a bordello. In the midst of the dispute Castro adopted a bold videogenic stratagem. He moved his delegation to the Hotel Theresa on 125th Street, in the heart of Harlem, a hotel seldom seen by white men. Khrushchev, whose arrival on the Soviet ship *Baltika* had been acknowledged only by a minor State Department protocol officer, at once left the Soviet Park Avenue quarters to visit Castro in Harlem. In view of far-flung television audiences, and surrounded by crowds of blacks on the Harlem sidewalk before the Hotel Theresa, the two leaders embraced, laughed, pummeled each other, talked eagerly, waved to the Harlem crowd and—via television—to blacks and whites of the world. Sandwiched between lily-white telefilms and toothpaste and deodorant commercials, the scene was dumbfounding—a strange intrusion into the accepted world of cowboys, detectives, and other heroes.

But the interruption did not last long. Khrushchev had played his cards. On this visit, flattering attentions of another day were replaced by official hostility and much unofficial truculence. The State Department went so far as to urge broadcasters to play down his reappearance. On the television series *Open End,* which risked official frowns by inviting him, Khrushchev showed startling patience under a juvenile belaboring by David Susskind—a generally astute interviewer far out of his depth. During the broadcast a note was passed to Khrushchev to inform him that the "station identification" breaks were being used for anti-Soviet messages on behalf of Radio Free Europe. Khrushchev remained calm in spite of this tricky ploy—by which the station, WNTA-TV, was presumably trying to show that it was on the side of the angels. But in subsequent public appearances, on and off the television screen, Khrushchev became more edgy; like a comedian who feels he is slipping, he overplayed. At a televised United Nations session he ranted and blustered; as rulings went against him he took off his shoe and pounded the desk. It was a spectacle that delighted his enemies.

In the end, many who had deplored giving Khrushchev a "platform" were delighted with the results. The Hearst reporters who collaborated on *Khrushchev and the Russian Challenge* noted with satisfaction that the public had come to know his "every twitch, every obscenity, every evasion, every declaration, determination, disarming gambit." And this, they felt, was good. The American media should not ignore Khrushchev, said the reporters, merely because of disagreement with him.[6]

Apparently wisdom of this sort was not to be applied to Castro. Castro, perhaps assuming that his United Nations appearance would give him a chance to tell the American audience something about United States-Cuba relations, prepared a long address. The television industry felt it could ignore Castro and the United Nations proceedings, and returned to profit-making. It perhaps felt virtuous in doing so, for the State Department—as in the case of Khrushchev—did not favor giving much attention to the bearded leader, whose extinction was already being rehearsed in the Guatemala mountains.

In not carrying the Castro speech, the networks missed a historic opportunity. Castro ranged over much Caribbean history from a Castroite point of view. It was often a diatribe and would have been unpalatable to many, but would also have given them some understanding of revolutionary pressures facing them in Latin America—a subject scarcely hinted at in prime-time television. It would also have given them a scoop, for Castro told of an invasion force being paid, armed, and trained at the expense of the American taxpayer, in violation of American law and treaty obligations. The speech was not telecast, scarcely even quoted. Its charges were denied by American spokesmen. In some newscasts they were mentioned as ravings of a madman.

Castro's journey to the United States, cold-shouldered by television, made hardly a dent in American consciousness. Meanwhile the Cubana airliner that had brought him was impounded by legal action of those

6. Hearst, Considine, and Conniff, *Khrushchev and the Russian Challenge*, pp. 249–50.

whose properties he had expropriated. He accepted the offer of a Russian plane to take him back to Cuba. Every aspect of the journey seemed to push him closer to Khrushchev.

For Khrushchev, the itinerant television spectacular was over. It had failed in its first purpose. The United States rapprochement had collapsed.

His final United Nations oratory made clear the complete redirection of his strategy. Praising communist China and "courageous Cuba," he denounced American interventionism. Only a few weeks earlier, gambling on American rapprochement, he had told his people of his trust in Eisenhower and the spirit of Camp David. But Eisenhower's statements during the U-2 incident had made a fool of Khrushchev and played into the hands of commissars of industry and defense who had, from the start, distrusted the idea of befriending America. Though Khrushchev might be simply a dictator in American newscasts, this was always an oversimplification; his power clearly rested on the support of various constituencies he dared not alienate. Khrushchev, railing on the television tube, was mending fences at home and in China—or trying to. So the cold war heated.

COMPLEX

There was a sadness about Eisenhower in the final days of his administration. He had never quite grappled with, or faced, the divisions in his nature, his cabinet, his administration, his country. In his final days he was moved by a strong desire to do *something* for peace, but in the end found himself propelled in other directions by forces he himself had set in motion. His problem, at the end, was not different from that of Khrushchev.

Every Eisenhower peace gesture had been hedged by moves that helped to doom its success. Yet somehow this had all seemed necessary, amid the alignment of forces surrounding him.

This alignment had been a central development of his administration. One of its most important components was the electronics industry.

Electronics had fathered broadcasting—and then much more. It exemplified serendipity at its most fantastic. While helping other industries expand into foreign markets via television and radio transmitters, receivers, telefilms, and recordings, it was also involved in each of the expanding industries through an electronic offshoot, the computer, which was revolutionizing the world of business. Meanwhile electronics was equally involved in the military bases that guarded the advance, and in the array of weaponry giving them muscle. Virtually every new weapon since World War II had involved electronics. Electronics triggered the falling bomb and brought the homing missile to target. It helped the submarine steer its course and find its target. Through the mushroom cloud at Bikini flew a pilotless plane reporting with television eyes. The advance into space—a growing military obsession—depended on electronics. The spaceship had to be tracked and guided by electronics. Scientists bounced radio waves off the moon to study and map its surface. Radio telescopes surveyed the mysteries of distant quasars. Electronics also turned inward. Its offshoot the transistor was producing a whole new industry of miniaturization. This was important to the spy and his "audio surveillance" or bugging. It put the transmitter in the martini olive. While involved in new weaponry, electronics was equally involved in the cold war as it crackled through the airwaves on all continents.

All this explains the development and mounting importance of a company like RCA. It had descended from British-controlled Marconi, which at birth attracted the interest of armies and navies of the world. The wresting of the American branch of Marconi from British control to create the Radio Corporation of America was in the nature of a nationalistic expropriation—with compensation. The Navy Department, which promoted this 1919 coup, had wanted a navy monopoly over radio; failing to get that, it worked for a private monopoly in congenial hands. RCA first took shape as a corporation controlled by four co-owners—General Electric, Westinghouse, AT&T, and United Fruit—with the navy represented on the board. In its formative years RCA had a general as president, to strengthen government ties. The eruption of broadcasting as a large source of revenue was something of a surprise and altered RCA's line of development for a time, but World War II and the cold war brought it back to its original direction. President David Sarnoff became General David Sarnoff; at RCA and NBC he was always "the General." In 1955, when the company entered the circle of the twenty-five largest corporations with income of a billion dollars—it had been a million in 1920—Senator Lyndon B. Johnson referred to RCA as "a key element in our defense structure."[1] Less than a quarter of its income now came from

1. *Broadcasting*, May 16, 1955.

broadcasting. In 1956 it formed a separate Defense Electronic Products Division. Throughout the Eisenhower administration General Sarnoff was deeply involved in government projects. He was at Fort Meade in Maryland, showing how battles could be coordinated by television, the new "eyes of the top command." He submitted to the administration *A Program for a Political Offensive Against Communism,* to be implemented through radio and television for swift and dramatic victories. He advised Radio Free Europe and Radio Liberation. Radio Liberation staff members had a training course at NBC studios. Sarnoff urged that "pockets of guerilla forces" be maintained in Poland and other communist countries; these pockets were to be supplied via radio with "information, slogans, and leadership." Sarnoff had a strong belief in slogans; at his suggestion the Voice of America adopted the sign-off, "This is the Voice of America, for freedom and peace." Sarnoff recommended that transistor radios be dropped from airplanes in communist areas; these should be made to receive only American transmitters, to shut out the totalitarian enemy. The plan was tried by American advisers in Vietnam. Sarnoff, supreme salesman of electronic hardware, slipped readily into the role of adviser and strategist.

The original RCA co-owners, though long since separated from it, had careers that continued to intertwine with RCA and its lines of interest. General Electric and Westinghouse remained in broadcasting as licensees and network affiliates, and AT&T as provider of network interconnection via cables and relay stations—with plans for relay satellites. For each of the electronic companies, broadcasting was overshadowed by other interests, including government contracts, through which they gravitated toward the aerospace industry. Selling military hardware and media hardware, the electronic companies rose in power along with all those interests—from automobiles to drug and food products—that rode the media into other lands.

United Fruit, briefly an RCA co-owner, was a pioneer in communication and transportation in many lands. When RCA first entered international broadcasting commercially in 1939, United Fruit was its first sponsor. The holdings of United Fruit involved it deeply in government affairs here and abroad. It had a division it considered its "state department." Its 1930 and 1936 agreements with Guatemala were drafted by the law office of John Foster Dulles. Allen Dulles was for a time a director of United Fruit. At the time of the Arbenz crisis, it had on its payroll

Spruille Braden, former Assistant Secretary of State specializing in Latin American affairs, who was considered an ardent interventionist.[2] The United Fruit vice president for public relations was Edmund S. Whitman, whose wife was confidential secretary to President Eisenhower throughout his presidency. After the CIA overthrow of Arbenz, as United Fruit returned to normal operation, Guatemala acquired American aid and commercial television—operated by the Guatemala RCA distributor. When the CIA in 1960 started Radio Swan to stir revolt in Cuba, the station acquired—like Radio Free Europe and Radio Liberty—a New York office and front organization. It was headed by Thomas Dudley Cabot, former president of United Fruit.[3] The lines between government and business were becoming strangely blurred.

In a vague but persistent way, all this seemed to worry Eisenhower, perhaps because he sensed he had abetted the trend. In his final presidential telecast he—or a speech-writer—gave a name to his uneasiness. The term passed quickly into the language of the time. He said:

> EISENHOWER: We must guard against the acquisition of unwarranted influence, whether sought or unsought, by the military-industrial complex. The potential for the dangerous rise of misplaced power exists and will persist.[4]

Much of the Eisenhower regime had had a confident air. But in his final television appearance he touched a new mood of anxiety. America was on the move—but where?

2. Gerassi, *The Great Fear in Latin America,* pp. 239–41.
3. Wise and Ross, *The Invisible Government,* p. 330.
4. January 17, 1961.

3 / FISSION

What happens to a dream deferred?

Does it dry up
Like a raisin in the sun?
Or fester like a sore—
And then run?

Does it stink like rotten meat?
Or crust and sugar over—
Like a syrupy sweet?

Maybe it just sags
Like a heavy load.

Or does it explode?

LANGSTON HUGHES[1]

fission: Reproduction by spontaneous division of the body into two parts, each of which becomes a complete organism.
Webster's New International Dictionary

Amid clusters of specials and spectaculars, television barreled on with telefilms. Night after night, squad car and horse thundered down the streets and byways of America—and elsewhere—drawing crowds and sponsors.

New arrivals often got bad reviews, but this mattered little. Press jealousy was assumed to play a part in the critical carping. Besides, critics were few and were almost drowned out by the publicity machinery of networks and film producers, which fed streams of newsy items and photos to newspapers and magazines. This free material—not criticism—dominated television pages, building audiences. If ratings were good and sponsors followed, all was well.

Special programs interrupting the telefilm parade included lavishly sponsored show-business spectaculars like *A Musical Bouquet from Maurice Chevalier* or productions of old Broadway plays like *Ninotchka*

1. Copyright 1951 by Langston Hughes. Reprinted from *The Panther and the Lash*, by Langston Hughes, by permission of Alfred A. Knopf, Inc.

or *Men in White,* star-studded but predictable.[2] They came also, increasingly, from news events, sometimes created for international purposes—the new diplomacy. These might cut into the profits of networks but were still welcomed by them to some extent. The events—such as the Khrushchev tour, Paris summit meeting, election campaign, or Olympic games—gave scope to the growing resources and skills of network news divisions. The deployment of cameras by scores, the fantastic feats of coordination, the knowledgeable and sometimes witty ad libbing of correspondents, displayed facets scarcely suggested in the ritual programming. News specials often brought hosannas from critics, politicians, and educators and therefore buttressed the system, helping to ward off interference with such things as telefilms—which were the meat and potatoes—and gravy—of the industry.

By 1960 the networks knew that explosions of special events could be a boon as well as burden. They had also reconciled themselves to the need for film documentaries, which likewise served to neutralize critics. The specials and documentaries *might* mean financial loss—but then again, they might not. CBS had persuaded Firestone to sponsor some of the Khrushchev documentaries. It seemed possible that such programs might not, in due time, be loss leaders after all, but sources of profit. Meanwhile they were a necessary service.

Thus, in the wake of the quiz scandals, a split-personality pattern was becoming the network norm. One part, the news division, was always preparing for, and welcoming, the grandstand interruptions that were its specialty. The other part, the real money-making part, was thereby enabled to plunge ahead with normal business, which had come to mean telefilms.

In 1960 one half of the split personality was preparing, among other matters, coverage of a presidential election that promised to be hard-fought. The other was locked in a battle of its own—an inter-network battle that was suddenly turning savage.

UNTOUCHABLE

The big 1960 telefilm news was the rise of *The Untouchables.* The series had started in October of the year before and by mid-April was pushing

2. These 1960 "spectaculars" were broadcast February 4, April 20, September 30.

into top position with a 35.1 Arbitron rating.[1] A surprising aspect of all this was that the program was on ABC-TV, the also-ran network. The industry noted that *The Untouchables,* depicting struggles against gangsters, was probably the most violent show on television.

ABC-TV had other rating successes to trumpet in trade-press advertising. Its third-place position had stemmed partly from its late arrival on the television scene. In some cities ABC-TV had no regular outlet and could only place occasional programs on affiliates of other networks—usually in unfavorable periods. But in cities that had three or more stations, and in which all networks were represented, ABC-TV had surprising successes. In these cities it was winning rating battles regularly—on Thursdays with *The Untouchables,* on Fridays with *77 Sunset Strip,* on Sundays with *The Rebel,* on Mondays with *Cheyenne,* on Tuesdays with *Rifleman,* on Wednesdays with *Hawaiian Eye.*

All of these were violent. ABC-TV did not mention this in trade-press advertising, but the other networks were aware of it—and responded in kind.

ABC-TV, ever since its breakthrough success with *Cheyenne,* had been devoted to "action-adventure"; now this policy was paying off in profits.

It happened to be a moment when executive shuffles brought ABC executives to top positions at both CBS and NBC. The quiz scandals had made telefilms the main battleground; CBS, anxious for a top man with telefilm credentials, turned to James Aubrey of ABC-TV and made him vice president and then president of CBS-TV—edging out Louis Cowan who, as packager of several quiz successes, had been similarly lionized only a year earlier. Aubrey, besides canceling big-money quizzes, scrapped a CBS anthology remnant, the occasionally impressive *Playhouse 90,* and put the accent on action telefilms.

Previously ABC president Robert Kintner had left that network and become president of NBC—where Robert Sarnoff had been felt to need strong help. At ABC Oliver Treyz, whose salesmanship had brought many telefilm sponsors to the network, became ABC-TV president.

Thus all three television networks were led by men who had had a part in the rise of ABC; all proceeded to push the successful formula.

Along with ratings, other factors contributed to the trend. The 1950's

1. Arbitron used, in a fixed sample of homes, electronic devices which measured the tune-in every 90 seconds and sent it to a central office of the American Research Bureau. The organization also issued "ARB ratings," which were based on interviews, using a roster of program titles to prompt the interviewee's recollection.

had seen new developments in "executive compensation" to counteract the rising tax bite. A much-desired benefit was the stock option. Thus Aubrey as CBS-TV president acquired options to buy a number of shares of CBS stock at a fixed price. A rise in the CBS stock quotation could put him in position to make instant money. With such a policy, an executive could hardly eliminate awareness of stock prices from programming decisions. Stock prices were becoming increasingly sensitive to successes like that of *The Untouchables.* Stocks could be nudged up or down, it seemed, by shifts in Arbitron, Nielsen, or Videodex ratings.

It also happened to be a moment when network chiefs acquired powers they had not had before. The quiz scandals had brought this on. Networks had promised to take firmer control of programming. CBS, said corporation president Frank Stanton, would be absolute master in its own house. No longer would sponsors control periods and determine how they would be used. The network would do the scheduling and let the sponsor know what was available.

To many critics of television, this sounded promising. Sponsors had seemed to be guided by merchandizing considerations; networks at least had split personalities.

But at the networks the decisions were now being made by Aubrey, Kintner, and Treyz.

In 1960, the year of the big shift in network control, film producers came to understand that series had to be contracted—"licensed"—to the network, not to the sponsor. The network would determine scheduling and deal with the sponsor.

As early as May 1960, *Broadcasting,* discussing the coming season, noted: "Four out of five shows in prime time will be licensed to the networks which carry them, and sold in turn to advertisers." It was a reversal of the practice of radio days, when most programs had been sponsor-controlled.

The new system, described by networks as an exercise of "responsibility," was also more profitable for them. A film producer soon found it difficult to get a series accepted by a network unless the network had "profit participation." *Broadcasting* continued:

> *Cut in the profits.* Going hand in hand with the increase in number of network-controlled shows is an increase in the number of shows in which networks have a profit participation. Authoritative but unofficial estimates indicate that ABC-TV, for example, had profit participa-

> tion in 49% of its nighttime programs this season, will boost them to 58% in the fall. CBS-TV's score was 52% this season, will go up to 68%. NBC-TV's was 46% this season, will go up to 60%.[2]

The power concentrated in the hands of three men made these levies possible. The networks, by paying for a pilot film, set up a plausible justification for the share in profits. It soon became a new source of wealth for the networks and their executives.

Controlling schedules, the three executives began to juggle them for advantage in the rating war. ABC-TV had developed the technique of "counter-programming." If CBS-TV and NBC-TV had competing comedies, ABC-TV felt it could "knock them off" with a sharply contrasting western. Action programs became the principal weapon in the struggle. In their orders to Hollywood studios the networks demanded more action. They never—or seldom—said "violence," but that is what they got.

The Untouchables was produced by Quinn Martin at the Desilu Studios under contract with ABC-TV. A network script supervisor in Hollywood, H. Austin Peterson, apparently combed through each script to make sure it was active enough, and relayed the information to ABC-TV vice president Thomas W. Moore in New York. An early script was described in reassuring terms:

> This is loaded with action. Many exciting scenes.
>
> Opens right up on a lot of action—a running gunfight between two cars of mobsters who crash, then continue to fight in the streets. Three killed. Six injured. Three killed are innocent bystanders. . . .
>
> There is a good action scene where the mail-truck is held up and driver killed.
>
> Colbeck suspicious it was Courtney and beats it out of Joe's henchman. Courtney is trapped in an alley and beaten unconscious and tossed in the river. . . .
>
> Colbeck wants Courtney's gal killed so she won't talk, but an irate landlady frustrates the killers. Colbeck pressures a police lieutenant who owes him a favor to pick up the gal and deliver her to a spot on a bridge where Colbeck's men will shoot her dead.
>
> Finally, Ness's men find Colbeck is going to send securities out of the country by airplane and that the plane is to land at the racetrack

2. *Broadcasting*, May 16, 1960.

that night. The Feds take over—land the plane themselves and catch the hoods red-handed. Colbeck is shot as he drives away.[3]

Scripts of insufficient vigor might bring complaints from network headquarters in New York to executive producer Quinn Martin in Hollywood:

> We have been advised that two of the recent episodes of *The Untouchables,* "Mexican Stake-Out" and "Ain't We Got Fun," lacked some of the dynamic excitement of the earlier episodes. . . . I hope that you will give careful attention to maintaining this action and suspense in future episodes. As you know, there has been a softening in the ratings, which may or may not be the result of this talkiness, but certainly we should watch it carefully.[4]

One such complaint led Quinn Martin to write a memorandum to an assistant:

> You'd better dictate some scenes of action in acts 1 and 2 before you mimeograph the treatment and send it out or we are all going to get clobbered.

Martin was obviously anxious to meet network requirements, but he was also anxious to avoid repetitiousness. He wrote to one writer:

> I wish you would come up with a different device than running the man down with a car, as we have done this now in three different shows. I like the idea of sadism, but I hope we can come up with another approach to it.

ABC was not leaving it all to Martin. It instituted a search for stock footage that might beef up action sequences. A memo reported:

> The following material was called in and run on Moviolas by librarians for possible exciting shots of gangsters, action, and violence—
>
> Desilu: Man-prison riot and break; woman-prison riot; 1933 police raid on Esco Creamery—very good gun battle.
>
> Fox: Good material of machine-gunning 10-gallon milk cans.
>
> Paramount: Four takes of barber-shop bombing—interior and exterior. . . .
>
> In accordance with your request, spectacular accidents and violence scenes of the years 1930-36 have been requested from all known sources of stock footage. You will be advised as material arrives.

3. *Juvenile Delinquency,* pp. 2328–9.
4. This and following memoranda are from *Television and Juvenile Delinquency,* pp. 25–6.

ABC had a continuity acceptance division, mentioned in quiz hearings as a guardian of network standards. On February 2, 1960, ABC-TV president Oliver Treyz said the division was "absolute boss" on policy matters and that producers could never use anything "that our continuity acceptance editors say take out." The continuity acceptance editors did not feel so all-powerful. Their notes of protest were often overruled and more often had minimal results. Their demands were, in any case, cautious. A continuity acceptance memorandum of January 11, 1960, requested:

> Page 33, scene 66, if more than one slash with the riding crop—only hit the face once.
>
> Page 59, scene 115, kill pigeon off camera please.
>
> Pages 60, 61, 62, 63, be careful in shooting the beating of Allison. As described it is much overdone and overdescribed. Cut down on the lashes—the blood, the "raw and bloody" face, the fall downstairs, etc. All this—plus the number of people killed in the closing scene—makes this a pretty horrible ten minutes.[5]

The Untouchables, heading rating lists, was estimated to reach 5,000,000 to 8,000,000 juvenile viewers with each program. Evidence of its popularity came from diverse sources. A juvenile gang in Cleveland called itself "the Untouchables" and its leader said he was the "second Al Capone." When some were rounded up on an assault-to-kill charge, one said: "We're untouchable—you can't do anything with us."[6] A Department of Justice survey at two correctional institutions—Ashland Youth Center and National Training School—found that almost all listed as their favorite programs *The Untouchables, Thriller, Route 66, Rebel, Have Gun—Will Travel.* Most had spent 3 to 5 hours a day watching television.

The ABC-TV concentration on action, which later became known in government circles as the "Treyz trend," was rivaled at CBS-TV by an "Aubrey dictum." It apparently stemmed from a staff discussion in November 1960 about the CBS-TV series *Route 66.* No transcript was made of this meeting, but the dictum was referred to in later CBS script memoranda as an item of canon law. One memo, criticizing a script for its deficiencies, said:

> You remember Jim Aubrey saying, "Put a sexy dame in each picture and make a 77 *Sunset Strip* if that is what is necessary, but give me sex and action."

5. *Juvenile Delinquency,* pp. 2331–2.
6. Cleveland *Plain Dealer,* May 19, 1961; quoted, *Juvenile Delinquency,* p. 2341.

Aubrey demanded action but wanted girls to be a part of it. One staff member summarized the dictum as "broads, bosoms, and fun."[7]

NBC-TV, referred to as "network Z" in ABC-TV trade advertising, apparently dared not leave the field to its rivals and this led, according to later government studies, to a "Kintner edict." Robert Kintner, former newspaper columnist, had a strong interest in news programming and was pushing NBC into a lead in that field. For telefilms he had a realist's contempt; they were "entertainment" and served their purpose if they won audiences and sponsors. So he too began a push for action. NBC-TV's chief supplier, MCA, was ready to provide it. One of its most violent contributions was *Whispering Smith,* the story of a soft-talking gunman.

A group in Los Angeles surveying nighttime television during one week—November 12-19, 1960—tabulated:

> 144 murders (scenes of mass murder not tabulated), 143 attempted murders, 52 justifiable killings, 14 cases of drugging, 12 jailbreaks, 36 robberies, 6 thefts, 13 kidnappings (1 of a small boy), 6 burglaries, 7 cases of torture, 6 extortion cases, 5 blackmail, 11 planned murders, 4 attempted lynchings, 1 massacre scene with hundreds killed, 1 mass murder of homesteaders, 1 planned mass murder by arson, 3 scenes of shooting between gangland posses, many killed, 1 other mass gun battle, 1 program with over 50 women kidnapped, this one including an hour of violence, kidnapping, murder, brutal fighting. These figures do not include the innumerable prolonged and brutal fights, the threats to kill, the sluggings or the many times when characters in the crime programs manhandled the victims, the forced confessions, and dynamiting to illegally destroy.

The group also noted that ABC-TV used film clips of its most violent scenes as promotion material at station breaks throughout the day.[8]

The precise effects of all this were not known. With juvenile delinquency rates still rising, some conjectured—as Senator Kefauver had done—that television was a factor in the situation; but others said that television was obviously mirroring an increasingly violent world. The debate received only limited attention in 1960.

While many watched the murder parade with fixed gaze, something else was more widely publicized. It was interruption time: the other half of the split personality was demanding attention. This was election year.

7. *Television and Juvenile Delinquency,* p. 30.
8. The National Association for Better Radio and Television (NAFBRAT) also made surveys in New York and Atlanta, with similar findings. *Juvenile Delinquency,* p. 1874.

PRIMARY

When forty-one-year-old Senator John F. Kennedy of Massachusetts, campaigning for the Democratic nomination for President, announced his intention to enter the Wisconsin primary, he received a flying visit from Robert Drew of Time, Inc. They had a talk in Detroit.[1]

Drew had a suggestion. He was a film enthusiast and had brought with him Richard Leacock, the cameraman who had done the magnificent photography for *Louisiana Story,* produced by Robert Flaherty.

Drew and "Ricky" Leacock wanted, during the primary, to be everywhere with Kennedy: in and out of cars, in hotel rooms, on speech platforms. Throughout the marathon they would never, at any time, ask him to *do* anything. They would never say, "Would you come through that door again, we missed that?" They wanted to be ignored, but wanted access.

Kennedy, who had grown up surrounded by reporters and cameramen and whose father, Ambassador Joseph P. Kennedy, had once been chairman of the board of Pathe, was aware that the arrangement could damage him. Leacock and Drew did not argue the point. Drew merely said, "You'll have to trust us." Kennedy had to fly back to Washington. The film makers got seats on the same plane and kept talking.

The Drew film obsession was, in a sense, a logical outgrowth of the candid photography that had established *Life*—where he had served briefly as assistant picture editor—and of the past glory of the *March of Time.* But Drew was also a television watcher. He was appalled at the sterility of most news programs, but had seen some telecasts he could not shake off.

He remembered an early *See It Now* program, which had shown Senator Robert A. Taft and Senator Leverett Saltonstall campaigning in a New England town. Normally the speeches, or tidbits of them, would have been the meat of newsfilm coverage. But the *See It Now* camera showed the end of each speech, then followed the speaker down the platform stairs. Senator Taft came to a group of small children with autograph books, and moved them all aside with sweeping gesture. Later Senator Saltonstall, coming down the same steps, noted the same group with amused interest and stopped to enjoy a moment with them. No narration commented on this. After seeing this program, Drew went to

1. Interview, Robert Drew.

Murrow and offered himself "body and soul" to work in the Murrow unit. Not finding a place with Murrow, Drew badgered Time, Inc., for funds for "candid" motion picture experiments. The first experiments pleased no one. An interval as a Nieman Fellow at Harvard gave him a chance to look into the reasons.

He began to restudy programs he had particularly admired, including "Toby and the Tall Corn," about a traveling tent show in Missouri, produced for *Omnibus* by Willard Van Dyke. Most people ascribed its impact to a narration by Russell Lynes, but Drew disagreed. The photography had been directed by Richard Leacock, and it seemed to Drew that Leacock's feeling for significant detail, mood, and juxtaposition had been the key elements. Drew sought out Leacock and became determined to work with him.

He also restudied a 1956 film called *Out of Darkness,* made at a mental hospital by Albert Wasserman, for the Irving Gitlin documentary unit of CBS News. Through material shot over many months, the film had followed a psychiatrist's patient work with a catatonic girl who had not spoken for months. The camera caught her first re-entry into the world of speech when, one day, she touched the psychiatrist's cuff-link and said quietly, "That's pretty. Is it a real pearl?" An intermittent commentary by Orson Welles had drawn rave notices after the first telecast, but Drew tried an experiment. Re-editing the footage, he omitted the commentary. The viewer was now thrown on his own resources and had to put two and two together without the guidance of an omniscient voice. To Drew the experience seemed more powerful—difficult, demanding, but in the end, more meaningful. This became the guiding principle of his film work at Time, Inc.

On the flight to Washington he said to John Kennedy: "Suppose you were elected, think what historic interest the film would have." Kennedy did not respond directly to this. But at the end of the trip he said abruptly: "If you don't hear from me, assume it's on."

The same arrangements were made with Senator Hubert Humphrey, who had also entered the Wisconsin primary.

A remarkable film team got to work. Besides Drew and Leacock, it included Donn Pennebaker, Terence McCartney-Filgate, and Albert Maysles. All were young and searching for their roles as film artists.

Pennebaker had collaborated with Leacock on other projects and had a strong technical background. Terence McCartney-Filgate was a Canadian in search of greener pastures. Albert Maysles was a former Boston University psychology teacher who had been on an early intercultural tour to Russia. Wanting to do something special with the opportunity, Maysles wrote to the Soviet government asking permission to bring a motion picture camera, to record his observations in Russian mental hospitals. To his surprise he received prompt approval. He went out and bought a 16mm camera. He later met Donn Pennebaker and got his advice and encouragement. Maysles's first footage was negligible, but he became obsessed with the same purpose that animated Drew—to capture significant moments, letting viewers experience them without *ex cathedra* guidance.[2]

This idea absorbed others and had various names. In France it was *cinéma vérité;* Maysles preferred *direct cinema.* The dislike for narration had philosophic as well as aesthetic aspects. The guiding commentary was regarded as limiting—an instrument of social control. These artists wanted to prod curiosity and sharpen vision—not harness it. So it was with a "movement" feeling that the Drew unit, as a division of Time, Inc., began to trail the Kennedy and Humphrey forces across Wisconsin.

They made a film of endless fascination—and historic interest. To public awareness of two leaders on the rise, and of the mechanics of a campaign, the film contributed a rich array of sights and sounds—informative, puzzling, disturbing, amusing, provocative. No previous film had so caught the pressure, noise, euphoria, and sweaty maneuvering. Detail and long-shot, noise and privacy, alternated rapidly.

Across a misty, rolling countryside, the viewer saw a lonely motorcade of half a dozen Humphrey cars en route to their next stop; at a reception, an extreme close-up of John Kennedy's hands in a long handshaking sequence, as citizen after citizen—unseen except for the hands—is propelled down the line; on a platform, during a speech by Kennedy, the neatly gloved hand of Jacqueline Kennedy twiddling behind her; in a barren room, Hubert Humphrey talking to a sprinkling of farmers, telling them the big boys back East don't like him much because they know his heart is with the farmer. Humphrey plugs his coming telethon over WEAU-TV, Lacrosse, urging people to phone in.

> HUMPHREY: Tell them you want to ask a question to that Humphrey fellow. And make them tough. That's what they like. They like to see me squirm.[3]

2. Interview, Albert Maysles.
3. *Primary,* Time, Inc., television stations, 1960.

Then the viewer is suddenly right behind Kennedy as he pushes through an endless, screaming crowd, acknowledging shouted questions, laughing, pushing on.

Film technology was not fully ready for such a sequence. Leacock and his camera were behind Kennedy; behind Leacock came Drew and a Perfectone tape recorder. To keep film and tape synchronized, a short length of cable had to connect the two. As Leacock moved, Drew squirmed along at his heels at the end of a four-foot umbilical cord. It was mad and impossible: they squeezed together through a revolving door, upstairs, downstairs, into limousines. When they began editing, they found disaster: the connection had been broken. Pennebaker spent weeks going over the material frame by frame, to put sound and picture *in sync.* One thing they knew: the umbilical cord had to go; some other synchronizing system must be found. In 1960 various groups were working desperately to that end.

On primary day as the returns came in, McCartney-Filgate was covering Humphrey, who at first seemed victorious. But in the end victory was in a Milwaukee hotel room where Leacock and Drew maneuvered around John and Jacqueline Kennedy and Bobby Kennedy and other relatives and co-workers, milling, grinning, planning.

Primary, an extraordinary achievement, was telecast on a few stations owned by Time, Inc.[4] It was rejected by all three networks. The rejection involved a newly developed—and perilous—network policy.

All the networks were now fostering their own documentary units, which might or might not prove financial assets. Each considered such a unit a government-relations necessity, but hoped to recoup costs through commercial sponsorship. Each was working on documentaries on a range of topics: space travel, election campaigns, Latin America, the cold war.

Early in 1960 an enterprising independent producer, David Wolper, completed a film titled *The Race For Space.* A sponsor, Shulton, contracted to sponsor it—on any network. But each network was developing similar projects, which would be less salable if the film were accepted. Each network announced that it would, as a matter of policy, carry only news documentaries produced by the network itself. Independent producers were stunned, dismayed. The network action seemed a move toward a documentary monopoly of three producing companies. Since

4. KLZ-TV, Denver; WFBM-TV, Indianapolis; WOOD-TV, Grand Rapids; WTCN-TV, Minneapolis.

network exposure influenced use in all other markets, the decision had a bearing on the whole documentary field.

Some loopholes remained. Shulton, having contracted to sponsor *The Race for Space,* bought time in many parts of the country, station by station. *The Race for Space* was thus, after all, able to reach a substantial audience, though under difficult circumstances, and often using undesirable periods. But documentary writers and directors were hardly reassured.

The networks explained the policy as a matter of responsibility. John Charles Daly, vice president for news at ABC-TV—where documentary was in a rudimentary stage—answered a Writers Guild of America protest by writing: "The standards of production and presentation which apply to a professional network news department would not necessarily apply to, for instance, an independent Hollywood producer." NBC-TV explained that its obligation "for objective, fair, and responsible presentation of news developments and public issues" required that it do all news documentaries itself. CBS-TV similarly ascribed its decision to a need for standards. "Since *The Race For Space* was not produced by CBS News and used a newsman not on the CBS News staff, our policy required its rejection."[5]

The film *Primary,* coming on the heels of such policy statements, again raised the issue. Film artists wondered whether Time, Inc., might perhaps lead a crusade against a policy they saw as dangerously monopolistic. But the possibility of such a crusade was averted by a compromise.

American Broadcasting-Paramount Theaters president Leonard Goldenson, after screening work of the Drew unit, was so impressed that he made a deal with Time, Inc. Its Drew unit would make a series of films for ABC-TV and function, in effect, as an ABC-TV documentary unit. The move pretended to sustain the new policy while at the same time evading it. John Charles Daly, feeling repudiated, resigned as ABC-TV vice president. His side-job as master of ceremonies on the CBS-TV series *What's My Line* earned him, in any case, a greater degree of celebrity as well as a considerable income.

The challenged policy apparently remained, but the Drew unit went on to higher things. Goldenson asked the unit to begin with a film on the Latin American turmoil. With no inkling of United States invasion prepa-

5. Files, Writers Guild of America, east. The letters were dated March 25 (ABC), 29 (CBS), 31 (NBC), 1960.

rations, Drew sent Albert Maysles to Cuba, accompanied by Negro reporter William Worthy, who had once angered the State Department by going to China. Worthy, like Drew, had been a Nieman Fellow at Harvard. Leacock was assigned to Venezuela and Costa Rica, and Drew went with him.

The Drew productions were to be telecast under the series title *Close-Up*. ABC-TV secured Bell & Howell as sponsor for a number of *Close-Up* programs. *Cinéma vérité* won a foothold in prime time.

I ASSUME YOU TWO GENTLEMEN KNOW EACH OTHER

The year 1960 was crucial for documentaries, but above all it was the year of the Great Debates.

Even before the Democrats at their Los Angeles convention nominated John F. Kennedy and Lyndon B. Johnson and the Republicans at Chicago nominated Richard M. Nixon and Henry Cabot Lodge, the networks were agitating for television confrontations between major candidates.

The idea had had a few previews. In the West Virginia primary—right after Wisconsin—Humphrey and Kennedy debated on television. Humphrey made a very "spirited" impression while Kennedy, in contrast, impressed viewers with the brevity and conciseness of his replies, an engaging wit, and apparent grasp of local issues. His debating skill clearly contributed to a smashing West Virginia victory.

This had led, during the Democratic convention, to another televised debate. By this time the two favored contenders were John F. Kennedy of Massachusetts and Lyndon B. Johnson of Texas. They were regarded as hostile to each other; Kennedy was pictured as a champion of liberalism while Johnson, a skilled parliamentarian, was considered too much a southerner to be acceptable to liberals. Challenges between their adherents led to a debate staged before a joint meeting of the Massachusetts and Texas delegations. The debate won network coverage as a news events.

In this debate Johnson described fulsomely his own achievements as Senate Majority Leader and heaped innuendo on "some people" who had been absent at quorum calls. Kennedy managed to puncture this neatly without antagonizing Johnson supporters. Since Johnson had not identified the "some people," Kennedy said, "I assume he was talking about some other candidate, not me." Then he heaped praise on Johnson and his "wonderful record answering those quorum calls," some of which he himself had had to miss.

> KENNEDY: So I come here today full of admiration for Senator Johnson, full of affection for him, strongly in support of him—for Majority Leader.

The debate had not been meaty but had provided illuminating glimpses of the candidates. Many observers felt that television should foster such contests and thus give new birth to the Lincoln-Douglas tradition.

The networks, in promoting the idea, had a special purpose of their own. They hoped to rid themselves of Section 315 of the Communications Act, which they regarded as tyrannous. In 1959 they had begun a concerted attack on it, led by CBS president Frank Stanton. This had already had results.

Section 315 provided that if a station licensee let a legally qualified candidate "use" the station, other candidates for the same office had to be given "equal opportunities" to use the station. This was assumed to mean that if one candidate got free time, rival candidates were entitled to free time; if one paid for time, the same rates had to apply to the rivals.

The law had not been difficult to interpret in regard to candidates' speeches, but campaigns invaded many other kinds of programming. Many a candidate was ready to dedicate a playground with a "nonpolitical" speech which he hoped stations would cover in newscasts; to appear on a quiz program; to visit a late-night variety program for a human-interest chat and a piano solo; to perform in a documentary. Such programs made licensees uncertain about their obligations, and the FCC had contributed to this uncertainty with confusing decisions.[1]

In 1959 Congress, pressed to clarify the situation, amended Section 315 by exempting "bona fide" newscasts, "bona fide" news interviews, "bona

1. Particularly the "Lar Daly decision" of February 18, 1959. The FCC ruled 4 to 3 that Lar Daly, who had announced himself as a candidate for mayor of Chicago, was entitled to free time to answer incumbent Mayor Richard Daley, who during the campaign had opened a March of Dimes drive on a television newscast, and had made several other ceremonial newscast appearances. President Eisenhower called the FCC ruling "absurd" and most observers concurred. The FCC decision reflected an awareness of how frequently "nonpolitical" appearances were used, especially during campaigns, for political ends.

fide" news documentaries, and broadcasts relating to political conventions.[2] It was this final provision that allowed the networks, without hesitation, to carry the Kennedy-Johnson debate from the Democratic convention. But problems remained.

The presidential debate proposal dramatized these. In 1960 sixteen parties chose candidates for the presidency. Would a debate between two major candidates violate the law? Would the networks have to devise a debate for sixteen candidates—obviously an absurdity? The solution urged by the networks was outright repeal of Section 315.

But to this there were strong objections. In several crises of American history new parties had risen to play decisive roles. The Republican Party had emerged from such a time of crisis. Was the nation ready to scotch the possibility of such innovations by giving the two dominant parties almost official standing?

In spite of such philosophic objections many Congressmen leaned toward the network plan. Immediately after the conventions the networks added pressure by wiring debate invitations to Nixon and Kennedy—subject to favorable legislative action.

Kennedy at once wired his acceptance; Nixon delayed a few days. He had little to gain, much to lose, by joint appearances with Kennedy, the lesser-known candidate. Eisenhower advised Nixon not to debate. But Nixon, with a reputation for "standing up to Khrushchev," could not afford to seem afraid of Kennedy. And Kennedy's acceptance speech, which Nixon had watched at home on television, had reassured him; it had seemed to Nixon "way over people's heads."[3] Nixon felt he would not make that mistake, and decided to accept. Spokesmen for the two candidates and the networks worked out a plan for four debates—still subject to favorable legislative action.

With the candidates in agreement, Congress finally acted. It did not go as far as the networks had hoped; it merely suspended Section 315 for the 1960 presidential drive. But the networks hoped, through success of the plan, to accomplish the final extinction of Section 315. President Eisenhower signed the bill on August 24, 1960.[4]

On September 26, at CBS station WBBM, Chicago, the candidates and their retinues converged for the first debate. It would, by agreement, deal

2. For text of amendment see Appendix B.
3. White, *The Making of the President 1960,* p. 214.
4. Public Law 86-677. See Appendix B.

with domestic issues. Each candidate would have an 8-minute opening, answer questions from a panel of correspondents, and have a 3 to 5 minute closing. Howard K. Smith would moderate. Directing the staff would be Don Hewitt of CBS.

Hewitt, one-time telephoto editor for Acme, a United Press subsidiary, had joined CBS-TV in 1948 to work on *Television News with Douglas Edwards*. He had also worked on *See It Now* and directed convention telecasts.

For days and hours before the first debate, tension hung over the premises. Nixon advisers, arriving in Chicago, found the studio background too pale. Their candidate's light-colored suit would merge with it. At their insistence it was repainted; but the new paint dried lighter than had been expected and an additional coat was required. The paint was still tacky at debate time. There was also argument—and suspicion—about lighting and camera positions. The Republicans did not want Nixon photographed from the left. This was agreeable to all concerned, including the Democrats.[5]

Hewitt was still juggling studio arrangements before program time when he suddenly found Nixon entering from one side, Kennedy from the other. The occasion seemed to call for initiative on his part so he said: "I assume you two gentlemen know each other?" The candidates shook hands.

To keep things moving, Hewitt asked Kennedy: "Do you want makeup?" Kennedy had been campaigning in California and looked tanned, incredibly vigorous, and in full bloom. He promptly said, "No."

Hewitt asked Nixon the same question. Nixon looked pale. He had made a vow to campaign in all fifty states and had been trying to carry it out. Besides, he had had a brief illness and had lost a few pounds; his collar looked loose around his neck. But after Kennedy's "no" he replied with an equally firm "no."[6]

Later his advisers, worried about his appearance, applied some Lazy-Shave, a product recommended for "five-o'clock shadow." Hewitt had a CBS makeup girl ready, but neither candidate used her services.

Nixon appears to have had conflicting advice about debate strategy. According to Theodore H. White, one adviser urged him to "come out swinging" and jolt Kennedy at the start. But Henry Cabot Lodge, in a

5. White, *The Making of the President 1960*, pp. 343–7.
6. Interview, Don Hewitt.

phone call, is said to have urged a conciliatory manner, to counteract a Nixon "assassin image." Nixon apparently tried to please both, beginning with elaborate courtesy, then growing more pugnacious.

The first debate was disastrous for Nixon. This had little to do with what was said, which on both sides consisted of almost ritualized campaign ploys and slogans. Douglass Cater called it a "paste-up job." What television audiences noted chiefly was the air of confidence, the nimbleness of mind that exuded from the young Kennedy. It emerged not only from crisp statements emphasized by sparse gestures, but also from glimpses of Kennedy not talking. Don Hewitt used occasional "reaction shots" showing each candidate listening to the other. A glimpse of the listening Kennedy showed him attentive, alert, with a suggestion of a smile on his lips. A Nixon glimpse showed him haggard; the lines on his face seemed like gashes and gave a fearful look. Toward the end, perspiration streaked the Lazy-Shave.

Edward A. ("Ted") Rogers, principal television adviser to Nixon, protested the reaction shots. But Hewitt said they were a normal television technique and that viewers would feel cheated without them.[7]

Such elements may have played a decisive part in the Nixon catastrophe. Among those who heard the first debate on radio, Nixon apparently held his own. Only on television had he seemed to lose.

In the second and third debates, which were limited to questions and answers—with each candidate having a chance to comment on the other's answers—Nixon fared better. The second debate, on October 7, was held in Washington. For the third, on October 13, Kennedy was in New York and Nixon in California; under this separate arrangement Nixon was more at ease and probably at his best. The fourth debate, on October 21, was held in New York. It used the same arrangement as the first, but dealt with foreign policy. Nixon, having rested and drunk much malted milk, looked better in all these debates, and his makeup was well handled.

All the debates had vast audiences.[8] Polls tended to show that Kennedy had won a larger share of previously undecided voters. Fence-sitting politicians swung to his side; he had clearly created a victory psychology.

7. Wyckoff, *The Image Candidates*, p. 45. Rogers later wrote *Face to Face*, a novel about a presidential debate.

8. The American Research Bureau estimated the television audiences of the four debates at 75,000,000, 61,000,000, 70,000,000, and 63,000,000. *Broadcasting*, November 7, 1960.

After the first debate increasingly large crowds gathered at his campaign stops, and the jumping, clutching, and screaming women produced scenes of hysteria.

Critics of the Great Debates stressed their superficiality. Like most newscasts, the programs were collections of tidbits. Major topics had to be tucked into capsules, each without context. The situation measured coolness and adroitness, not wisdom. The Great Debates, said historian Henry Steele Commager, glorified traits having no relationship to the presidency; he was sure George Washington would have lost a television Great Debate.[9] But enthusiasts saw in the public ordeal of the debates a relevance to leadership in an age of instant crisis and instant communication. They also pointed out that the Great Debates caused Republicans to listen to a Democrat, and Democrats to listen to a Republican, whereas separate speeches had always garnered audiences consisting mainly of the faithful.

Time periods for the Great Debates were made available free by networks and stations, but the debates were only one element in a furiously contested campaign. Both parties also bought time—in each case, more than in any previous campaign. The Republicans spent $7,558,809, the Democrats $6,204,986, on television and radio.[10]

The Democrats, perhaps remembering their 1956 difficulties in finding an advertising agency, made the unusual move of selecting a San Francisco-based agency, Guild, Bascom & Bonfigli—thirty-fifth in standing among advertising agencies.[11] But Robert F. Kennedy, thirty-four-year-old brother of the candidate, was definitely in charge of strategy, abetted by the still younger Ted Kennedy and other Kennedy relatives, friends, and associates.

For the Republicans, the television and radio campaign was again in the charge of Batten, Barton, Durstine & Osborn personnel, but Nixon wanted to avoid a Madison Avenue stigma, so BBD&O executive Carroll Newton left his Madison Avenue quarters for a Vanderbilt Avenue office a block away, where a temporary agency, Campaign Associates, was created, with recruits from a number of organizations.

Among them was Gene Wyckoff of NBC, who completed a film from stock footage of Nixon's travels. He added a connecting thread of planes

9. New York *Times,* October 30, 1960.
10. FCC statistics.
11. *Broadcasting,* November 11, 1960.

"slashing through clouds," and called the film *Ambassador of Friendship.* It began:

> *Banks of cumulus clouds rolling toward the camera.*
> (*Sound: slow fade in, airplane motors*)
>
> NARRATOR: This is the story of one man who traveled over two hundred thousand miles through sixty countries of the world to ease international problems and foster good will for the United States.
> *Main title zooms forward to fill the screen, superimposed over the clouds:*
>
> AMBASSADOR
> OF
> FRIENDSHIP
>
> (*Sound: plane motors stronger*)
> *The Nixon plane suddenly slashes through the clouds, sunlight flashing on its wings.*

The "slashing" plane was not actually the Nixon plane—the editors used surplus footage from an airplane sequence in the Jerry Lewis feature *Geisha Boy,* produced by Paramount. When Nixon was shown making a speech, they used reaction shots of people supposedly listening to Nixon but actually gathered from footage libraries. When Nixon was seen speaking in a Moscow television studio, the editor cut to a shot moving up an antenna which, as Wyckoff later explained, "could pass for a Russian antenna," and close-ups of people "who looked as if they might be Russian and who looked as if they might be listening to radio or watching TV." The faces came from various sources. The antenna that looked suitably Russian was from a vintage low-budget Hollywood feature about a radio patrol. Not everything was precisely what it purported to be, but this apparently caused no concern, even at CBS stations. The film was tried out in California, and seemed so effective that Carroll Newton scheduled it for the final Sunday of the campaign, immediately after the Ed Sullivan program on CBS-TV. However, in a late switch in plans Nixon decided on a fireside-chat format for that occasion.

The Campaign Associates film personnel lost their early enthusiasm for Henry Cabot Lodge. He looked statesmanlike, but when he spoke something else came through—a patrician condescension. When a Lodge campaign film was planned, his aides insisted: "Just let him sit and talk to the people." But the result was considered a disaster, and the film was scrapped. Instead Wyckoff assembled a 4:15-minute film—to be used as a hitchhiker—composed largely of still photos from Lodge family albums. Stills used with camera motion had been an avant-garde preoccupation for some years and had been especially popularized by the Canadian film *City of Gold,* a 1957 Oscar winner. The Lodge campaign film helped to make the technique a fixture in politics. A similar film was made for Nixon. For both, narration was pieced together from President Eisenhower's press conferences, using laudatory references to Lodge and Nixon.[12]

For Kennedy the most important campaign film dealt with the religious issue. Attacks on him as a Catholic had been widespread, and he knew he must, at some stage, meet them head-on. An invitation from Texas to appear before the Houston Ministerial Association provided a springboard. He confronted the ministers on September 12, read a short statement, and answered questions freely. The session was broadcast live throughout Texas and also filmed—for later rebroadcast. Kennedy was at his best.

> KENNEDY: I believe in an America where the separation of Church and State is absolute—where no Catholic would tell the President (should he be a Catholic) how to act, and no Protestant minister would tell his parishioners for whom to vote—where no church or church school is granted any public funds or political preference. . . .

Last-minute staff research culled from names of those who had died at the Alamo a few who *might* have been Catholic—although no one could find out if they were. This resulted in the astute lines:

> KENNEDY: . . . side by side with Bowie and Crockett died McCafferty and Bailey and Carey, but no one knows if they were Catholics or not. For there was no religious test at the Alamo.

Seasoned Texas politician Sam Rayburn considered the session a triumph and said that Kennedy "ate 'em blood raw."[13] Filmed highlights of the confrontation became a basic campaign weapon. Time was bought on television stations across the country, with special concentration where the message seemed most needed.

If this film and the Great Debates were crucial to the outcome, so was one other element.

12. Wyckoff, *The Image Candidates,* pp. 21–5, 50–51.
13. Sorensen, *Kennedy,* pp. 189–93.

On October 19 Martin Luther King, already world famous through his nonviolent campaigns for social justice, was arrested in Atlanta with other Negroes for refusing to leave a table in the restaurant of Rich's department store. A few days later all except King were released; he was sentenced to four months of hard labor and spirited off to the state penitentiary. Mrs. King, who was six months pregnant, was unable to communicate with him there. Many Negroes thought he would not emerge alive.

To the Nixon and Kennedy campaigns, the crisis brought sudden challenge. Nixon was campaigning for votes of northern blacks, but more assiduously for those of southern whites. A statement from him as Vice President might have carried weight, but he said—and did—nothing.

Kennedy, when he heard the news, picked up the phone and called Mrs. Martin Luther King at her home in Atlanta. He merely assured her of his interest and concern—if necessary, his intervention. Next morning Robert Kennedy telephoned a plea to the Georgia judge who had sentenced King.

The Kennedy forces did not trumpet these small moves. But Mrs. King told friends, and word spread quickly through the Negro leadership and from them into newspapers and newscasts. Dr. King's father, a Baptist minister who had said he would support the Republican ticket, switched because Kennedy had been "willing to wipe the tears from my daughter's eyes." He urged others likewise to support Kennedy. Meanwhile King was released on bail—unharmed.

In scores of cities where Negro voters were beginning to wield political power, news items about Kennedy's call to Mrs. King may have been more potent than any oratory.

The campaign had suddenly thrust a new personality on the consciousness of television audiences. Like Lucy and Van Doren and the *Bonanza* group and *The Untouchables,* John Kennedy had caught on suddenly with a spectacular rise in ratings. He did not belong to the regularly scheduled world but seemed as professional as a regular. He had wit and drama. He went after an adversary with style. He said: "Mr. Nixon may be very experienced in kitchen debates, but so are a great many other married men I know." When Nixon accused him of a "barefaced lie," Kennedy replied that he could not accuse Nixon of anything bare-faced because "I've seen him in a television studio, with his makeup on." With utmost brevity, he sometimes made statements that rang out. Concerning the developing world he said: "More energy is released by the awakening of these new nations than by the fission of the atom itself."

He had an air of confidence that to some people, including Adlai Stevenson and Edward R. Murrow, had an element of arrogance. He seemed to have a lot of information on many problems, and ready statistics on them. Perhaps he reacted to them as fields for political action rather than as deeply felt personal concerns; yet he was aware, alert, growing. In the fall of 1960 his vitality crackled from the television tube.

Reviewing a videotape of one of his own television appearances, John F. Kennedy commented: "We wouldn't have had a prayer without that gadget."

The final television debate produced a historic moment that scarcely anyone in its vast audience could appreciate, because the facts were hidden. Kennedy did not know—Congress did not know—that the United States was training and equipping an invasion army in Central America to crush the Castro government. But Kennedy felt sympathy for Cuban refugees. So his office had issued a policy statement—actually, Kennedy himself had not seen it before it was released[14]—that the United States should try to strengthen "democratic anti-Castro forces in exile, and in Cuba itself, who offer eventual hope of overthrowing Castro."

Nixon, who knew of the invasion plan and supported it, was perhaps alarmed that Kennedy had wind of it and was bringing it into the open. So in the fourth Great Debate he denounced the idea with vehemence as "dangerously irresponsible" and "shockingly reckless" and a violation of American treaty obligations and the United Nations Charter. Nixon warned: "It would be an open invitation to Mr. Khrushchev to come in, to come into Latin America."

The Nixon outburst was an elaborate stratagem. He did not deny the existence of the invasion force about which Cuba was complaining but, without lying, he accomplished the same thing. The denunciation made it seem impossible that such a thing was being planned—much less, was approaching the action stage.

Nixon also said that a far better procedure was the one used in Guatemala—where, he said, our quarantine had produced a "spontaneous revolt against communism." Nixon knew there had been nothing spontaneous about it, and Kennedy was by now also aware of that fact; but the debate hardly seemed a time to say so.[15]

14. Schlesinger, *A Thousand Days,* p. 73. Schlesinger says "it did not happen again."
15. Sorensen, *Kennedy,* p. 206.

The television campaign went on till the last moment. On the final day Nixon conducted a four-hour nationwide telethon from Detroit at a cost of at least $200,000—the Democrats estimated the cost at $500,000. Kennedy, after six speeches in five states, closed that night with a telecast from historic Faneuil Hall in Boston, with film excerpts of campaign travels, words from Jacqueline Kennedy in Hyannisport, from Lyndon B. Johnson in Austin, and from Kennedy himself. Next day Kennedy was elected by a popular vote of 34,221,463 to 34,108,582, and an electoral vote of 303 to 219. Crucial states had been won by a hair.

FORBIDDEN FRUITS

The Radio Liberty staff felt most assured when discussing Russia. Here the commentators, through past experience, had a basis for evaluating news items obtained via Soviet broadcasts, newspapers, travelers, and defectors.

But Radio Liberty also, at times, took up items from other parts of the world, including America. The purpose, as its spokesman Howland Sargeant explained it, was to give the Soviet listener "forbidden fruits"—knowledge and pleasures denied him by his own government.[1]

Among the forbidden fruits sent to him during 1960 via twenty-nine transmitters were items about Guatemala, Nicaragua, the American consumer, local politics in the United States, and the American Indian.

The Indian received attention on May 2, 1960 on Radio Liberty's *Panorama of the Week*. A news item from the United States was seen as a chance to correct Soviet impressions that the United States had not dealt generously with its aborigines. With saturation coverage, the presentation began:

> (*Music: American Indian music*)
>
> VOICE 2: The leaders of eleven Indian tribes recently met for a council of war in the state of Oklahoma—in the United States—and decided to declare war on their paleface brothers. However, do not let your imagination run away with you, for they were not dressed in buffalo skins and eagle feathers, but in everyday jackets. They were not sitting around a campfire, but in an office. And at this meeting they decided to file a complaint with the United States Congress about American TV programs, in which red Indians are shown as bloodthirsty savages. But in point of fact. . . .

1. Lecture, The New School for Social Research, April 4, 1957.

VOICE 3: . . . but in point of fact young Indians go to school, college, and universities, and even the older generation has long ago "buried the hatchet." The council of war was attended by doctors, lawyers, engineers, and businessmen: half a million red Indians now live peacefully on the income from their 20 million hectares of land.

The glowing account went on to tell of the Navajo tribe of Arizona acquiring a computer to aid in management of its jewelry, wool, and leather industries and its oil and mining properties. The modern Indian "wigwam," said one of the Voices, had modern furniture and television. The only problem of its occupant seemed to be that his image on the television set was not quite satisfactory.

On July 13, 1960, Radio Liberty commentator Natasha Zhukova dealt with democracy in Stamford, Connecticut, which through a slip-up became "Stanford" in her broadcast. With the usual multi-transmitter coverage, she told her audience that the mayors of many American cities maintain lively contact with their citizens.

ZHUKOVA: Take the town of Stanford, in the State of Connecticut, for example. There, every Thursday afternoon, the Mayor of Stanford, Walter Kennedy, interrupts his affairs for two hours in order to listen to complaints, statements, proposals, and requests put to him by his citizens. . . . And the people go to him concerning the most varied things. One house-owner, for example, complains that his neighbor's children smash his windows when they play baseball. A lady house-owner complains that the noise coming from a near-by tavern at night prevents her from sleeping. A shop-owner is concerned that he will lose his source of income due to the planned reconstruction of the section of the town where his shop is situated.

And Kennedy finds a solution for everyone who comes to see him, the Mayor of the town of Stanford.

An item broadcast May 11 took its inspiration from *Izvestia*, which published a complaint from a Russian woman that she had bought a pair of stockings which lasted only two weeks, and a more expensive pair that lasted three days. A Radio Liberty economic commentator, Georgy Osokin, explained why such things could not happen in the United States.

OSOKIN: In a free market the problem of quality is resolved by consumers not buying poor goods, having an opportunity to choose better ones from the goods of various firms.

Even where monopoly existed, said Osokin, it did not usually lead to a lowering of quality, although it might increase the price.

Such program items were what one might expect from hard-pressed writers dealing with unfamiliar matters but faithfully following an official line. Those who were paying for all this, American taxpayers, might have been astonished at the content, but of course it was only for Russians.

The items on Nicaragua and Guatemala were on a somewhat different level.

In mid-November, in the United States, the magazine *The Nation* published an editorial titled "Are We Training Cuban Guerillas?" It was based on information from a scholar returning from Guatemala and an item in a Guatemala newspaper. *The Nation* mentioned talk in Guatemala of a secret American airstrip high in the mountains, and suggested: "Fidel Castro may have a sounder basis for his expressed fears . . . than most of us realized." It urged the administration, if the reports were false, to scotch them.[2]

Proofs of this editorial were circulated some days in advance and brought several reactions. In an Associated Press dispatch from Guatemala City, dated November 17, President Ydigoras denounced the report as "false" and "a lot of lies." Next day Radio Liberty—with saturation Russian coverage—took up the subject. Without mentioning the article in *The Nation,* it spoke of a complaint by President Ydigoras that Castro was sending airplanes to rebels in Guatemala, which might therefore, said Radio Liberty, "have to declare war on Cuba." Later Radio Liberty said that President Eisenhower, because of the threat to Guatemala and Nicaragua, was sending them troops and planes.

> VOICE: According to a report from Washington, United States marines and navy planes have occupied strategic positions to prevent foreign intervention in Guatemala's and Nicaragua's domestic affairs.

Thus the CIA ingeniously explained the military traffic to Guatemala and Nicaragua and laid an international smoke screen around its operation.

Those who dutifully read these dispatches over Radio Liberty had no way of assessing their truth. They were being used. The material came from higher echelons.

2. *The Nation,* November 18, 1960.

Among its forbidden fruits Radio Liberty offered sabbath programs in a spiritual vein. One such program—December 4, 1960—posed the question: "Does the end justify the means?" Any societies not recognizing the importance of this question, said the commentator, "will sooner or later reveal their antihumanitarian essence."

Another broadcast quoted Chekhov: "I hate lies and violence in all their forms!"[3]

HIGH PRIORITY

On November 18 President-elect Kennedy, then at Palm Beach, was briefed by CIA director Allen Dulles on activities of his agency. Kennedy now learned for the first time of the Cuban invasion plan—in very general terms. The information was kept even from close campaign associates. Theodore Sorensen, often referred to as Kennedy's chief policy aide and "alter ego"—his appointment as special counsel to the incoming President was the first appointment to be announced—was told nothing.[1]

The Nation of November 18 was just reaching its limited list of subscribers. The question-raising editorial had already been counteracted by official statements from strategic points.

Visitors to the Kennedy Palm Beach headquarters during following days included Robert Drew of the *Time* film unit. He was invited to show *Primary,* which Kennedy had not yet seen. He also brought "Yanki No!", the film on Latin American tensions just shot for the ABC-TV *Close-Up* series.

John and Jacqueline Kennedy expressed delight with *Primary.* The President-elect spoke of having his presidential administration documented in similar fashion. On the following evening they looked at "Yanki No!"—this time with two other visitors to the Palm Beach quarters, Senator and Mrs. J. William Fulbright.

Shot *cinéma-vérité* style in Cuba, Costa Rica, and Venezuela, "Yanki No!" announced itself as a warning, to the United States, of a widening gulf. The viewer had glimpses of a Costa Rica meeting at which the United States, strongly backed by various military regimes, pushed through a Declaration of San José aimed at building an economic wall

3. Radio Liberty, *Papers.* The broadcasts quoted in the preceding pages were selected by Radio Liberty as examples of its work.
1. Sorensen, *Kennedy,* p. 295.

against Castro. The foreign minister of Venezuela was shown opposing it. Leacock and his camera then followed him back to Venezuela.

We see the foreign minister entering the presidential palace, and then emerging. Deeply disturbed, confident of popular support for his views, he tells the camera he has been summarily fired. The viewer is given the impression of an elite, fortified by United States support, holding against wide unrest. The scene shifts to Cuba, where Maysles's camera follows Castro through surging, delirious, admiring throngs. Here the dam has burst.

At the end of the film Kennedy paced up and down restlessly. He and Fulbright appeared to respect the film and its disturbing portrayal. Kennedy walked up to Drew, and with his face three inches away, said, "So what do we do about it?"

Drew was nonplussed at the question. "All I can do," he said, "is try to show the problem."[2]

About this time Edward R. Murrow at CBS had a visit from a journalist he had known during World War II. The man said he had details about a coming invasion of Cuba, which he insisted was being organized by the United States. Murrow discussed it with Fred Friendly as a possible *CBS Reports* subject. Friendly could scarcely believe the story, already dismissed and denied on various occasions. Murrow seemed to have more faith in it, but they hardly knew how to proceed. "It just seemed to us we were getting mixed up with something we couldn't handle," said Friendly later.[3] Soon afterwards Murrow was offered the directorship of the United States Information Agency, and accepted; he and Friendly had no further discussions on Cuba.

The Cuban government still talked about a threatened invasion. In January 1960 President Eisenhower, saying that United States endurance was at an end, severed diplomatic relations with Cuba. It was two weeks before the end of his presidency.

At about the same time a spokesman for refugee groups in Miami flatly denied the invasion rumor.[4]

At the isolated CIA training camp in Guatemala, trainees were splintering into factions. Some were former Batista soldiers; others former Castro followers who had defected. These groups were antagonistic to each

2. Interview, Robert Drew.
3. Friendly, *Interview*, pp. 10–13.
4. Johnson, *The Bay of Pigs*, p. 58.

other, and both were resented by many younger recruits. The CIA mistrusted those with liberal leanings—they might turn out to be land-reform enthusiasts—and kept the leadership in hands regarded as safe. This aggravated tensions and disputes. In January 1960, to prevent leaks, a number of dissidents were transferred by the CIA to a jungle prison accessible only by helicopter.[5]

The CIA was thus managing to keep its activity well submerged. In the United States only two periodicals, *The Hispanic-American Report* and *The Nation*, both with small circulation, had treated the rumors with any seriousness. Their effect had been negligible. Television and radio had been kept almost spotless. Allen Dulles, in later memoirs, explained his policy toward the mass media.

> I always considered, first, how the operation could be kept from the opponent and, second, how it could be kept from the press. Often the priority was reversed.[6]

The high-priority task of keeping it from the press—and the public—was being performed with impressive success as the new administration prepared for office.

GOLDEN DRAGON

With the election interruption out of the way the industry—the main part, the regularly scheduled part—could resume full force.

There was no let-up in the boom. Evening schedules were practically sold out. New advertisers were coming in.

ABC-TV, for added profit, decided to stretch the 30-second break between programs to a 40-second break. This would allow four instead of three 10-second commercials at station-break time, and mean millions in additional revenue. The stations of the network were pressing for the innovation.

There was some anxiety that it would be denounced as commercialism. "Why don't we explain," said one ABC-TV executive, "that it will seem *less* commercial? Wherever possible there will be two 20-second commercials, not four 10-second commercials."

"Why don't we explain," said vice president Giraud Chester, who some-

5. *Ibid.*, p. 238.
6. Dulles, *The Craft of Intelligence*, p. 238.

times made irreverent suggestions, "that we want to make more money? That will strengthen capitalism and be a blow to communism."

In 1960 a new institution made a flourishing start: a film festival for commercials. It was so successful that a thousand entries were expected for the following year. Entries would be eligible for election to a Commercial Classics Hall of Fame and for awards in numerous categories—apparel, automobiles, cake mixes, cigarettes and cigars, coffees and teas, cosmetics and toiletries, laundry soaps and detergents, paper products and wraps, pet foods, pharmaceuticals, soft drinks, and twenty other categories.

Commercials were becoming lavish. While budgets in the telefilm field were generally about $2000 per minute, budgets for commercials were running to $10,000 and $20,000 per minute and were still climbing. Major Hollywood studios, including Metro-Goldwyn-Mayer, were competing for a share of this business.[1] Animated commercials were especially costly but avoided residual payments to on-camera actors and the problem of obsolescence of clothing and hair styles. For lower costs, some agencies were having animated sequences done in Japan, Spain, and other countries where costs were low. Impressive talents were being drawn into the making of commercials, which sometimes—in the opinion of some viewers—outshone the entertainment.

The 1960–61 season brought its new crop of telefilm series. *The Roaring Twenties* was somewhat like *The Untouchables, Surfside 6* like *77 Sunset Strip, The Aquanauts* like *Sea Hunt, The Man From Interpol* like *A Man Called X.*

CBS executive William Morwood once explained his technique for persuading an advertising agency or sponsor of the virtues of a script. He always began by saying, "This is something that has never been done before. . . ." At this point he would identify the novel element—a setting or occupation or plot device. Then he would add: "And it's *exactly* like. . . ." Here he would point out the similarity to a well-known success.

Pierre Berton, the Canadian writer, was surprised but flattered when he was whisked to Hollywood as a well-paid consultant for a telefilm series to be called *Klondike.* Berton had known the Klondike from childhood, had written a book on the subject, and narrated the memorable Canadian film *City of Gold.* He was pleased to find the production team

1. Among MGM's salesmen was a former first violinist for the Toscanini NBC Symphony orchestra.

fanatically interested in details. Clothing and furniture had to be authentic. But the rest could apparently be nonsense. The series, shot on a Hollywood lot, premiered in October 1960. "The last time I looked at *Klondike*," wrote Berton in his Canadian column a few weeks later, "I felt completely detached from it; the action might be taking place in Madagascar or on the moon, but not in the land I knew and wrote about." A fact of particular interest to Berton was that even at the height of the gold rush no guns were carried in the Klondike; it was extraordinarily law-abiding. But in the *Klondike* series gunplay was incessant. "The idea that every story on TV must end with a fight or a gun battle depresses and annoys me. . . ."[2]

With America on the move throughout the world, interest in foreign settings was increasing. Thus *Hong Kong* made a glittering October 1960 debut as a one-hour ABC-TV series produced by Twentieth Century-Fox and sponsored by Kaiser Industries. There was early sponsor and network discontent over scripts, but things were soon under control: an actress called Mai Tai Sing was added to the cast as owner of the nightclub, The Golden Dragon, a scene of international intrigue and treachery. There was exotic dancing and murder in back rooms and alleys. Lloyd Bochner was the Hong Kong police inspector, and Rod Taylor was an American correspondent, presumably covering Red China from this vantage.

Business was good but the industry was wary of the new administration; budgets for news, special events, and documentaries were pushed up.

CAMELOT

The 1960–61 season brought to the documentary field a sense of a beginning renascence. The documentary spurt, which owed some of its impetus to the quiz scandals, had been further stimulated by disasters like the U-2 affair, which brought foreign policy under more earnest public scrutiny. The new critical atmosphere prompted NBC, for example, to undertake a documentary on the U-2 incident and to schedule it in the closing weeks of the Eisenhower administration—behavior that seemed uncharacteristic of NBC.

But the feeling of renascence acquired more positive momentum from

2. Toronto *Daily Star*, January 30, 1961; Berton, *Fast Fast Fast Relief*, p. 114.

the tone set by the Kennedy administration as it gathered the reins of government. That tone emerged from interviews, appointments and—unforgettably—from the Kennedy inaugural.

In interviews Kennedy said he did not plan fireside chats in the Roosevelt tradition but believed strongly in documentaries. He felt that the motion picture camera could help keep citizens informed about activities of government and should be allowed to do so.[1] Documentary producers sensed they had a spokesman in the White House—a place where the western had held a place of honor.

The appointment of Edward R. Murrow to head the U. S. Information Agency confirmed their sense of representation. Plans for the inaugural further encouraged expectations of a new dawn. Kennedy wanted "The Star Spangled Banner" sung by Marian Anderson, who had once been barred from singing in Constitution Hall by the Daughters of the American Revolution. Whereas President Eisenhower had asked the Negro to be patient, his successor clearly wanted to move forward.

The inauguration, held on a bitterly cold day in a city all but paralyzed by eight inches of snow, produced warm and memorable moments on the television screen.

There was the eighty-five-year-old Robert Frost, who had been asked to read a poem—again, a symbolic moment, since poets had not recently been noticed in high places. He began to read words he had written for the occasion.

> FROST: . . . a Golden Age of poetry and power,
> Of which this noonday's the beginning hour . . .

Blinded by the glare of sun and snow, he could hardly read the words, and faltered briefly. But those words were only a prologue to an older poem he knew well. He held his head high and the words rang out.

During the prayer by Richard Cardinal Cushing a short circuit under the lectern smoldered. As the long prayer rumbled on, incoming and outgoing Presidents peered furtively but anxiously toward the smoke. J. Leonard Reinsch of the Cox Broadcasting Corporation, television adviser to Kennedy, crept around the lectern in his cutaway, trying to be invisible but not succeeding. Over NBC-TV David Brinkley murmured quietly: "Leonard Reinsch seems to be the chief fireman."

Striking was the contrast between two men—one, the oldest to serve as

1. *Broadcasting*, April 3, 1961.

President; the other, the youngest ever to be elected. The contrast gave emphasis to a passage in the inaugural address:

> KENNEDY: Let the word go forth from this time and place, to friend and foe alike, that the torch has been passed to a new generation of Americans, born in this century, tempered by war, disciplined by a hard and bitter peace, proud of our ancient heritage, and unwilling to witness or permit the slow undoing of those human rights to which this nation has always been committed, and to which we are committed today at home and around the world.

What actions were foreshadowed by these ringing words? No one could even guess, but their rhythm and eloquence, the verve of their delivery, won extraordinary response from the young. Kennedy's address included standard cold-war language, but also departed from it. He did not speak of Russia and its allies as "enemies"; at Walter Lippmann's suggestion, he used the word "adversary."

> KENNEDY: Finally, to those nations who would make themselves our adversary, we offer not a pledge but a request: that both sides begin anew the quest for peace, before the dark powers of destruction unleashed by science engulf all humanity in planned or accidental self-destruction.

He dared to use a term John Foster Dulles had considered equivalent to *appease—negotiate.*

> KENNEDY: Let us never negotiate out of fear, but let us never fear to negotiate.

Finally, to a nation that had been obsessed with the acquisition of consumer goods—and, at higher levels, with capital gains, stock options, inside deals—the young President spoke an invitation.

> KENNEDY: And so, my fellow Americans, ask not what your country can do for you; ask what you can do for your country.

Among the listening hundreds of millions, Archibald MacLeish heard it by short wave in the Windward Islands. He wrote: "It left me proud and hopeful to be an American—something I have not felt for almost twenty years."[2]

The whole day was a dazzling series of telecasts—a glorious interruption. Next day, for most of the industry, it was time to get back to regular

2. Schlesinger, *A Thousand Days*, p. 732.

schedules. But many in news and documentary units felt they had an agenda; they had heard some sort of call. The Kennedy aura seemed to encourage unthinkable thoughts and impossible dreams. Kennedy enthusiasts were reminded of Camelot.[3]

The rise of the documentary was now stimulated by three-network rivalry. No longer did CBS, as in the *See It Now* years, totally dominate the field. NBC president Kintner had lured Irving Gitlin from CBS to head the "creative projects" unit in NBC News, and Gitlin brought with him Albert Wasserman—of *Out of Darkness*—and gathered other vigorous young talents. They began a series of specials titled *NBC White Paper,* the first of which was "The U-2 Affair"—broadcast November 29, 1960—which did not hesitate to state uncomfortable facts and ask painful questions.

That same week—on November 25—*CBS Reports* came up with "Harvest of Shame," produced by David Lowe under Fred Friendly's supervision, and narrated by Edward R. Murrow shortly before his departure from CBS. It portrayed the plight of migrant workers—so vividly that many people simply rejected its truth. Such poverty and human erosion could not easily be fitted into the world as seen in prime time. This reaction became a familiar one to documentary producers.

A week later, on December 7, ABC-TV's *Close-Up* series presented "Yanki No!" and followed it a few weeks later with "The Children Were Watching," which gave viewers a sense of how it felt to be a six-year-old black child attending the first integrated school in New Orleans. The competition was keen—and produced moments of brilliance.

USIA, too, began to experience a feeling of renascence. Murrow felt that candid reporting of the American scene would, in the long run, do honor to the United States; thus he considered integrity, not salesmanship, to be at the heart of good propaganda. His arrival brought an enormous lift to USIA morale.[4]

It was not yet a golden age. The network news divisions—which had jurisdiction over the network documentary activity—were aware of their

3. The songs of *Camelot,* a current musical success, were said to be favorites of the Kennedy family.

4. Murrow's first days at USIA were, however, marred by a lapse of judgment. He asked BBC director general Hugh Carlton Greene, as a "personal favor," not to broadcast "Harvest of Shame," which the BBC had purchased. Intensive badgering at the Senate foreign relations committee confirmation hearing, over his alleged overemphasis on the "seamy" side of American life, apparently led to this action by Murrow; he later regretted and deplored it. The BBC refused the request.

second-class citizenship. Network news, at this critical juncture of American history, still had as its main achievement an early-evening 15-minute telecast of newsfilm items threaded by one or more anchor-men. Its scheduling made it a preliminary to the real business of the evening. As for network documentaries, they *might* get into prime time, but this usually depended on Aubrey, Kintner, Treyz. Documentaries were made in news divisions but scheduled by executives outside the news divisions.[5] The scheduling decisions often seemed to depend on how things looked in Washington. In 1960–61 the omens were good for documentaries.

In March 1961 NBC telecast one of its finest documentary achievements—*The Real West*, from the unit that had made *Victory at Sea*. The unit's early projects had been concerned with historic newsfilm. Under Donald Hyatt it turned to historic photographs, and in *The Real West* produced a film of extraordinary scope. Its researcher, Daniel Jones, ransacked historical societies and attics in a score of states for photographs of the westward push of the late nineteenth century; a script by Philip Reisman, Jr., matched them in poignance, drama, comedy, irony, tragedy. Gary Cooper, already aware that he had cancer, heard about the project and wanted to narrate the film. He seemed oppressed by the thought of his own contributions to a western mythology. Coming to the NBC office repeatedly to pore over the fantastic assemblage of photographs, he said he wanted to help put the record straight. His usual fee—one of the highest in Hollywood—was not an obstacle. Whatever was usual in these documentary things would be fine. So Cooper came to play his last role. He felt and seemed somewhat out of place in the documentary context, but his participation insured prime time, commercial sponsorship, and a massive audience. The one-hour epic at once became a television classic.

The final section of *The Real West* was one of the most moving. It pictured briefly what the westward tide had done to the Indian—a tragic, bitter story. In his search for photographs, Daniel Jones discovered so many of Indians of the period that NBC decided on another film, to tell the same story from the point of view of the Indian. Long, patient work on the sequel was begun.

Another product of brilliance and patience erupted on CBS-TV in 1961. This was "Biography of a Bookie Joint," seen on *CBS Reports* on

5. Videotape was introducing a new hazard. It was easy for affiliate stations to videotape a prime-time documentary and use it in a fringe period; this was becoming frequent.

November 30. It was chiefly the work of Jay McMullen, who was becoming known as an "investigative reporter." CBS had become so confident of his talents that it sometimes permitted him to disappear for months without telling anyone exactly what he was up to. The results were usually good. In 1960 he began photographing the entrance to a Boston bookie parlor—from a window across the street.

The bookie joint pretended to be a key shop, but it had as many as a thousand visitors a day, including members of the Boston police, who sometimes left their patrol cars double-parked during their visits. McMullen, with help from Palmer Williams, developed a lunch box with a concealed 8mm camera, and completed sequences inside the bookie parlor. The film provided a detailed look at a gambling operation, its customers, and its protective machinery. It led to the resignation of a police commissioner.

Meanwhile the *NBC White Paper* series had triumphs of its own. Its "Sit-In" featured newsfilm of Negro resistance to restaurant and department store segregation in Nashville, and the resulting violence. But the film makers—Albert Wasserman and Robert Young—used this material merely as a starting point. Robert Young sought out identifiable participants, then filmed and recorded their accounts of the incidents. The two strands of material, edited together, gave an illuminating picture of the trickiness of human testimony. Again and again a participant could be seen reshaping an episode to fit his own biases, purposes, and self-image.

The film made Robert Young something of a hero in the black community, and led to another *NBC White Paper*. In northern Angola in West Africa—still in colonial status—the Portuguese were attempting to subdue a native uprising. Portuguese bulletins spoke of native savagery and ingratitude, but foreign newsmen were not allowed to enter Angola for direct observation. Robert Young persuaded NBC to let him—along with black cameraman Charles Dorkins—go to the Congo to try to enter the Angola war area from there. Armed with recommendations from American black leaders they walked 300 miles through jungle—carrying Arriflex camera, film, and tape—and returned with footage that became "Angola: Journey to a War." The war they had seen and photographed was one of atrocities on both sides. They were especially disturbed to find, in a native village demolished from the air, napalm bomb fragments with instructions in the English language. They photographed the evidence and took a bomb nose-cone with them. En route home via the

Congo, they met an Englishman who was sure the bomb was of American manufacture. "How do you know?" he was asked. He pointed to the English instructions. "Split infinitive." In the United States, American manufacture of the napalm bomb was confirmed. Was it part of American military aid to Portugal? Or was it NATO materiel, supplied for protection against communist aggression? The issue remained unresolved, but NBC made a decision. The bomb sequence was removed from the film because, said Gitlin, the Russians would "use it against us."[6] "Angola: Journey to a War" won awards, but the film makers felt wounded.

Increasing numbers of film makers wanted to go to far places and bring back the truth. What they often brought back on film was their own preconceptions. This happened even with the best. In 1960 the able Reuven Frank of NBC news was dispatched to Hong Kong with David Brinkley to produce the documentary *Our Man in Hong Kong*, and he later published a journal of hectic days of shooting sandwiched between fittings for Hong Kong suits. They shot shacks of hillside squatters, busy streets, maritime police at work in the teeming harbor. On the final days they assessed what they had achieved. Frank writes:

> December 21. Last minute rush. Not enough clichés. Send crew out to film clichés. Sequence of slit skirts. . . . Very difficult to film. Finally hire four dance hall girls who do as told, into and out of rickshaw, up and down streets for different angles. Girl in cheogsam getting out of sports car—best of both civilizations. Tea for dance hall girls. How to list on expense accounts? Suit doesn't fit. Brinkley's do.[7]

There was a feeling, perhaps, that it should resemble the Twentieth Century-Fox *Hong Kong*—so that people would believe it.

Film makers were becoming acutely aware of the ambiguity of the film image. It had the ability to arouse strong emotions, but these emotions could be steered in various directions. Narration could, to some extent, control what the viewer saw and how he responded. In 1960 the San Francisco hearings of the House committee on un-American activities were picketed by students protesting the committee's procedures. The students were driven off with fire hoses, and many were beaten. San Francisco television stations covered these events; the footage was later subpoenaed by the

6. Interview, Robert Young.
7. Frank, "Life With Brinkley," in Barrett (ed.), *Journalists in Action*, p. 300.

House committee and turned over to film producer Fulton Lewis III, who used it to make *Operation Abolition.* Its narration tells a story of violent students manipulated by a communist conspiracy and posing a threat to law and order. The film had wide use during 1960–61 among conservative groups and on television stations. But another film maker took the same footage and wrote another narration, making it a case history of police brutality and the stifling of democratic processes.

An element in the documentary upsurge was the development of new equipment. The 16mm camera was becoming standard for television news and documentaries. Whereas *See It Now* had used bulky 35mm equipment, *CBS Reports* used 16mm. NBC News had also made the transition from 35mm to 16mm, gaining maneuverability.

For the documentary producer who wanted to record synchronized sound on location—as the Drew unit, against all odds, had done in *Primary*—there was an especially notable advance. In 1960 the umbilical cord between camera and recorder became obsolete with the invention of methods for synchronizing them without wire connection.[8] The wire between microphone and recording equipment could likewise be abolished by use of the wireless microphone, which communicated its signals to the recording equipment via miniature transmitter. Murrow's *Person to Person* series had pioneered the use of this device in television. Now the performer with his microphone, the cameraman with his camera, the sound engineer with his recording equipment, could all be free agents. To the *cinéma vérité* movement all this was especially important.

Kennedy, pursuing his policy of accessibility to the camera, made an historic innovation in presidential press conferences. They could be filmed or telecast live, without strings. Pre-release scrutiny was abolished. Kennedy at once proved fantastically nimble and effective—and sometimes witty—in his ad-lib replies and exchanges. He prepared carefully, letting aides hurl questions at him on every conceivable current question. They enjoyed mimicking prominent newsmen in anticipated questions.

The Kennedy innovation was a smashing success for everyone but the newspaper correspondent, whose role was further reduced by the change. There seemed little point in phoning a report afterwards to the newspaper—the office could get it all on television. The newspaperman survived partly as a publicity agent for his paper: he could get up, be seen on tele-

8. The first such system was based on the use of tuning forks, but these were soon superseded by crystal-controlled motors.

vision, and ask a question—after mentioning his name and that of his newspaper.

President Kennedy also pursued promptly his interest in the documentary. He invited Drew and his group to document the beginnings of the regime, in a film that acquired the title *Adventures on the New Frontier.* Drew, with Pennebaker at the camera, even followed Kennedy into a meeting with the Joint Chiefs of Staff. This was the President's idea. He would let them know, he said, when it was time to leave. Entering the meeting, he explained to the Chiefs, "These men are with me." There was brief organizational talk; then came the real agenda. It was time, someone said, to discuss "Caribbean maneuvers." The President nodded to Drew to indicate that it was time to leave.

At this point, government withdrew from public scrutiny. Soon the sense of a golden age was harshly shattered.

ATTENTION, HAVANA!

Published accounts of the invasion of Cuba at the Bay of Pigs have dealt mainly with reasons for its failure. Military difficulties have been reviewed in detail. But the invasion deserves study as a classic case history of deception—of government leaders by government leaders, of news media by government leaders, of the public by government leaders, and of the public by news media.

The plan developed by the CIA was built around a series of lies, or cover stories—the preferred term gave a sense of professionalism. The planners were not troubled by the deceptions, which were always considered a necessary part of saving the free world, but the cover stories inevitably ensnared others, often unwittingly at first—in the White House, the Defense Department, the State Department, and the news media. The extent to which these, when called on to participate consciously, were able to rationalize and overcome their scruples is illuminating. Some—including intellectuals—seem to have felt exhilarated by finding themselves in an atmosphere of *Realpolitik.* Others, according to Arthur Schlesinger, Jr.—who may have been talking, to some extent, about himself—wanted to show the CIA and the Joint Chiefs "that they were not soft-headed idealists but tough guys, too."[1]

In going along with the cover stories, many were not aware of the scope

1. Schlesinger, *A Thousand Days,* p. 256.

of the deceptions until they themselves were deeply involved. And most apparently felt sure the discomforts would be short.

According to the plan, landings on the beaches would cause thousands within Cuba to rise and overthrow Castro. To assist these uprisings, CIA planes began dropping arms and ammunition into Cuba late in 1960, during the Eisenhower administration. The invasion would have the active support, according to CIA estimates, of at least a quarter of the Cuban people.[2] Under these circumstances, the United States role in recruiting and training the invasion force of some 1500 men and providing it with money, arms, ammunition, ships, tanks, airplanes, bombs, napalm, and a United States navy escort, would soon be forgotten in the ecstasy of victory. The action would become a movement of millions rather than 1500. The lies, the puzzling discrepancies, would shrink into minor historical footnotes.

By 1960 approximately half the people in the United States depended for their news primarily on television; many others depended primarily on radio. The broadcasting media thus played a crucial role in shaping public ideas about world events.

Few viewers and listeners were aware that broadcasters had only thin channels of information concerning many foreign events, such as those in Cuba and Guatemala. For public relations as well as human reasons, broadcasters seldom discussed this problem on the air; besides, they had developed a confidence that they could put two and two together. The networks during the invasion period did not have regular correspondents in Cuba or Guatemala. They depended heavily on wire services, especially the Associated Press. The AP stringer in Guatemala, when asked to check stories about the training camp rumors, merely consulted the Ydigoras government and relayed its denials.

Thus the network news items stemmed largely from Washington representatives of networks and wire services, who got their information to a large extent from people in the White House, the State Department, and the Defense Department—via press conferences, releases, background sessions. In the Cuban affair these government spokesmen gave out data that had come, in the first place, from the CIA.

Other sources of information available to reporters included two of particular interest. The various Cuban exile groups had set up a Cuban Revolutionary Council, which was calling on the people of Cuba to overthrow

2. *Ibid.*, p. 247.

The Virginian. At left, James Whitmore, Harper Flaherty, James Drury, Doug McClure. At right, Jeannette Nolan, Sara Lane.

NBC photos

Bonanza. Extreme right, Lorne Green; second from left, Connie Hines.

Drew Associates photos

Yanki No! Produced by Drew unit of Time, Inc. for ABC, 1960. Below, Fidel Castro.

Crisis: Behind a Presidential Commitment. ABC special, 1963.

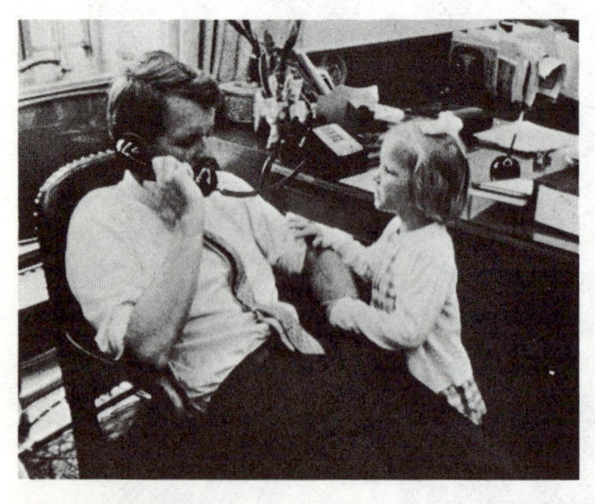

U.S. Attorney General Robert F. Kennedy and daughter Carrie.

Deputy Attorney General Nicholas Katzenbach.

Wide World

Shot seen round the world: Jack Ruby (right) shoots Lee Harvey Oswald, November 24, 1963.

United Press International

Meanwhile at the Capitol: the body of John F. Kennedy is moved to the rotunda. Mrs. John F. Kennedy with children and Robert Kennedy; President and Mrs. Lyndon B. Johnson.

CAMPAIGN SPOTS

Charles Guggenheim, campaign film specialist. Below, his crews at work.

Guggenheim

Black Star

Making a campaign film for Senator George McGovern.

T. Mike Fletcher

On location with Representative John J. Gilligan.

Castro. It was considered the genesis of a government in exile. It issued bulletins through the New York public relations firm of Lem Jones, who had formerly done public relations work for Wendell Willkie and for Twentieth Century-Fox. Though these bulletins were all issued in the name of the Cuban Revolutionary Council, all were dictated to Jones over the phone by the CIA, which actually employed him at the expense of the American taxpayer—a fact not revealed until much later. The CIA had organized the Cuban Revolutionary Council by bringing together exile representatives it felt it could trust, but the CIA role was concealed.[3]

Another source was Radio Swan, the 50,000-watt transmitter broadcasting from Swan Island in the Caribbean and also claiming to represent Cuban exiles. It also—as was later revealed—was a covert CIA project, under absolute CIA control. Correspondents covering the invasion story sometimes checked it for information, without knowing its true status.

As the Cuban situation grew hotter, these various voices—a well-orchestrated ensemble—gave a coherent impression. Their story was passed on to the American people via television, radio, and the press, and seemed to be "the facts."

One false note sounded through the music. Cuban exiles in Miami heard rumors that did not agree with the handouts. The efforts of reporters—including network reporters—to track these rumors down eventually gained headway.

On April 6, 1961, CBS broadcast a short statement from Stuart Novins in Miami, suggesting that the United States was, in spite of denials, deeply involved in preparations for a Cuban invasion and that the invasion was "imminent." He had worked with Tad Szulc of the New York *Times,* who sent the *Times* a similar dispatch. The *Times* was frightened by the "security" implications and toned the dispatch down, omitting "imminent"; as toned down, it appeared on April 7. Five days later President Kennedy seemed to sweep the story aside, telling the cameras and microphones that there would not be, "under any condition, an intervention in Cuba by the United States."[4] His words seemed to repudiate Novins and Szulc. The *Times* felt its caution had been wise and that Szulc had gone overboard.[5]

3. The Lem Jones role and relationship to the CIA are discussed in detail in Schlesinger, *A Thousand Days,* pp. 275, 284–5; and Wise and Ross, *The Invisible Government,* pp. 29–72, which provides the texts of bulletins quoted in the following.
4. Press conference, April 12, 1961.
5. Szulc, "The New York *Times* and the Bay of Pigs," in Brown and Bruner (eds.), *How I Got That Story,* pp. 321–3.

President Kennedy had actually—it was later revealed—approved the invasion plan two days before the Novins bulletin. He had stipulated that no United States "armed forces" should take part in the landings, and perhaps felt that this stipulation reduced the American role to less than "intervention." The stipulation had been accepted by the CIA.

Among the President's top-level advisers, only Senator J. William Fulbright appears to have protested the plan, calling it "of a piece with the hypocrisy and cynicism for which the United States is constantly denouncing the Soviet Union in the United Nations and elsewhere." Allen Dulles was supremely sure of results—even surer, he said, than he had been of the Guatemala intervention. Secretary of State Dean Rusk and Secretary of Defense Robert McNamara approved the plan, along with the Joint Chiefs of Staff. USIA chief Edward R. Murrow, who would have the task of explaining it all to the world, was not consulted. He is said to have been aghast when he learned the facts.

Kennedy faced the choice of approving—or scuttling—a plan in preparation for a year and approved by President Eisenhower, the most celebrated military leader of the age. Kennedy was told delay would be fatal because Cuba would soon get Soviet MIGs and pilots trained in Czechoslovakia, and because the rainy season would soon turn the Guatemalan training sites into a sea of mud. Besides, the bitterness that would result from cancellation was portrayed as unimaginable. The trainees could not be left in Guatemala; they would have to be brought to the United States. As Dulles put it, "We have a disposal problem."[6]

In the end it seemed easiest to proceed. And it might just turn out to be grand and glorious. So far, Kennedy had been lucky.

The invasion force was transferred by plane from Guatemala to an embarkation point in Nicaragua. There the CIA had assembled a fleet of seven ships. President Luis Somoza of Nicaragua, surrounded by bodyguards and with his face well powdered, came to wave goodbye. He said, "Bring me a couple of hairs from Castro's beard!" As the ships sailed late on April 14, trainees were comforted by the sight of U. S. destroyers and one or two larger ships. A U. S. submarine was seen circling. The men were told by their CIA trainers that they would have "protection by sea, by air, and even from under the sea," and that marines would be near by if needed.[7]

6. Schlesinger, *A Thousand Days*, pp. 242, 251, 258–9.
7. Details of the invasion are based on accounts by participants in Johnson, *The Bay of Pigs;* other sources as noted.

The CIA also had thirty-six planes available and had trained exile pilots to fly them. Before dawn on April 15, nine planes took off—B-26's of World War II design. They had been chosen by the CIA because many nations had such planes and their source would be difficult to establish. More importantly, Castro's own air force included B-26's. The CIA planes had been repainted with the insignia of Castro's FAR—Fuerza Aerea Revolucionaria.

Eight of the disguised planes headed for Cuba. Their mission was to bomb Castro's air force into extinction by the first light of dawn—before the FAR planes could get off the ground.

The ninth plane, which had already been provided with photogenic bullet holes, headed for Florida. A cover story was unfolding.

Its pilot, Mario Zuniga, who had been recruited a few weeks earlier in Miami, landed at Miami International Airport and announced that he was one of a number of FAR pilots who had revolted and were devastating Castro's air bases. He himself had been hit during these operations, he said, but had managed to reach Miami. Press photographers took pictures of the plane and bullet holes. Then Zuniga was hurried into seclusion. Story and pictures went out over the press wires. An April 14 Associated Press bulletin began: "Pilots of Prime Minister Fidel Castro's air force revolted today and attacked three of the Castro regime's key air bases with bombs and rockets." The bulletin was broadcast by stations throughout the nation and the world, including those of the Voice of America—which assumed it was the truth.

The Cuban Revolutionary Council in New York released a statement acclaiming the "heroic blow for Cuban freedom." The council said it was not surprised because it had been in touch with the brave pilots in Cuba and had encouraged the move. The statement was quoted on radio and television.

White House press secretary Pierre Salinger said he knew nothing about it and that the United States was trying to get information. He was duly quoted.

That same afternoon at 3 p.m. Ambassador Raul Roa of Cuba spoke in the General Assembly of the United Nations. He was angry. He said the United States had launched a surprise attack on Cuba with mercenaries trained by "experts of the Pentagon and the Central Intelligence Agency," and that seven people had been killed. The United States, he said, was scandalously passing this off as an attack by FAR defectors.

U. S. Ambassador Adlai Stevenson, who did not know the truth but thought he did, rose in shocked indignation. Television cameras were on him as he spoke.

> STEVENSON: No United States personnel participated. No United States government airplanes of any kind participated.

He held up a photograph of United Press International.

> STEVENSON: I have here a picture of one of these planes. It has the markings of the Castro air force right on the tail, which everyone can see for himself. The Cuban star and initials FAR—Fuerza Aerea Revolucionaria—are clearly visible. Let me read the statement which has just arrived over the wire from the pilot who landed in Miami. . . .

Stevenson repeated the full CIA fabrication. That evening and on the following day—Sunday, April 16—his firm rejection of "wild" Cuban charges was widely reported and applauded on the air and in newspapers. Kennedy is said to have been aghast at the realization that the credibility of the principal American spokesman at the United Nations was being placed in jeopardy before the world. But in hope of quick developments in Cuba, Stevenson was still allowed to remain in ignorance.

On Monday Raul Roa again rose in the General Assembly. He stated that the CIA had organized the invasion, that it was using Opa Locka air field, that a CIA agent with the cover name of "Bender" was in charge there, and that the CIA had poured $500,000 into invasion preparations. He apparently underestimated the cost; otherwise, his statements were correct.[8] But for the second time in forty-eight hours Ambassador Stevenson stood up before the cameras and delegates and flatly denied the charges.

Lem Jones, on CIA instructions telephoned the night before, was meanwhile releasing a bulletin to all news media.

> CUBAN REVOLUTIONARY COUNCIL
>
> VIA: Lem Jones Associates, Inc.
> 280 Madison Avenue
> New York, New York
>
> Bulletin No. 1
>
> The following statement was issued this morning by Dr. Jose Miro Cardona, president of the Cuban Revolutionary Council.

8. The CIA agent "Bender" figures in every participant's account. Schlesinger gives his real name as Droller and says he was a German refugee. The Guatemala air base, built by an American contractor with offices in the Chrysler building, cost $1,200,000, according to Wise and Ross, *The Invisible Government*, p. 28.

"Before dawn Cuban patriots in the cities and in the hills began the battle to liberate our homeland. . . ."

The bulletin, marked *for immediate release,* was dated April 17, 1961.

Dr. Miro Cardona, in whose name this was issued, actually knew nothing about it. But he heard it broadcast over a Florida radio station, and was furious. He was in a small building near Opa Locka, held incommunicado. The CIA had whisked him from New York the day before so that he could be flown to Cuba as soon as a beachhead was secured. Then his government was to be recognized at once as a sovereign nation; this, it was understood, would remove obstacles to overt military help. But meanwhile he was neither informed nor consulted; he was virtually held prisoner.

The same No. 1 bulletin was also being broadcast by Radio Swan, along with mysterious action calls: "Alert, Alert! . . . the fish will rise . . . the fish is red!"—apparently code signals to an underground. Then Radio Swan broadcast further prefabricated items.

VOICE: Forces loyal to the Revolutionary Council have carried out a general uprising on a large scale on the island of Cuba . . . the militia in which Castro placed his confidence appears to be possessed by a state of panic . . . An army of liberation is in the island of Cuba to fight with you against the communist tyranny of the unbalanced Fidel Castro . . . attack the Fidelista wherever he may be found. Listen for instructions on the radio, comply with them and communicate your actions by radio.

To victory, Cubans!

The invasion had in fact begun, but not on the lines of the pre-planned CIA bulletins. In violation of the President's orders, the first man ashore at each of the two main landing places was a United States frogman. Two of the ships were quickly sunk by the FAR air force, which had not been knocked out. Two other ships, carrying supplies for the landing parties, fled south. They were later intercepted and turned back by the U. S. navy. Meanwhile the landing parties were in peril. Six CIA planes were shot down the first day; the second air strike was called off, then reinstated. The invaders had been told that the defenders would not have tanks, but this proved incorrect; the CIA tanks of the invaders clashed with defending tanks on the beaches. Falsely marked CIA planes caused confusion and unforgettable bitterness. A convoy of Castro militiamen waved caps at a plane with familiar FAR markings. As they waved, machine guns and rockets hit them full. An ambulance exploded. In some encounters the invaders used United States phosphorus grenades. "Everything was on fire," an invader said later. The screaming of the defenders was "just like hell."

In Washington a State Department spokesman, Joseph Reap, who perhaps did not know he was lying, announced: "The State Department is unaware of any invasion."

White House press secretary Pierre Salinger, who may also have been ignorant, was likewise uninformative. An AP bulletin quoted him saying: "All we know about Cuba is what we read on the wire services."

Secretary of State Rusk was more explicitly deceptive, telling a press conference: "The American people are entitled to know whether we are intervening in Cuba or intend to do so in the future. The answer to that question is no."[9]

Over the phone the CIA dictated to Lem Jones a second bulletin, announcing a successful landing; then a third, proclaiming "a wave of sabotage and rebellion."

Meanwhile the CIA's tax-supported Radio Swan exhorted Cubans:

> VOICE: Attention! . . . take up strategic positions that control roads and railroads! Make prisoners of or shoot those who refuse to obey your orders!

It urged Cuban pilots:

> VOICE: See that no Fidelist plane takes off. Destroy its radio; destroy its tail; break its instruments; and puncture its fuel tanks!

Lem Jones's Bulletin No. 4, again dictated by the CIA, claimed that the landing parties were being attacked by "heavy Soviet tanks and MIG aircraft which have destroyed sizable amounts of medical supplies and equipment." The statement about the MIGs was unfounded; no MIGs were seen during the invasion.

With a sense of showmanship, Castro gave many of these American bulletins added circulation. A funeral for those killed in the first raid was broadcast throughout Cuba and heard in Miami. Castro himself made the funeral address, calling the attack "cunning" and "cowardly" and com-

9. Press conference, April 17, 1961.

paring it to Pearl Harbor. He read various CIA-inspired news bulletins, calling them "pure fantasy. . . . Even Hollywood would not try to film such a story." Without question, the occasion consolidated his position in Cuba.

On Tuesday, April 18, Soviet Premier Nikita Khrushchev repeated the charge that the United States had armed and trained the exiles. He said the Soviet Union would give Cuba "all necessary assistance" unless Washington halted the invasion. In reply, Kennedy warned Khrushchev to stay out of Western Hemisphere affairs. Kennedy said he did not intend to be lectured "by those whose character was stamped for all time on the bloody streets of Budapest." It was a statement curiously reminiscent of Khrushchev's retort: "We also have some dead cats we can throw at you."

On April 18 the exile air force was exhausted and rebellious. In violation of President Kennedy's orders, United States instructors began to fly bombing missions from Nicaragua. By the end of the day ten Cuban and four American pilots had been killed. Catastrophe was near. It was the day of the President's formal reception for congressmen. In white tie and tails he danced with Jacqueline Kennedy as the band played "Mr. Wonderful." Vice President Johnson danced with Lady Bird; then they switched partners. At midnight, still in white tie and tails, the President met with the CIA and Joint Chiefs, who pleaded for permission to send U. S. naval and air power into immediate action to save the invasion. Kennedy finally approved a one-hour flight by *unmarked* navy planes, which were to fly between the CIA B-26's and any opposing aircraft, and could fire if fired on. This absurd compromise—apparently intended to provide some thin basis for diplomatic justification of resulting action—was accepted by the military leaders, but due to a mix-up the navy planes arrived at the wrong time; the B-26's were using Nicaragua time.[10] On April 19 the invasion collapsed. One by one the invaders were hunted down. Some escaped via rescue ships and planes.

In Cuba foreign reporters were allowed to interview prisoners. A tall blonde French girl asked a prisoner why he had come. "To fight communists." She suggested he had been "brainwashed," but he insisted, "I'm glad I came." She said, "You're crazy," and walked away.

While government spokesmen had stuck through four bitter days to the story of an uprising of patriots in which the United States had no part, the story was coming apart. The photo of the flier Zuniga and his plane, which

10. Sorensen, *Kennedy*, p. 299.

Stevenson had held up as evidence in the United Nations, was causing trouble. In the Miami Cuban colony, New York *Times* reporter Tad Szulc found some who recognized Zuniga as an exile. The trail led Szulc to Zuniga's wife and child, living in Miami.[11] Zuniga himself had been spirited back to Nicaragua for more flights. Such findings were causing the cover story to unravel. The fury of some survivors, who said they had been led to expect United States "air cover" and early Marine landings, brought further information to light.[12]

Reactions to the collapse and the revelations that followed were varied. To a Fair Play for Cuba Rally sociologist C. Wright Mills telegraphed: "I feel a desperate shame for my country." While he attacked American policy as a moral disgrace, network newsmen generally used the term "fiasco." In busy discussion of planning errors, the horrifying implications of the Bay of Pigs venture were shunted aside. There was little discussion of the precariousness of our news sources, and the demonstrated effectiveness with which the CIA, to conceal its activities, could corrupt them and maneuver other government agencies—and the news media—into helping them do so. Discussion of the illegality of the venture and its violation of treaty obligations—clearly indicated by Nixon in the Great Debates—was also minimal. The very term "Bay of Pigs fiasco," used repeatedly on the air, was a face-saving term. It made the failure, not the action, the subject of discussion.

In this spirit ex-President Eisenhower advised at Gettysburg: "Don't go back and rake over the ashes. . . . To say you're going into methods and practices of the administration—I would say the last thing you want is to have a full investigation and lay all this out on the record."[13]

A few days later all records of the Guatemala camp were put in a freshly dug hole and bulldozed over.[14]

A committee appointed by Kennedy to find out "what went wrong" came up with the answer that it was "a shortage of ammunition." This preposterous four-man committee of inquiry included two people who had had a leading part in planning the venture and its web of cover stories. One was Allen Dulles.

Among some people the reaction was: "Next time let's do it right."

11. Szulc, "The New York *Times* and the Bay of Pigs," in Brown and Bruner (eds.), *How I Got That Story*, p. 325.
12. Johnson, *The Bay of Pigs*, pp. 74, 81.
13. Press conference, May 1, 1961.
14. Johnson, *The Bay of Pigs*, p. 350.

It soon appeared unlikely that facts slowly filtering out through various media would ever succeed in amending the version experienced via newscast and television press conference. In May the former Cuban television executive Goar Mestre, who had transferred his commercial television activity to South America—mainly Argentina—came to Washington for an NAB meeting and addressed his American colleagues, with whom he shared many interests. He said it was certainly not true, as Castro claimed, that the United States was "interventionist." That might have been true before 1934, he said, but not now. Mestre urged American broadcasters to say, whenever quoting Castro, that his tirades were not true.[15] Most broadcasters had, of course, been doing exactly that.

If the invasion had gone according to plan, how much of the truth would ever have emerged? Would history have been based on the official record—the cover stories? This raised another question: how many other events were going into the books in terms of televised cover stories?

If the nation learned little from the Bay of Pigs invasion, it may have been partly because the mass media, discomfited at the extent to which they had been used to serve conspiratorial ends, exposed it reluctantly. As a result, the disaster brought few changes. Allen Dulles was replaced, but the machinery that had made it all possible, including the extraordinary Central Intelligence Agency Act—a charter of deception—remained intact, available for other ventures.

Radio Swan remained on the air, changing its name to Radio Americas—to the annoyance of the Voice of America, which felt, with reason, that its own uncertain credibility would be weakened.

Kennedy, under stress, said that the mass media should have refrained "in the national interest" from discussing the imminence of the invasion—as Novins had done. Later Kennedy felt that the national interest might have been better served if the news media—in spite of official obfuscation and deception—had ferreted out the facts and made them known. Perhaps, he thought, they might have prevented the invasion.

The events were a severe setback for the Kennedy leadership. Liberals blamed him for having approved the invasion; right-wing elements, for failing to give it more military muscle.

Liberals were quicker to come back. They did so because of Kennedy's actions in a number of directions including the Peace Corps, civil rights—and television.

15. *Broadcasting*, May 15, 1961.

WASTELAND

When President Kennedy appointed Newton N. Minow chairman of the Federal Communications Commission, few broadcasters knew the name. They wondered who had recommended him.

Investigation brought out that it was Kennedy's own idea and that he had met Minow at the 1956 Democratic convention when Minow was an aide to Adlai Stevenson.

Chairman Minow at once made clear that he wanted to strengthen noncommercial television—"for real diversification of program fare"—and favored federal financial help for the purpose. He also proved himself a phrase-maker, and gave the industry an unnerving sample of the talent at the 1961 Washington convention of the National Association of Broadcasters.

The session was, on the whole, a dismaying experience for the assembled executives. Chief speakers were President Kennedy; ex-Governor LeRoy Collins of Florida, newly selected president of the NAB; and Newton Minow, new FCC chairman. The only one who was kind to his audience was Kennedy.

He scored a political coup by bringing with him Commander Alan B. Shepard, "America's astronaut," who had just been lofted from Cape Canaveral and come down in the near-by Atlantic. It was a modest space achievement in comparison to one the Russians had just completed—a manned orbit of the earth, by Yuri Gagarin. But Kennedy made a point that struck home. The Gagarin flight had been made in secret and announced only after success was assured. Shepard's flight had been launched in full view of television cameras—as had humiliating failures that preceded it. Kennedy was able to exploit both success and failure.

> The essence of free communication must be that our failures as well as our successes will be broadcast around the world. . . . That is why I am here with you today. For the flow of ideas, the capacity to make informed choices, the ability to criticize, all the assumptions on which political democracy rests, depend largely upon communication. And you are the guardians of the most powerful and effective means of communication ever designed.

The broadcasters had, through trade-press speculation, been led to expect a move toward censorship, to prevent disclosure of future operations like the Cuban landings. But Kennedy moved directly in an opposite direction, and brought cheers.

Then came Governor LeRoy Collins. Most NAB members looked on the NAB presidency as a lobbying job. Its occupant was supposed to keep Washington off their necks. But Collins seemed to feel he should lead them to higher levels of aspiration. He said they should equate their interests with the public interest and not just strive "to make every dollar possible." He recommended self-discipline. Many NAB members were dumbfounded. They wondered, said *Broadcasting*, whether they had hired "a president or a proctor." In the corridors there was talk of resigning from the NAB.

On top of that came Minow. He did not *look* menacing, but mild-mannered and clerkish. He began with words of admiration.

> Yours is a most honorable profession. Anyone who is in the broadcasting business has a tough row to hoe. You earn your bread by using public property. When you work in broadcasting you volunteer for public service, public pressure, and public regulation. . . .
>
> I can think of easier ways to make a living. . . .
>
> I admire your courage—but that doesn't mean I would make life easier for you.

He was happy to find their "health" good. A 1960 gross revenue of $1,268,000,000 had given broadcasters a profit, he noted, of $243,900,000 before taxes—a return of 19.2 per cent. "For your investors the price has indeed been right."

He said television had had great achievements and delightful moments, and mentioned some in the course of his speech. Except for the fantasy of *Peter Pan* and *Twilight Zone,* his choices were oriented toward reality. They included *Project Twenty, Victory At Sea, See It Now, CBS Reports,* the Army-McCarthy hearings, convention and campaign broadcasts, the Great Debates, *Kraft Theater, Studio One, Playhouse 90*. When television was good, he said, nothing was better.

> But when television is bad, nothing is worse. I invite you to sit down in front of your television set when your station goes on the air and stay there without a book, magazine, newspaper, profit and loss sheet or rating book to distract you—and keep your eyes glued to that set until the station signs off. I can assure you that you will observe a vast wasteland.

> You will see a procession of game shows, violence, audience participation shows, formula comedies about totally unbelievable families, blood and thunder, mayhem, violence, sadism, murder, western badmen, western good men, private eyes, gangsters, more violence, and cartoons. And endlessly, commercials—many screaming, cajoling, and offending. And most of all, boredom. True, you will see a few things you will enjoy. But they will be very, very few. And if you think I exaggerate, try it.
>
> Is there one person in this room who claims that broadcasting can't do better? . . .
>
> Gentlemen, your trust accounting with your beneficiaries is overdue. Never have so few owed so much to so many.

Near the end he said:

> I understand that many people feel that in the past licenses were often renewed *pro forma.* I say to you now: renewal will not be *pro forma* in the future. There is nothing permanent or sacred about a broadcast license.

In the corridors, reporters for *Broadcasting* sampled reactions. Most NAB members agreed that there had been nothing like this since James Lawrence Fly compared them with a dead mackerel in the moonlight, shining and stinking. "A young smart alec." "He certainly speaks well—but not of us." "An unrealistic bureaucrat." "A mixture of arrogance and ignorance." "I think he's bucking for a bigger government job." "I have to admit he's got guts—or else he's pretty naïve." "Much of what he said was true but he could have used more tact." "A naïve young man who has read all the books but hasn't had to meet a payroll." "I can watch any TV station all day long and enjoy it." "It's a sneaky kind of censorship. . . ."

Many NAB members even forgot momentarily to be angry at Governor LeRoy Collins, their new president. Collins's own reaction was: "I think he overstated his case, to a degree."[1]

The Minow attack received wide praise from newspapers. He became a favorite bogeyman for much of the broadcasting and advertising trade press.

Minow proved to be more than a phrase-maker. His concern for noncommercial broadcasting came to an early test, in which he showed himself a dexterous bureaucrat.

1. *Broadcasting,* May 15, 1961.

Noncommercial television was clearly in a bad plight. Producing on starvation budgets, it still lacked outlets in New York, Los Angeles, Washington. It had scarcely begun to make an impact on the national life.

In 1961 National Telefilm Associates, licensee for WNTA-TV—operating on channel 13 in the New York metropolitan area—announced that it would sell its facilities and transfer the license. It was offered $6.6 million by a group headed by David Susskind and backed by Paramount Pictures. Another group, headed by Ely Landau, offered $8 million for WNTA-TV and a related radio station, WNTA. A spirited auction seemed to have begun.

National Educational Television was desperately anxious to have channel 13 as New York outlet for its noncommercial programming. Pledges from foundations scraped together an offer of $4 million, but it was quickly rejected by National Telefilm Associates.

Chairman Minow was equally anxious for channel 13 to go noncommercial, but his hands appeared tied. In 1952 an ingenious amendment had been grafted onto the Communications Act. It *forbade* the FCC, when acting on transfer proposals, to consider

> . . . whether the public interest, convenience, and necessity might be served by the transfer, assignment, and disposal of the permit or license to a person other than the proposed transferee or assignee.[2]

National Telefilm Associates was thus in a position to hold its auction, submit the name of the winning bidder to the FCC, and insist that his qualifications be assessed without reference to any other question. Judged by precedent, Susskind and Landau were qualified.

But the auction spectacle was a disgrace. The station facilities were modest, perhaps worth a half million. What was being auctioned was the ear of a world metropolis. Theoretically the channel was public property; under the law the licensee, National Telefilm Associates, had no ownership in it.[3] Yet it was auctioning the channel. The 1952 amendment made the spectacle possible.

This apparently blocked any possible action on behalf of noncom-

2. Communications Act of 1934, Section 310 (b) as amended by Public Law No. 554, July 16, 1952, 82nd Congress.
3. "It is the purpose of this Act . . . to provide for the use of such channels, but not the ownership thereof, by persons for limited periods of time, under licenses granted by Federal authority, and no such license shall be construed to create any right, beyond the terms, conditions, and periods of the license." Communications Act of 1934, Section 301.

mercial television, but Minow saw one possible weapon. National Telefilm Associates was anxious for cash and wanted a quick deal.

Minow now persuaded the FCC to schedule hearings—an "inquiry"[4]—on the desirability of securing noncommercial outlets in New York and Los Angeles. During the inquiry applications for license transfers in those cities would be held in abeyance.

National Telefilm Associates and the commercial bidders fumed. Minow was offering them a prospect of months—perhaps more than a year—of delay. Funds might have to be held in escrow awaiting FCC action. Bids began to be withdrawn. National Telefilm Associates finally asked the FCC to approve transfer to the noncommercial group.

Minow had to face one more challenge to his maneuver. Because the channel was technically allocated to Newark, N.J., Governor Robert Meyner of New Jersey was enlisted in efforts to block the transfer. Intensive diplomacy, with a guarantee of New Jersey-oriented programs, resolved this obstacle.

The price finally paid to make channel 13 noncommercial was $5.75 million. The licensees for all competing New York television stations—WABC-TV, WCBS-TV, WNBC-TV, WNEW-TV, WOR-TV, WPIX—contributed. All gained advantage from a reduction in the number of local competitors for advertising revenue. Before contributing, they therefore asked for letters from the U. S. Department of Justice stating that the gifts would not be considered moves in restraint of trade. Such letters were provided. After months of intensive moves and countermoves, Minow had the satisfaction of seeing channel 13 transferred to educational television.[5]

Educational television received another important boost in 1962 through legislation advocated by Minow. It authorized federal grants—for station construction, not programming.[6]

Later Minow moves had an anti-monopoly flavor. Network affiliation contracts had long contained clauses giving the network virtual control over blocks of time on affiliate stations. Some lawyers had argued that these "network option" clauses were equivalent to the block booking through which major Hollywood studios had controlled theaters, and which the U.S. Supreme Court had outlawed in 1948 in *U.S. v. Para-*

4. FCC Docket No. 14006.
5. Boekemeier, *The Genesis of WNDT*, pp. 1-32.
6. Educational Television Facilities Act of 1962.

mount et al.[7] In 1963 the FCC, led by Minow, banned "network option" clauses as an improper surrender of licensee responsibility. Although strongly opposed by the networks—which made prophecies of disaster—the move had only minimal effects on schedules and profits.

In another move, Minow persuaded Congress to require that sets manufactured after January, 1963, be equipped to receive UHF as well as VHF channels. Set manufacturers, many of whom had VHF stations, had been in no hurry to spread the competition. The move was of crucial importance to noncommercial television.

Kennedy is said to have given Minow continual encouragement, telling him, "You keep this up! This is one of the really important things."[8] This firm support for Minow kept the industry in a state of uneasiness.

One other public figure seized the spotlight in 1961 as a warrior on the battlefield of television: Senator Thomas J. Dodd of Connecticut. In this case the outcome was different—and mysterious.

Thomas J. Dodd had entered the Senate in January 1959 with a mixed background and reputation. An impressive figure with classic profile, he had considered a career as actor or priest before turning to law. He served a year as FBI agent, then became a prosecutor for the Department of Justice, working on espionage cases. He also handled some civil rights cases, which enabled him later to enter politics with a civil liberties aura. But he once told an aide that he wanted, above all, to become director of the FBI or the CIA.

He began his Senate career with cold-war speeches but leaped into national prominence when, as chairman of the Senate subcommittee on juvenile delinquency, he took up a subject that had troubled Senator Kefauver—television violence.

Clearly the intervening years—and thousands of telefilms—had given the subject new attention value. As Dodd held hearings he began to win banner headlines, and letters from aroused parents, educators, clergy, police, psychiatrists, and social workers.

Researchers were beginning to have strands of evidence on the role of dramatized violence. There was testimony on the findings of Professor Richard H. Walters of the University of Waterloo in Ontario, to the effect

7. The issue was exhaustively discussed in the 1958 "Barrow Report," by a committee under Dean Roscoe L. Barrow of the University of Cincinnati Law School. See *Network Broadcasting*, pp. 279–400.
8. Schlesinger, *A Thousand Days*, p. 736.

that male hospital attendants, after seeing a filmed knife-fight, were much harsher in their handling of patients than a comparable group that had seen an innocuous film. Similarly Dr. Albert Bandura, professor of psychology at Stanford University, testified that children, after seeing a violent film, played more aggressively than comparable groups who had not seen the film. The findings suggested that the filmed action, instead of "purging" hostile feelings, might stimulate them, or at least bring them to the surface. Dr. Bandura also found that children copied the *kinds* of aggressive action they had seen, regardless of whether the film story had or had not rewarded the aggressor. The findings cast doubt on the value of "crime does not pay" endings stressed by code administrators.[9]

But when Dodd turned to industry practices favoring the cult of violence, he really hit pay dirt. Subpoenaing voluminous files from networks and producers, his committee uncovered the phenomena which it named the "Treyz trend," the "Aubrey dictum," and the "Kintner edict."

Dodd was in his glory. The life roles to which he had aspired—priest, actor, spy, lawyer, prosecutor—all seemed to unite in the crusading Senator Dodd. His handsome head, prematurely white, became familiar in front-page photos and occasionally in television film clips. He was a national figure, mentioned as vice presidential material.

After triumphant hearings of June-July 1961, Dodd was so busy that he left things more and more to the subcommittee staff. At his direction it prepared for further hearings and drafted a summary of committee findings.[10]

But things were changing, especially with Dodd. As a member of important committees he received many blandishments and gifts, to which he responded warmly, telling an aide: "Generous-hearted friends!" He had entered politics with modest means, but he soon converted his Connecticut farm into a baronial estate—with private road, artificial lake, waterfall, stables, guest house—and lived on a lavish scale. Company planes took him to vacation resorts. As his tastes mounted, so did his financial needs, and he became anxious for "campaign contributions," which often arrived in bundles of cash. His staff—once dedicated and even hero-worshiping—began to feel uneasy about him.

He was on friendly terms with president John Kluge of Metromedia, a

9. *Television and Juvenile Delinquency,* pp. 60, 65.
10. The following is based on Boyd, *Above the Law,* pp. 192–3; other sources as noted.

company that had suddenly risen to power with valuable frequencies, including channel 5 in New York (WNEW-TV) and channel 5 in Washington (WTTG-TV).[11] Senator Dodd was a frequent Kluge guest; Senator and family were invited to the Kluge home in Beverly Hills, recently acquired from Frank Sinatra, and the Dodd daughters had a chance to sit in Frank Sinatra's barber chair. Dodd received gifts of imported champagne and campaign contributions and testimonials from Metromedia officials and friends. Metromedia executive Florence Lowe sent him a television set. Dodd wrote her: "The picture is so beautifully clear."

Meanwhile the subcommittee staff was planning further hearings. They noted that many network series mentioned in earlier testimony as especially violent were being syndicated, and shown on independent stations throughout the country. One committee aide observed: "It's as if they used our 1961 hearings as a shopping list!" Many of the programs were scheduled at earlier hours than before, and were reaching younger audiences. The staff felt that the extension of the violence wave through syndication was worthy of study. In Washington they considered channel 5 a glaring example; they suggested this as a starting point for new inquiries.

But Dodd said abruptly: "Channel 5 is out." To the astonishment of the committee, he appointed Mrs. Lowe's son to the subcommittee staff.[12]

Meanwhile there were other surprises. In November 1961 Senator Dodd arranged with NBC that its chairman, Robert Sarnoff, should testify in secret. On the day of the hearings Dodd summoned Paul Laskin, subcommittee counsel, to his office. Laskin found Sarnoff and his attorney already there. In front of Sarnoff, whom Laskin was presumably to cross-examine, he found himself berated by Dodd for pressing the investigation too hard. Such aggressiveness, said Dodd in a booming voice, would no longer be tolerated. After a brief hearing, Dodd ordered the transcript of the session locked up.[13] There were further hearings, but *Variety*—May 16, 1962—commented on their "hot and cold running nature"; Dodd seemed a "reluctant dragon." No report on his violence probe was ever

11. Also television stations KOVR, Stockton, Calif.; WTVH, Peoria, Ill.; WTVP, Decatur, Ill.; radio stations WHK, Cleveland, Ohio; WIP, Philadelphia, Pa.; WNEW, New York, N. Y.; short-wave station WRUL, Scituate, Mass. The company had grossed $40,000,000 in 1960 and was expanding rapidly. *Broadcasting,* January 23, 1961.

12. Pearson and Anderson, *The Case Against Congress,* pp. 65, 80–81.

13. A copy was leaked, much later, to Jack Anderson. See the Pearson-Anderson column, *Washington Merry-Go-Round,* May 20, 22, 1967.

published. An interim report was mimeographed in a watered-down version for subcommittee members, but never released to the public. According to James Boyd, long-time aide and speech-writer to Dodd, staff members "bitterly deplored it as a sell-out."[14]

By the end of 1962 the broadcasting industry, while still nervous about Minow, felt easier about the Dodd menace. Hollywood people who testified under subpoena at his Washington hearings felt mainly contempt for him. They were convinced his interest was in political dividends the hearings offered—a judgment his staff had come to share. Besides, film people held to the conviction that they themselves were only producing "entertainment," no doubt mirroring a violent society but not shaping it.

Network executives were likewise unconvinced of any "causal connection," but took defensive measures in the form of more varied programming. Some was in fringe periods and included informational children's series like *Exploring* and *Discovery*. But the pressure also affected prime time, bringing a rash of new comedy telefilms. The 1961-62 arrivals also included more meaty drama. Notable was *The Defenders*, a series with legal background created by Reginald Rose of *Twelve Angry Men*. With stories that often touched current issues, it stirred memories of anthology drama. There were also impressive series with a medical background including *Ben Casey, Dr. Kildare,* and *The Nurses*. To the apparent surprise of network officialdom, these were extremely successful in terms of ratings. The success of *Ben Casey* and *Dr. Kildare* threatened a television stampede to the operating table. They encouraged similar series around schoolteachers and social workers—*Mr. Novak* and *East Side/West Side*.

Among the comedy series, many seemed unrelentingly though cheerfully familiar—*The Dick Van Dyke Show, The Bob Cummings Show, Mrs. G. Goes to College, The Father of the Bride, Hazel, The Lucy Show.*[15] But these were overwhelmed by two other phenomena. One was the clattering arrival, in September 1962, of *The Beverly Hillbillies,* which by the end of the year headed all rating lists and diverted indignation from the subject of violence. The series concerned a mountain clan from the Ozarks that had struck oil and moved to Beverly Hills on an old flatbed truck loaded with jugs of corn liquor and $25,000,000. The family re-

14. The unpublished report, titled *Television and Juvenile Delinquency: interim report,* is available in the FCC library.

15. *The Lucy Show* was a new Lucille Ball program following her divorce from Desi Arnaz and the split-up of their $20,000,000 television empire. Reruns of *I Love Lucy* continued.

mained unchanged by the new environment, and this was the source of humor.

The series popped jokes with abandon. "Do you like Kipling?" "I don't know—I ain't never kippled." The son of a Beverly Hills banker calling on Elly May asked, "Is Elly May ready?" Granny answered, "She shore is! She's been ready since she was fourteen!" Many critics were enraged at such humor, seeing it as exploitation of old hillbilly stereotypes. To their consternation the series won admirers among sophisticates, who saw it rather as a lampoon on a money-oriented society represented by Beverly Hills. The unchanging ways of the Clampett clan seemed a kind of incorruptibility. In an early episode Granny, returning to the thirty-two-room mansion, went on a rampage.

> GRANNY: I been all through these "Beverly Hills" and let me tell you it ain't easy! You got t' climb fences an' walls, jump over hedges, git around them cement ponds, where they's usually a bunch o' half-nekkid people smearin' theirselves with oil, an' yellin' at you to git out! This place is full of the laziest, greasiest, unfriendliest mess o' people I ever laid eyes on!

One of the Clampetts feeds on a bowl of wax fruit; complaining of the taste, he is told perhaps it wasn't meant to be eaten. This sets Granny off again. "Ain't s'posed t' keep cows er chickens . . . ain't 'sposed t'fire up th' still and make a little moonshine whiskey!" However, defying local mores, she pulls her rocker to the side of the swimming pool to fish, and soon catches a plastic toy fish. "There's Beverly Hills fer you! All flashy an' show on th' outside, but nothin' inside where it counts."

The Beverly Hillbillies director Richard Whorf, a former Shakespearian actor, saw another virtue in the series: "You know that no one will be killed, no one will have a brain tumor."[16]

The Beverly Hillbillies was a creation of Filmways, which had started as producer of commercials and, after earning a small fortune, turned to telefilms as a step toward feature-film production. With money gushing in from *The Beverly Hillbillies,* Filmways began offering a series about a jet-set beauty settling on a farm—a completely new idea, and reassuringly like *The Beverly Hillbillies.* It reached the air as the Eva Gabor-Eddie Albert series *Green Acres.*

The other telefilm phenomenon of 1961-62 was the animated telefilm.

16. Quotations from Hano, "The G. A. P. Loves the Hillbillies," New York *Times,* November 17, 1963.

The animators who had flocked to television had become a sizable industry of small and large units. It harbored extraordinarily gifted artists—John Hubley of Storyboard, Inc., and Ernest Pintoff of Pintoff Productions made occasional award-winning theatrical shorts—but the main work of the industry was to depict the rapid effect of decongestants on the sinuses, and similar actions. Some units survived on low-budget children's cartoons, shown mainly on Saturday morning and usually violent in a humorless way, although here and there a zany spirit emerged, as in the mock-heroic *Crusader Rabbit*, whose creators seemed determined to please themselves as well as children. In a typical episode the two heroes, Crusader and Rags, encounter the fire-breathing dragon Arson Sterno, who has two heads and therefore a split personality. The heroes tell the monster that he is serving evil. "We know," the heads say in unison, "but a job's a job, and think of the residuals we'll get." Later they are in the deadly clutches of a python. Music can perhaps save them, for the python has a cobra fondness for tunes and longs to sway in a basket. "Quick, Rags," says Crusader, "whistle a chorus of Dixie!" But Rags is afraid it hasn't been cleared. "Is it in public domain?"[17]

For years frustrated animators had banged at the doors of prime time with one proposal after another. The series that finally made the leap was *The Flintstones*. Launched late in 1960, it won high ratings during 1961 and became a Friday night leader. By 1962 it had established animation as a prime-time commodity.

Like *The Beverly Hillbillies, The Flintstones* capitalized on cultural contrasts and was aggressively plebeian. It dealt with a stone-age family which was at the same time controlled by current middle-class mores and loaded with modern artifacts. The series began with hints of social satire but settled for easier forms of comedy.

The Flintstones was the work of William Hanna and Joseph Barbera, descendants of the old Metro-Goldwyn-Mayer animation unit. The series was a bitter disappointment to many animators. The Hanna-Barbera team set up a rigorous assembly line that solved financial problems of animation but in ways that seemed to sacrifice the medium. Reducing lip movements to standardized cycles—vowel to vowel, ignoring consonants entirely—they made speech easy and cheap. This encouraged a heavy reliance on dialogue and developed a form of drama that, in spite of fanciful settings and time relationships, resembled acted telefilms.

17. "Gullible's Travels," *Crusader Rabbit* series.

By 1962 the Hanna-Barbera organization was said to be reaping an annual profit of over a million dollars from merchandising tie-ups—toys and other articles using animated characters. It was ready to launch *The Jetsons,* the story of a space-age family—a completely new idea, and reassuringly like *The Flintstones.*

During 1961–62 action series did not go into limbo. They were too deeply ingrained in industry habits and were still winning ratings, and residuals from syndication at home and abroad. There were even new arrivals—*The Gunslinger, Two Faces West, Whiplash* (1961), and *Brenner, Sam Benedict, The Virginian* (1962).

The broadcasting industry was fighting off interference not by changing money-making ways but largely by trying to protect them with increased services. A mounting schedule of news, special events, and documentaries was among the benefits. In 1962 a new factor began to contribute to this. Television entered the space age.

IN ORBIT

In 1962 the United States lunged forward in space development—and thrust television into a new era.

The United States, President Kennedy had said a few months earlier, should determine to "put a man on the moon" by the end of the decade. His aim was to "beat the Soviets," and for this purpose space research budgets were sharply increased. The President's statement seemed at first a flight of rhetoric, but in 1962 developments came so fast that the dream began to seem plausible.

In February 1962, when Lieutenant Colonel John H. Glenn, Jr., became the first American to be shot into orbit around the earth, the United States again won world prestige by permitting live television coverage. Within months Lieutenant Commander M. Scott Carpenter repeated the achievement. Again television viewers had front-row seats, both at blast-off and in recovery operations. Films were flown to television stations and theaters throughout the world—by networks, newsreels, and the USIA.

Within weeks another Cape Canaveral blast-off involved television more directly. In July 1962 Telstar I, a communication satellite, was boosted into orbit. Culminating several earlier experiments, it was the first communication satellite that could relay all forms of communication, including television. The event was thus comparable to the laying of the first Atlantic cable or the sending of the first radio signal across the Atlantic. David Brinkley, honoring the new era with a telecast from Paris, indulged his puckish humor by announcing solemnly "via Telstar" that there was no important news. But the possibility of transmitting events "live" to and from all parts of the globe clearly introduced momentous vistas.

The event contributed to an important decision by each network—to expand its 15-minute evening newscast to 30 minutes in 1963. It was scarcely the millennium, but seemed to news divisions a forward stride of great promise.

Telstar I was launched by the National Aeronautics and Space Administration "for AT&T" and was described as "paid for" by AT&T. But its existence was of course made possible by space experimentation that had cost billions in public funds. That public investments should thus be channeled into a private preserve agitated some observers, although it was hardly discussed on television.

Since the beginnings of network broadcasting, the relaying of programs from station to station had been the province of AT&T, and AT&T now appeared set to play a comparable role in the relaying of programs from continent to continent. But delicate issues were involved. Who would decide what programs would flow—and on what terms—from continent to continent? A new communication highway had been opened—and with it, a new tollgate. The operation of the tollgate involved relations between nations.

In spite of the touchiness of the issues the Kennedy administration wanted to push ahead as rapidly as possible with satellite communication, as a key to international prestige and power. It was anxious to set up a functioning organization that could press ahead, and most congressmen shared the sense of urgency. The result was the Communications Satellite Act of 1962, which became law a month after Telstar I.

The bill placed international satellite communication firmly in the private sector. It authorized the creation of COMSAT—the Communications Satellite Corporation—a private corporation. Half the stock would be offered to the general public. The other half would be owned by AT&T and other major communications companies, with the proviso that no company could own a majority.

The law called for a fifteen-man board of directors—three to be chosen by the President, six by the public stockholders, six by the communications companies investing in COMSAT. However, no one company was to have more than two representatives on the board. The arrangement was reminiscent of the way RCA had been established in 1919 under navy auspices, with provision for government representation on the board but with control lodged in four companies—among them, AT&T. In that case, as in the case of COMSAT, the result was an accommodation of existing power groups.

By February 1963 COMSAT was organized. On the board AT&T was joined by representatives of RCA, Western Union International, and the International Telephone and Telegraph Company.

COMSAT stock, when offered to the public, was sold overnight and quickly doubled in value. To many observers it was a triumphant demonstration of private investment serving a public purpose. Others were uneasy that so prominent a role in international affairs would revolve around the profit motive.

One function of COMSAT was to negotiate with foreign governments concerning ground stations, to implement their participation in a global system. Such international diplomacy was at once begun, but the private status of COMSAT proved an obstacle. This was solved by another compromise: a consortium was created—INTELSAT, the International Telecommunications Satellite Consortium—which would be titular owner of the evolving satellite system. But all management functions were delegated to COMSAT.

Telstar I, though highly successful, was already obsolete when it went into orbit around the globe. It could link only areas which were, at any moment, in its line of sight. As early as 1945 Arthur C. Clarke, British scientist and science fiction writer, had outlined—in an article in *Wireless World*—the possibility of a different kind of satellite—one that would move in an orbit so synchronized with the earth's rotation that it would seem to hover in a fixed spot.[1] In 1962, as Telstar I began its career, a *synchronous* satellite was already being built by Hughes Aircraft Corporation, and early in 1963 it was blasted into space. Its electronic equipment failed, but a second synchronous satellite, lofted a few months later, functioned perfectly. Synchronous satellites became the focus of COMSAT planning—for global television and other communication.

The whole development was being pushed forward with staggering

1. Clarke, "Extra-terrestrial Relays," *Wireless World*, October 1945.

speed. One of COMSAT's organizers, contemplating its tasks, said: "It is like being ordered to organize a worldwide airline six months after the Wright brothers first flew."[2]

Even as COMSAT plunged ahead, great uncertainties surrounded it. Its synchronous satellites would be linked to participating countries by ground stations. A standard ground station, using a dish approximately 100 feet in diameter, cost between three and seven million dollars. To some countries this seemed a modest entry fee; to others it was formidable. But the Kennedy administration was anxious that many countries join the system as promptly as possible, for there might soon be competing systems sponsored by rival powers using other technology. Thus United States economic aid as well as military aid were invoked to help developing countries join up. Though COMSAT was private, Kennedy policy aimed to ensure its quick success.

But the thoughts of Arthur Clarke, who had led scientists to the synchronous satellite, were already outrunning the COMSAT system. In the September 1959 issue of *Holiday* he outlined ideas on "how to conquer the world without anyone noticing." He considered the means available to the United States and the Soviet Union alike. He pointed out that ground stations were not really necessary. The Soviet Union might, for example, put a synchronous satellite high over Asia, reaching the entire continent. If through Soviet trade missions it could then flood the continent—perhaps at a slight profit—with low-cost sets designed to receive the satellite directly, the ground stations would pass into limbo. The same technique could be applied to Africa and elsewhere. From the satellite would flow exciting drama, sports events, quiz programs, brisk newscasts—everything to enthrall nations in the way that "even ostensibly educated nations have been unable to resist." First prize on quiz programs would always be a free trip to the Soviet Union. Before long, uncommitted nations would become committed. Priority in establishing such a system, said Arthur Clarke, "may determine whether, fifty years from now, Russian or English is the main language of mankind."[3]

Officials at COMSAT did not talk that way; it was not designated as a propaganda agency. It was private and its assignment was to perform a service at a profit. But such thoughts were in the air.

In 1963, for the first time, a majority of people told Roper researchers

2. Quoted, Dizard, *Television: A World View*, p. 266.
3. Clarke, "Faces From the Sky," *Holiday*, September 1959.

that their chief source of news was television rather than newspapers.[4] Spectacular blast-offs from Cape Canaveral, news bulletins "by satellite," news specials on the age of the missile, must have helped to produce such a result.

In 1962 American television presented bizarre juxtapositions. There were *The Beverly Hillbillies* and other Nielsen pacemakers; and there were news specials that seemed to come from another world. The two worlds often seemed incompatible. They represented the two worlds into which television had fissioned. Within television they were interdependent, but antagonistic.

There was no doubt which commanded the chief loyalty of audiences. Such series as *Bonanza, The Red Skelton Hour, Ben Casey, The Andy Griffith Show, The Flintstones, Gunsmoke* hovered near the top of most rating lists. But now and then the other world broke in.

In October the most sensational interruption featured John F. Kennedy. It was an international ultimatum delivered by television—about missiles in Cuba.

TV ULTIMATUM

While Kennedy was campaigning in the midwest, boosting Democrats in state and local elections, it was announced that he had a slight upper respiratory infection and was returning to Washington. Some bulletins hinted it was a cover story for more important developments.

Shortly afterwards he asked for television time on all networks for Monday, October 22, 1962, at 7 p.m. eastern time.

The CIA had photographic evidence that missile sites were being built in Cuba with Russian help, not only for surface-to-air missiles for use against aircraft, but also for missiles of longer range, able to reach cities on the continent. There was no evidence of nuclear warheads, but it was supposed they might be on hand or en route.

In days of feverish behind-the-scenes debate, alternative plans for United States action were proposed, and developed by small groups. A

4. People were asked from what source "you get most of your news about what's going on in the world today—from the newspapers or radio or television or magazines or talking to people or where?" Some people named two or more media. In November 1963 the answers ran: television 55, newspapers 53, radio 29, magazines 6, people 4, don't know 3. In November 1961 the answers had been: television 52, newspapers 57, radio 34, magazines 9, people 5, don't know 3. Roper and Associates, *New Trends in the Public's Measure of Television and Other Media,* p. 2.

central role for television was so much taken for granted that each group recommended not only what should be done, but what President Kennedy should say on the air. Participants referred to the plans as "scenarios."[1]

The Joint Chiefs of Staff all felt that the United States should at once demolish the missile bases with a massive air-strike. They assumed the communists would use atomic weapons and therefore felt the United States should go ahead with atomic weapons. Robert Kennedy later commented that the recommendations of the military leaders always had the advantage that if they proved mistaken, no one would be around to know.[2]

An alternative plan, which John Kennedy, Robert Kennedy, and Secretary of Defense McNamara favored, was a blockade of Cuba. They argued that an attack would still be possible if the blockade proved unable to remove the missiles.

The blockade scenario—using the term "quarantine" because a "blockade" is an act of war—was adopted. The quarantine would be coupled with an ultimatum to be addressed not to Cuba but to the Soviet Union. Kennedy would demand that the Soviet Union remove its missiles from Cuba. Failure to do so would bring United States military action against Cuba. Meanwhile Soviet ships en route to Cuba would be stopped and searched for military supplies.

To give the ultimatum maximum force, it would be delivered via television—not merely through diplomatic channels. The televised commitment, relayed throughout the world by satellite, would create a situation from which retreat would appear impossible.

To the group around Kennedy, the impossibility of retreat was obvious for many reasons. One was the current election campaign. The Bay of Pigs debacle had given Kennedy, in many eyes, a "soft on communism" image which was proving a factor in the present campaign. He had tried to neutralize this with increased military deployments—for example, he had sharply increased the number of so-called military advisers in Vietnam—but these moves made little public impact. It was assumed by those around Kennedy that any half-way resolution of the missile situation would destroy Kennedy politically. During scenario discussions an air-

1. The following is based on Schlesinger, *A Thousand Days,* pp. 794–841; other sources as noted.
2. Kennedy, *Thirteen Days,* p. 48.

strike advocate passed a note to Theodore Sorensen suggesting that if the missiles were not eliminated, "the next House of Representatives is likely to have a Republican majority."[3]

Kennedy desperately needed an atmosphere of victory. Ironically Khrushchev, by his perilous gamble, had offered a chance for a victory. Kennedy saw the chance.

The ultimatum telecast of October 22 and the moves surrounding it showed Kennedy a brilliant technician in the consolidation and use of political power. He checked all details with minute precision; he was fully in control. Meanwhile his coolness astonished observers. A few minutes before the telecast NBC correspondent Robert Goralski was in the office of Evelyn Lincoln, Kennedy's secretary, to check some information. Through the open door of the washroom he saw Kennedy calmly combing his hair; makeup was already on.

A minute or two later Kennedy, speaking on television in his rapid, clipped style, was preparing his worldwide audience for atomic war. Referring to the bases under construction, he said:

> KENNEDY: The purpose of these bases can be none other than to provide a nuclear strike capability against the Western Hemisphere. . . . Several of them include medium-range ballistic missiles capable of carrying a nuclear warhead for a distance of more than 1000 nautical miles. Each of these missiles, in short, is capable of striking Washington, D.C., the Panama Canal, Cape Canaveral, Mexico City, or any other city in the southeastern part of the United States, in Central America, or in the Caribbean area. Additional sites not yet completed appear to be designed for intermediate-range ballistic missiles capable of traveling more than twice as far—and thus capable of striking most of the major cities of the Western hemisphere, ranging as far north as Hudson Bay, Canada, and as far south as Lima, Peru. In addition, jet bombers, capable of carrying nuclear weapons, are now being uncrated and assembled in Cuba, while the necessary air bases are being prepared.

Using the word "nuclear" eleven times, Kennedy drew a panorama of devastation enveloping the whole hemisphere. The moves that had made such things possible, said Kennedy, could not be accepted by the United States "if our courage and our commitments are ever to be trusted again by either friend or foe." He asserted:

3. Sorensen, *Kennedy*, p. 688.

> We will not prematurely or unnecessarily risk the costs of worldwide nuclear war in which the fruits of victory would be ashes in our mouth—but neither will we shrink from that risk at any time it must be faced.

Then he turned to Khrushchev:

> I call upon Chairman Khrushchev to halt and eliminate this clandestine, reckless, and provocative threat to world peace and to stable relations between our two nations. I call upon him further to abandon this course of world domination and to join in an historic effort to end the perilous arms race and transform the history of man. He has an opportunity now to move the world back from the abyss of destruction. . . .

He told the American people:

> Many months of sacrifice and self-discipline lie ahead.

About an hour before the President's speech, managers of ten powerful radio stations—mainly in the South—received telephone calls from Pierre Salinger at the White House to say that the federal government wished to take over their facilities for an indefinite duration. The managers were astonished to learn that AT&T had the arrangements all ready. At 7 p.m. the connection was made and the stations became a Spanish-language network of the Voice of America beaming the President's speech, with translations and commentaries, to Cuba.[4] The material was also carried by Radio Americas, formerly Radio Swan.

Although Kennedy prepared his audience for a long crisis, it proved short. So effective had been Kennedy's bludgeoning attack, coupled with vast moves of navy, army, and air force units and intensive diplomatic maneuvers, that the crisis came quickly. Soviet ships en route to Cuba halted in the Atlantic. A series of Soviet messages to the United States included at least two that offered a basis for settlement. In two weeks the crisis was largely over. Dismantling of the missile sites was begun, and the blockade was removed. Kennedy gave assurance that Cuba would not be invaded.

The crisis left some footnotes in television history. Among the messages

4. The stations were WCKR, WGBS, and WMIE, Miami; WKWF, Key West; WSB, Atlanta; WWL, New Orleans; WCKY, Cincinnati; KAAY, Little Rock; and shortwave stations WRUL, Scituate, Mass., and KGEI, San Carlos, Calif. The stations were operated as a government network for three weeks. New York *Times*, December 5, 1962.

from Khrushchev, one came via a network newsman. On October 26 at 1:30 p.m. John Scali, ABC diplomatic correspondent, was eating a bologna sandwich in his cubicle at the State Department when he received an urgent phone call from Aleksander S. Fomin of the Soviet embassy, asking Scali to meet him at once at the Occidental Restaurant. Although Fomin had the title of "counselor" at the embassy, newsmen suspected him of being a Soviet intelligence official, and this undoubtedly added to Scali's curiosity about the call. At the restaurant Fomin seemed highly agitated, saying that "something must be done!" Would the State Department, he asked Scali, be interested in settling the missile crisis on these terms: (1) the missile sites to be dismantled and shipped back to the Soviet Union; (2) such weapons not to be reintroduced into Cuba; (3) the United States to pledge not to invade Cuba. When Scali said he did not know, Fomin begged him to find out, and gave him a home telephone number to call. Scali conveyed the message to the State Department, and was given an encouraging letter to take back. This exchange and a personal letter from Khrushchev to Kennedy set the lines of the eventual settlement.[5]

For Khrushchev, who had a penchant for unorthodox channels of communication, the choice of a television newsman was perhaps not strange. At a time when diplomats were being shadowed relentlessly, the comings and goings of newsmen were taken for granted.

Although Kennedy in his television ultimatum showed a complete sense of assurance about the United States position and its moral rightness, his own feelings were not so unclouded. He had been furious to learn, during the intensive writing of scenarios, that the United States still maintained missile bases in Turkey, ready to rain atomic destruction on major Russian cities. He had ordered Rusk months earlier to withdraw those missiles, considering them an unnecessary and ultimately useless provocation. But Rusk, delaying, had not done so. Kennedy was furious because he considered those missiles, in the words of Assistant Secretary of State Roger Hilsman, a "political albatross."[6] They raised such questions as: if American missiles in Turkey were "defensive," why were Russian missiles in Cuba not also "defensive?"

Khrushchev at first justified missiles in Cuba on the ground that the United States had attacked Cuba, and that many United States leaders

5. Abel, *The Missile Crisis*, pp. 155–6.
6. Hilsman, *To Move a Nation*, p. 222.

were calling for a new attack. The Soviet Union had not attacked Turkey.

The comparative situations of Cuba and Turkey troubled Kennedy. They also worried Adlai Stevenson, who this time was brought into administration councils.[7] Stevenson, approving the quarantine plan, suggested that withdrawal of missiles based in Turkey—and also in Italy—could be elements in a settlement. Kennedy agreed with this but did not wish to mention such concessions in his ultimatum. Thus all troublesome details were swept under the rug to make the television address as unified and powerful as possible. It proved so powerful that the Turkish and Italian missile bases were ignored in the speedy resolution.

Television audiences were thus left with the picture of a good guy/bad guy crisis. A villain had been caught in a fiendish plan and been stopped by a good guy. It was an oversimplification—a defect not uncommon in television messages.

But few were inclined to criticize Kennedy on these grounds, for his strategy, brilliantly executed, had apparently accomplished miracles. Holding the Joint Chiefs in check, it had removed the Cuban missiles without atomic attack. For this all liberals were thankful. At the same time he had, through his anti-communist success, neutralized attacks from the right.

The end of the crisis was not marked by any television event. Kennedy was insistent that he wanted no "victory" statement, no gloating. Humiliation of Khrushchev and the Soviet Union would, he feared, stimulate the arms race, and he hoped to prevent that.

Of course he could not. The events had doomed Khrushchev. His misadventure opened him to attack from the Soviet military-industrial complex and the rulers of China. Khrushchev, in self-defense, tried desperately for a rapprochement with the United States. He halted all jamming; in return the United States curtailed, then stopped, its ten-year encroachment on the Moscow frequency of 173 kc. The nuclear test-ban treaty was another move toward better relations. But Khrushchev could no longer control events. In the end the arms race intensified.

Meanwhile Kennedy had achieved an extraordinary unification. The sense of Camelot returned. Kennedy, feeling more secure internationally, could hammer at domestic issues on which he wanted action—such as civil rights. His rapport with artists and writers increased. He was sur-

7. Also invited to confer at an early stage was Edward R. Murrow, but he was ill, apparently already suffering from the lung cancer to which he later succumbed.

rounded with an aura of history in the making. He seemed to have America "moving again." All this had its impact on television.

NEWS 1962–63

Conscious of new dignity, the network news divisions expanded. Their staffs did not inquire what had given them a warmer place in the sun. They were vaguely aware that Kennedy activism, international rivalries, and network tremors over Washington contributed to a changed situation. Meanwhile it was enough that budgets rose and scheduling was more favorable—sometimes.

The new affluence gave rise to a fascinating diversity of projects—among them *The Tunnel* on NBC-TV.

Reuven Frank, an executive producer for NBC News, wanted to do a film about East German escapees, who eluded the communist-built Berlin wall by jumping across roofs, wading through sewers, and digging tunnels. During a European trip he asked the NBC Berlin staff to shoot anything interesting on escapees.

In a sense he was trying to capitalize on a limitation. NBC had no representatives in East Germany. So, as Frank later wrote—

> We set out to do a study of what was going on inside East Germany—inside the Communist Empire—without going there ourselves. We were looking for effects, the sign, the small indicator that you have to find, of what was building up inside.[1]

It was somewhat like observing China from Hong Kong, or Cuba from Miami. However, Frank's drama developed in other directions.

In June 1962 Piers Anderton of the NBC Berlin Bureau came to New York to get married, and during the festivities backed Frank into a corner and said they had to have a meeting at once.

> I thought that was a little strange—hardly the occasion for business—but he insisted. We got back to the office and closed the door at his insistence, and he said, "I've got a tunnel."

A group of West Berlin students had started a tunnel to help friends or relatives in East Berlin escape. The students were willing to have cameramen join them. It would be a grueling, probably risky task.

1. Frank, "The Making of the Tunnel," *Television Quarterly*, Fall 1963. The following is based on this account, and other sources as noted.

A sticky problem was that the tunnelers—the three with whom the NBC Berlin staff had negotiated—wanted $50,000. Reuven Frank said, "They're crazy!" but he agreed to a smaller sum, later reported as $7500.

Only a few at NBC were told of the project, and these carried on in cloak-and-dagger fashion. Funds were disbursed "outside the NBC channels." An agreement signed with the diggers was not seen by an NBC lawyer. The project was never mentioned on the phone. Reuven Frank, though in charge, did not know the exact location of the tunnel. He stayed away until the digging was finished, because his presence in Berlin might cause speculation. No film left Berlin until it was all over. Processing was done in a local Berlin laboratory considered politically trustworthy. Two young Berliners, often lying flat on their backs, did the shooting for NBC. A tape recorder in the tunnel could pick up footsteps fifteen feet above them in the street.

The film was not pure *cinéma vérité*. Under Anderton's direction the group leaders reenacted planning phases for the NBC cameras.

On September 13 Reuven Frank, alerted by Anderton, flew to Berlin with a film editor. The following day the tunnel was completed and the escape of twenty-six people—including five babies—was filmed. The NBC group rushed the editing of twenty hours of accumulated film. By phone to William McAndrew, head of NBC News, Frank said cautiously, "I think we need ninety minutes." Carrying the film to New York as hand-luggage, he put it behind his rear-row seat. Then he was asked to move so that Mayor Willy Brandt of West Berlin could sit with associates. Throughout the trip Willy Brandt unknowingly shielded the film.

A surprise message from McAndrew awaited Reuven Frank at Idlewild international airport in New York. He was to fly on to Pittsburgh. There in a hotel room he met with representatives of Gulf Oil to tell them about *The Tunnel,* and they agreed to sponsor it, assuring prime time.

Although originally announced for October 31, *The Tunnel* was not shown until December 10. The missile crisis and the imminence of elections played a part in the postponement, along with crises over the film itself. The State Department, learning of the project, criticized it as imperiling international relations. The $7500 payment to the tunnel leaders had become known in East Germany and had given it valuable propaganda ammunition. Some of the tunnelers demanded that telecast plans be stopped as a danger to them and their relatives. They said they had assumed the film was for private use; they said they had been told nothing about television plans and had received no money. They had only been told about Americans giving money for tunneling equipment. In the United States some critics charged that NBC by providing funds had in fact financed the completion of the tunnel.

Jack Gould of the New York *Times*, probably the most influential of television critics, felt that NBC's initiative had been understandable, but he regretted that "the stamp of commercialism" had been put on the project. He added: "With peace hanging by a thread it is no time for adventurous laymen to turn up in the front lines of world tension."[2]

But such objections were largely forgotten when *The Tunnel* was shown on NBC-TV on December 10, 8:30–10 p.m. eastern time, sponsored by Gulf. Audiences found it gripping drama, and it matched *The Lucy Show* in ratings. It won awards. Amid the hosannas the State Department decided it was useful cold-war propaganda and permitted USIA to show it abroad. Questions about private cold-war initiatives were no less valid than before, but were left hanging.[3]

Television was often accused of timidity but became less timid in 1962–63. Along with the adventurous air of the Kennedy regime, a dramatic legal event contributed to the new spirit. In June 1962 the lawsuit of John Henry Faulk against Aware, Inc., Vincent Hartnett, and Syracuse supermarket executive Laurence Johnson reached its climax in a New York court. A parade of witnesses had laid bare methods by which self-styled patriots had conducted a purge of the industry, with much help from within the industry. Executives who had at first taken the "security" claims seriously, but had since sickened of the operation, testified in illuminating detail. As a climax Louis Nizer, attorney for Faulk, sought testimony from defendant Laurence Johnson, who through legal maneuvers and medical bulletins had staved off appearance in court. As the case was ready to go to the jury, word arrived that Laurence Johnson

2. New York *Times*, October 12, 14, 22, 1962.

3. The kudos won by *The Tunnel* may have encouraged the later involvement of CBS in a comparable venture—a film about arms smuggling, which gradually turned into a film about clandestine preparations for a Florida-based invasion of Haiti, designed to overthrow President François Duvalier. CBS investigative reporter Jay McMullen apparently arranged small payments to participants for "television rights." As in the case of *The Tunnel*, the payments laid the network open to the charge of having helped to finance the venture itself. The project and the film were both halted, but remained for many years a subject of rumor. *Scanlan's Monthly*, March 1970.

had been found dead in a Bronx motel. He had apparently taken barbiturates, vomited, and choked to death. The court ordered that the estate of Laurence Johnson be substituted for the deceased as a defendant in the case. A few hours later the jury awarded unprecedented damages of $3,500,000 to John Henry Faulk.

It was an extraordinary vindication for Faulk, who had given six years of his life to clear his name; also for those who, like Edward R. Murrow, had not hesitated to help him finance his suit. The verdict was upheld at every level, although the award was eventually scaled down.[4]

Thousands in the television industry breathed a sigh of relief. The blacklist machinery appeared to be disintegrating. Many an artist emerged from long obscurity. Topics that would have been considered too controversial a year or two earlier were now welcomed. Treatment became less fearful. "Meet Comrade Student," a *Close-Up* program produced by John Secondari and written by Robert Lewis Shayon, examined Russian education with a minimum of standard polemics; it noted that Russian schools had their successes and failures and that it was more important for Americans to understand the successes than the failures. Increasing courage also erupted in local programming, where the documentary was winning a foothold. Surveying local documentaries of this period, the writer William Bluem found many worthy of praise, among them *Suspect* (KING-TV, Seattle), an examination of a right-wing extremist group; *The Wasted Years* (WBBM-TV, Chicago), on the erosion of prison life; *Superfluous People* (WCBS-TV, New York), on New York's staggering welfare problems; and *Block Busting–Atlanta Style* (WSB-TV, Atlanta), on segregationist tactics.[5]

Race issues were ever-present and lurked behind many other problems. The year 1963 produced moments that seemed to offer extraordinary promise. After a sequence of violence in Birmingham, Martin Luther King was able to negotiate a desegregation agreement with white Birmingham leaders. Because prominent businessmen participated, the resolution gave hope of a new era, especially when King—by now often seen in television statements—told his followers: "We must not see the present victory as a victory for the Negro. It is a victory for democracy. . . . We must not be overbearing or haughty in spirit."

But uglier currents were at work. They were revealed occasionally in

4. Faulk, *Fear On Trial*, pp. 366–98; Nizer, *The Jury Returns*, pp. 387–438.
5. Bluem, *Documentary in American Television*, pp. 221–30.

network news programs, more tellingly and extensively on non-network stations. In June 1963 WRVR-FM of New York's Riverside Church broadcast a series of recordings that included a Ku Klux Klan meeting held near Birmingham, presided over by Imperial Wizard Robert Shelton. The recordings, later broadcast over many other stations, let listeners hear oratory about "communists . . . atheists . . . so-called ministers of the nigger race." Then more direct appeals by Klan leaders.

VOICE: If you are standing for God, you're our brother. Tonight we are facing the greatest darkness that this nation has ever faced. Tonight we as God-fearing men and women can turn Alabama upside down for God. . . . Don't you love your children? Wanna see 'em mongrelized? . . .

ANOTHER VOICE: King will be met with force. King and Kennedy are worse than Castro. We need to go back to the old-time religious time and the old-time Klan time . . . in Atlanta they're puttin' on a show about Cleopatra, and Cleopatra is a black girl and she kisses a white boy. . . . You have as much civil rights as any communist. There will possibly be bloodshed in every state.

As the meeting breaks up we hear:

VOICE (*on loudspeaker*): Please drive out carefully. Obey all traffic laws.

ANOTHER VOICE (*off mike*): Run over any niggers you see!

That same evening two buildings used by Negro leaders are bombed and rioting erupts. We hear the desperate efforts of King and other leaders to keep their drive nonviolent.[6]

In June 1963 television brought viewers glimpses of the funeral of Medgar Evers, black civil rights leader murdered in Mississippi. National Educational Television carried long interviews with Martin Luther King and with Malcolm X, a far more militant voice. King, assailed from one direction as "communist," was increasingly denounced by others as too moderate.

Amid such tensions, the announcement that Martin Luther King was planning a mammoth march on Washington to be held in August, with people coming from all states to petition Congress for equal rights, posed elements of a crisis. Some government leaders felt the plan must be halted as a threat to law and order. Memories of the Hoover era,

6. *Birmingham: A Testament of Non-Violence.* Six programs produced by Jack D. Summerfield for WRVR-FM.

when a march of veterans demanding a bonus had been driven back by mounted troops, stirred deep fears. But Kennedy decided to encourage the march. The result was one of the most inspiring of television spectaculars. An incredibly disciplined migration had its climax on August 28 when Martin Luther King spoke to 200,000 people at the Lincoln Memorial in Washington. He told them—and millions of others via television and radio:

> KING: I have a dream. It is a dream deeply rooted in the American Dream. I have a dream that one day this nation will rise up and live out the true meaning of its creed: "We hold these truths to be self-evident, that all men are created equal."
>
> I have a dream that one day on the red hills of Georgia sons of former slaves and sons of former slave owners will be able to sit down together at the table of brotherhood. I have a dream that even the state of Mississippi, a state sweltering with the heat of injustice, sweltering with the heat of oppression, will be transformed into an oasis of freedom and justice. . . .

The networks—television and radio—had no doubts about the necessity of relaying these events, in whole or part, to the American public. Through live and delayed coverage, millions shared the experience.

The USIA had its own decision to make. The story of a protest march required some word on what was being protested. Former USIA regimes, shell-shocked by the McCarthy attacks, would have handled such an event in gingerly style, reporting it as a triumph of democracy while saying as little as possible about issues involved. The New Frontier atmosphere permitted a more venturesome approach.

George Stevens, Jr., USIA film chief under Edward R. Murrow, had a policy of encouraging young film makers, and entrusted film coverage of the march to James Blue, a University of Southern California film graduate who had later studied in Paris at the Institut des Hautes Études Cinématographiques, and made a film about the Algerian revolt. In his hands, the event became a major USIA film—*The March.*

The March began with shots of hordes en route to Washington by all conceivable means of travel.

> NARRATOR: They came from Los Angeles and San Francisco or about the distance from Moscow to Bombay. They came from Cleveland, from Chicago or about the distance from Buenos Aires

> to Rio de Janeiro. They came from Birmingham, Alabama, from Jackson, Mississippi, or about the distance from Johannesburg to Dar-es-Salaam.
>
> He had been insulted, beaten, jailed, drenched with water, chased by dogs, but he was coming to Washington, he said, to swallow up hatred with love, to overcome violence by peaceful protest.

There were glimpses of fantastic preparations—the making of 80,000 cheese sandwiches, the network of walkie-talkies, the distribution of pins: "I MARCH FOR JOBS AND FREEDOM." At the end came the King speech, and "We Shall Overcome" sung by a sea of humanity, and final words by the narrator.

> NARRATOR: There were many who praised this day and said that there had been a new awakening in the conscience of the nation. Others called it a disgrace. In the wake of this day, more violence was to come, more hatred, but in the long history of man's cruelty to man, this was a day of hope.[7]

Some congressmen who previewed the film were determined to block its distribution. But test showings in India and Africa yielded eloquent evidence. Here was a nation, said viewers in astonishment, that could admit its errors, discuss them openly, and move to correct them. After delays the film went into wide distribution and won recognition as one of USIA's finest hours.

The changing climate of 1963 put pressure on producers of television drama and commercials to amend their largely lily-white world. There was response but also nimble foot-dragging. The daytime serial remained almost untouched. Progress was more noticeable in other areas. Inclusion of a Negro or two in crowd scenes was becoming standard. Some series went further. *The Defenders,* often breaking new ground, featured black actor Ossie Davis as a prosecutor. Medical shows occasionally had a black doctor or nurse. The Jackie Gleason program engaged a black dancer—granddaughter of Duke Ellington—for its formerly all-white chorus line. The Ed Sullivan program regularly welcomed Negro acts. *Mr. Novak* and *East Side/West Side* used stories about Negroes. But resistance appeared constantly. CBS found two of its southern affiliates refusing to carry an *East Side/West Side* episode because of its Negro roles. When a *Bonanza*

7. *The March.* Produced by James Blue for U.S. Information Agency, 1963.

episode introduced a Negro character, General Motors considered withdrawing its commercial—but was dissuaded by NBC.[8]

A National Urban League study revealed that the ten leading New York advertising agencies had, among 23,600 employees, only eleven Negroes. Four years of persuasive efforts by the League had produced one new hiring—in the research division of Foote, Cone & Belding.[9] On the other hand, a Manufacturers Hanover Trust Company commercial featuring the beautiful Diahann Carroll created enough stir to encourage bandwagon jumping. Statistics on Negro purchasing power got increasing attention from advertisers and their agencies. It was beginning to be argued that integration might turn out to be good business. The box-office success of theatrical films on interracial themes encouraged the idea.

However, when NBC-TV scheduled a three-hour documentary on civil rights as a 1963 Labor Day special it found sponsors reluctant. Gulf Oil, which had been ready to back *The Tunnel,* was not interested. Such problems had to be solved by selling participations at bargain rates. ABC-TV had similar trouble with a series of five films on civil rights scheduled late Sunday evenings.[10]

Meanwhile each network had acquired a black on-the-air newsman. It seemed "tokenism" to some, a revolution to others. When CBS president Frank Stanton testified at a Senate communications subcommittee hearing in July 1963, Senator Strom Thurmond of South Carolina asked him, "Don't you care about white people?" Thurmond, outraged at attentions given to Martin Luther King, accused the networks of "following the NAACP line."[11] When NBC-TV decided to cancel its telecast of the Blue-Gray football game, an annual event from which Negro players were barred, Governor George C. Wallace of Alabama denounced the network decision as "tragic and irresponsible." But the sponsors, Gillette and Chrysler, concurred in the cancellation.[12]

President Kennedy, pressing hard for civil rights legislation, kept nudging the whole process forward. In October he was seen in another—somewhat bizarre—*cinéma vérité* production. The Alabama racial crisis was the subject, and Governor George C. Wallace was co-star.

8. New York *Times,* November 5, 12, 1963; *Variety,* April 1, 1964.
9. New York *Times,* April 16, 1963.
10. *Variety,* August 14, 1963.
11. The reference was to the National Association for the Advancement of Colored People. *Variety,* July 3, 1963.
12. New York *Times,* November 14, 1963.

In 1963 the University of Alabama, the last to remain all-white, was under federal court order to integrate. For its summer session starting June 10, its admissions office accepted two black applicants, Vivian Malone and James Hood. But Governor Wallace intervened, telling a news conference: "I am the embodiment of the sovereignty of this state and I will be present to bar the entrance of any Negro who attempts to enroll at the University of Alabama." A scrappy politician and one-time Golden Gloves bantamweight champion, he had been elected to the governorship in 1962 proclaiming "segregation now, segregation tomorrow, segregation forever!"[13]

In June 1963, as a crisis approached, President Kennedy invited Robert Drew to document its development on film. Donn Pennebaker and his camera began to cover strategy sessions at the White House and at the home of Attorney General Robert F. Kennedy. Meanwhile, as Deputy Attorney General Nicholas Katzenbach was rushed to Alabama—with three thousand army troops kept ready near by—Alabama photography was organized by George Shuker of the Drew unit. George C. Wallace proved as ready as the Kennedys to take part; camera on shoulder, Richard Leacock and James Lipscomb were soon following him about the Alabama executive mansion, the governor's office, and the University of Alabama campus in Tuscaloosa. As events moved forward, with much telephoning between leaders in Washington and Alabama, both ends were covered by camera.

Clearly the Drew unit had scored an extraordinary coup. But the film that emerged, while full of fascinating moments, brought into focus some limitations of *cinéma vérité*. The events in *Crisis: Behind a Presidential Commitment* were a far cry from those photographed in *Primary*. In the earlier film, candidate John Kennedy, pushing through Wisconsin crowds and coping with mass meetings, tea parties, and press conferences, had been so absorbed in problems at hand that the camera could be truly forgotten—by him and audiences alike. Not so in *Crisis: Behind a Presidential Commitment,* in which the camera was often covering small groups: two or three White House advisers in shirt sleeves pacing about, discussing strategy while President Kennedy nodded thoughtfully; Robert Kennedy on the phone, instructing aides in Alabama. Television critics found themselves unable to forget the camera or to accept the notion that the participants—performers—could forget it. New York

13. Lewis, *Portrait of a Decade,* pp. 165–6.

Times critic Jack Gould called it "managed newsfilm . . . a melodramatic peepshow." Columnist Harriet Van Horne found it "oddly depressing."

Even the unfolding of events gave a managed feeling. As Katzenbach and the two students approached the administration building, Governor Wallace, as promised, "stood in the door" and delivered his statement defying federal authority. The students were then escorted elsewhere—to dormitory rooms already assigned to them—while Wallace capitulated off camera. He had his moment of on-camera glory, while the Kennedys had their victory. Both apparently had ammunition for future election campaigns.

A *quid pro quo* aspect touched other elements of the film. Robert Kennedy's small daughter Carrie, for some reason visiting her father's office during this crisis, was allowed to say hello on the phone to Deputy Attorney General Katzenbach in Tuscaloosa. "Hello, Uncle Nick." Perhaps to balance things, Governor George Wallace was shown hugging his small daughter. There were glimpses of his ante-bellum Southern mansion to balance shots of the Robert Kennedy home in Virginia.

There were memorable moments. Governor Wallace receiving praises from thin-lipped, elderly ladies: "We're proud of what you're doing—for Alabama. . . . Bless your heart!" The students being advised by Katzenbach: "Dress as though you are going to church—modestly—neatly." We see the students having their pictures taken for the covers of *Newsweek* and *Time*. The students are so handsome, so firmly armed with quiet humor, one wonders whether they were "cast" for these breakthrough roles.

During the film Kennedy is considering going on television, to put the whole issue on a moral basis. There is so much talk about what to say on television, and how people will react, that the film develops a Pirandello-ish atmosphere. Is it a television program about a television program about—about what? Do Wallace and Alabama actually exist?

In the film the President goes on television and says words he actually said on television. Although the context has somehow been reduced to ritual, the words are still powerful.

> KENNEDY: We preach freedom around the world, and we mean it. And we cherish our freedom here at home. But are we to say to the world—and much more importantly to each other—that this is the land of the free, except for Negroes; that we have no second-class citizens, except Negroes; that we have no class or

> caste system, no ghettos, no master race, except with respect to Negroes? . . .
>
> We face, therefore, a moral crisis as a country and a people. It cannot be met by repressive police action. It cannot be left to increased demonstrations in the streets. It cannot be quieted by token moves or talk. It is a time to act in the Congress, in your state and local legislative body, and, above all, in all of our daily lives.

These events took place in June, but the film *Crisis: Behind a Presidential Commitment* did not reach the air until four months later. Some thought the air showing was timed to help the unresolved struggle for civil rights legislation, but apparently the real reason for delay was that ABC-TV had trouble finding a sponsor. The one-hour film was finally telecast on October 21.

Around the Kennedy family, action and entertainment were unceasing. Relations with television correspondents always had sparkle and style. During a presidential plane ride from Hyannisport, reporters saw a puppy waddling along the corridor. Learning it was a new family acquisition, a gift from father Joseph P. Kennedy, they sent a solemn questionnaire to the Kennedy compartment. Jacqueline Kennedy filled it out by hand. One question was: "What do you feed the dog?" Jacqueline wrote: "Reporters."[14] Kennedy's mental agility and coolness constantly amazed correspondents. They saw him as a virile political animal whose decisions were often concerned with organizing and using power, but they saw him grow. With insatiable curiosity, he could overwhelm them with streams of questions, digging rapidly into almost any issue. Actress Jean Seberg, a White House visitor with her French husband, author-producer Romain Gary, recalled a dinner with Kennedy. "He asked Romain a million questions about de Gaulle—he was like an incredible IBM machine, digesting *everything*. I swiped a menu and wrote Malraux a long letter about it."[15]

A few days later Kennedy was the central figure in the most moving spectacular ever broadcast. It lasted four days.

DALLAS TO ARLINGTON

There was never anything like it. At times nine out of ten Americans were watching, along with throngs in Europe, Asia, Africa, Australia—

14. Interview, Robert Goralski.
15. New York *Times*, August 11, 1968.

who often watched at the same time via satellite. A New York critic wrote:

> This was not viewing. This was total involvement . . . I stayed before the set, knowing—as millions knew—that I must give myself over entirely to an appalling tragedy, and that to evade it was a treason of the spirit.[1]

The cast of characters in the four-day telecast exceeded anything ever seen before. And it included the first "live" murder on television, the first murder ever witnessed by millions.

It all began staccato at 12:30 Dallas time, November 22, 1963.[2] The sixth and final car in a presidential motorcade winding through Dallas contained reporters. When shots rang out and the motorcade, after momentary hesitation, careened ahead amid screams and pandemonium, Merriman Smith of United Press International grabbed the car's radiophone—he was in the front seat and nearest to it—and called the UPI Dallas bureau. At 12:34—four minutes after the shots—the bulletin he dictated was being sent out and began clacking out of UPI teletype machines at television stations, radio stations, newspapers.

> DALLAS, NOV 22 (UPI)—THREE SHOTS WERE FIRED AT PRESIDENT KENNEDY'S MOTORCADE TODAY IN DOWNTOWN DALLAS.

At 12:36 Don Gardiner at ABC Radio in New York went on the network with this announcement—six minutes after the shots.

At the same moment President Kennedy was being carried into Parkland hospital. Arriving immediately afterwards, Merriman Smith saw the limp body drenched in blood, and raced to a phone at the hospital cashier's booth. He called the UPI Dallas bureau again and dictated another bulletin, which at 12:39 began spilling out of UPI teletype machines everywhere.

> FLASH
> KENNEDY SERIOUSLY WOUNDED PERHAPS SERIOUSLY PERHAPS FATALLY
> BY ASSASSINS BULLET

One minute later in New York, on the basis of the two bulletins, Walter Cronkite broke into the serial *As the World Turns* and appeared on CBS-TV screens.

1. Mannes, "The Long Vigil," *The Reporter,* December 19, 1963.
2. The time sequence is based on *Four Days;* other sources as noted.

CRONKITE: In Dallas, Texas, three shots were fired at President Kennedy's motorcade. The first reports say the President was "seriously wounded."

The serial continued.

In Washington CBS bureau chief William J. Small was busy with others going over a list of invitees for a Christmas party when someone rushed in with the news. Small later recalled: "I can remember running out of my office yelling, 'We want remotes at the White House, Lyndon Johnson's house, and the Hill!'" Checking the continuing wire service bulletins, he began phoning other CBS stations—Baltimore, Philadelphia, New York, and others—to borrow equipment and personnel he felt would be needed.[3]

Within minutes radio and television audiences were growing fantastically. At 1:35 they learned that Kennedy was dead and that Lyndon Johnson, who had been in the motorcade, had left Parkland Hospital for Love Air Field.

Even those at the center of events found themselves linked to developments by television. Mrs. Lyndon ("Lady Bird") Johnson was with her husband speeding to Love Field. There, boarding Air Force One, the presidential plane, they found a television set on. A commentator was referring to "Lyndon B. Johnson, now President of the United States." The commentator also said there was a suspect.

A fantastic kaleidoscope began to reach television screens: interviews that networks were gathering with their remote units on Capitol Hill and elsewhere—"we kept throwing them on the air," William Small later recalled; a shot of Jacqueline Kennedy accompanying the dead President's body from Parkland Hospital, leaving for Love Field; information about the arrested suspect, identified as Lee Harvey Oswald; details about Lyndon Johnson taking the oath of office at 2:38 p.m. Amid all this, videotapes of events *before* the shooting.

Kennedy's visit to Texas had been covered by local stations, not by networks; now these videotaped events, juxtaposed with new developments, had almost unbearable impact. Suddenly John Kennedy and a radiant Jacqueline Kennedy were on the screen, waving to crowds from their limousine. Then—more grisly—a Chamber of Commerce breakfast in Fort Worth that morning, attended by President Kennedy and Vice

3. Small, *Interview*, p. 1.

President Johnson. Jacqueline was expected but, said Kennedy, was still "organizing herself." In fine mood, he tells breakfasters:

> KENNEDY: Two years ago I introduced myself in Paris by saying that I was the man who had accompanied Mrs. Kennedy to Paris. I'm getting somewhat the same sensation as I travel around Texas. Nobody wonders what Lyndon and I wear.

The breakfasters have gifts for the President. One is to protect him "from local enemies"—a pair of Texas boots, a shield against rattlesnakes. Each gift is presented with a local advertising plug, like a quiz prize.

In Washington, in the gloomy twilight at Andrews Field, Air Force One comes down and moves in strange stillness into a circle of light. Before scores of cameras the casket emerges and descends by hydraulic lift. As Jacqueline Kennedy joins the casket in the navy ambulance, the telescopic close-up shows that her suit and stockings are caked with blood. After she has left, President Johnson descends from the plane and makes his first public statement. It ends:

> JOHNSON: . . . I will do my best. That is all I can do. I ask your help—and God's.

Late Friday a shrill note enters pickups from Dallas. It continues throughout Saturday, growing in stridency. The Dallas police, who house the accused assassin, are not merely cooperative toward television; they have completely surrendered to its demands and whims. Determined to vindicate the honor of Dallas, whose right-wing extremists are blamed for an atmosphere of hate—an anti-Kennedy leaflet headed *Wanted for Treason* was being distributed before the visit—the police display their zeal to the world. Each new clue is shown on television. For better visibility the manacled Lee Harvey Oswald is exhibited on a platform for cameramen and reporters. The case is "cinched," an officer tells the world. A rifle is brandished triumphantly: "The murder weapon." A spokesman says, "I figure we have sufficient evidence to convict him." There are said to be "approximately fifteen witnesses." Asked if Oswald has a lawyer, the spokesman replies, "I don't know whether he has one or not."

Deputy Attorney General Katzenbach, watching all this on television in Washington, became troubled. The chance of finding an acceptable jury for an Oswald trial was diminishing rapidly. Oswald, without defense counsel, was being tried on television. Katzenbach phoned U. S. Attorney H. B. Sanders, Jr., in Dallas to express his dismay. It would be a nightmare, he suggested, if the Supreme Court eventually threw out an Oswald conviction because of what was happening. "That would be about as low as we could sink," said Katzenbach. Sanders tried without success to restrain the television production at police headquarters. The Dallas police were so eager to perform that the ordinary procedure of checking credentials was forgotten. The writer William Manchester, who later reviewed videotapes of the scene, found it "an MGM mob scene." Viewers may have assumed the milling crowd consisted of policemen and newsmen, but this was scarcely so. The "extras" included

> . . . complainants and witnesses who happened to be in the building, prisoners, relatives of prisoners, relatives of police officers, known criminals, and drunks who wandered in off Harwood Street.

The prisoner was to be transferred from city jail to county jail. This was planned for the middle of the night but rescheduled for Sunday at noon, mainly to accommodate broadcasters. Justice Department officials continued to urge a secret night transfer, but Dallas Police Chief Jesse Curry was reported to be unwilling to "put anything over on" the news media.[4]

Among those disturbed by the Dallas performance was Richard Salant, president of CBS News, who was himself a lawyer. CBS-TV, like NBC-TV and ABC-TV, was cutting back and forth between Dallas and Washington. Salant wondered whether, in the public interest, it would be better to de-emphasize Dallas and stay with more somber events in Washington. But NBC-TV was keeping Dallas well covered and was getting the major share of audience. This apparently reflected the "news sense" of NBC president Robert Kintner. CBS continued with a similar policy.

Meanwhile Washington, in spite of drizzly weather, became a scene of unparalleled pageantry. On Saturday President Johnson received ex-President Truman and ex-President Eisenhower. Here was the drama of constitutional succession, of unity before the world. Simultaneously, planeloads of the great began arriving from abroad, and television cameras watched them: the towering Charles de Gaulle of France, the diminutive Emperor Haile Selassie of Ethiopia, President Chung Hee Park of South Korea, King Baudouin of Belgium, Chancellor Erhard of West Germany, Queen Frederika of Greece, President Eamon de Valera of

4. Manchester, *The Death of a President*, pp. 378, 599.

Ireland, Prince Bernhard of the Netherlands, Foreign Minister Golda Meir of Israel—and countless others.

The funeral was to be on Monday. After midnight Friday and again Saturday, long inter-network meetings were held to plan joint coverage of coming events. To cover essential locations and routes they would need about fifty television cameras. These would provide a "pool coverage" on which each network could draw at any time, while adding its own commentary. In addition each network kept cameras in reserve for added features of its own. Extra equipment and personnel were being flown in from as far as California. CBS was chosen to coordinate the pool coverage. ABC felt unable to handle it; the choice between NBC and CBS was made by the toss of a coin. Art Kane was chosen to direct for CBS.[5]

By Saturday the audience had much information—and occasional misinformation—about Lee Harvey Oswald. Police kept supplying details; networks and wire services dug up additional details. He had been born in New Orleans. His father had died before Lee Oswald was born, and he had been dominated by an overpowering mother. He had lived for a time in New York, where a school psychiatrist considered him a potentially dangerous schizophrenic. He had entered high school near Dallas, left for the marines, won a marksmanship award, was court-martialed twice, went to Russia, tried to gain Russian citizenship, got disgusted, came back with a Russian wife and a U. S. government loan, tried without success to go to Cuba, bought a mail-order rifle, got a job in Dallas at the Texas school-book depository, broke up with his wife, tried to win her back. He was a loner, truculent, self-pitying, paranoid. His flirtation with communism was often mentioned; it seemed to exonerate Dallas.

On Sunday cameras in Washington were ready for the solemn transfer of the casket from White House to Capitol, where a river of humanity would wind slowly past it. Cameras were ready in Dallas for another transfer. They were lined up in basement and garage of police headquarters for the removal of Oswald to the county jail.

Policemen were stationed around the scene. A television cameraman, finding himself photographing the back of a police uniform, spoke to the policeman. Television viewers saw the policeman turn to hear, then obligingly move aside to provide a clear gap.

At 12:21 eastern time ABC-TV and CBS-TV were carrying the pool

5. Small, *Interview*, p. 4.

coverage from Washington, where the caisson was about to leave the White House. All around the White House, and along the route the caisson would go, were weeping men and women. The weather was clear.

NBC-TV was taking Dallas, live. Its cameras showed Oswald coming into view, escorted by detectives, moving from the basement to the garage and toward an armored vehicle. Then a figure with a hat suddenly pushes into the picture. He thrusts a pistol toward Oswald and fires a shot. Oswald twists and falls out of sight. A policeman shouts, "Jack, you son of a bitch!" NBC's Tom Pettit shouts, "He's been shot! Lee Oswald has been shot! There is panic and pandemonium . . . !" Amid wild scuffling and shouting, the assailant is overpowered. Oswald is hauled into the armored car with limp arm dragging.

Only NBC-TV has shown it live, but the scene is available to the other networks via videotape, and within seconds they leave the Washington scene for a Dallas replay. It is shown repeatedly, intercutting with continuing pandemonium at headquarters, as announcers frantically seek information about the assailant.

In front of television sets throughout the country, diverse reactions. In Washington young Joanie Douglas, as the shot rings out, leaps up and shouts: "Good! Give it to him again!" Later when her husband, Supreme Court Justice William O. Douglas, arrives home, she tells him the news. He takes a deep breath and murmurs softly, "Well, it's just unbelievable." Later Mrs. Douglas wonders about her own "barbarism." Elsewhere George Reedy, assistant to President Lyndon Johnson, is busy on a telephone, watching television out of the corner of an eye; he thinks that, unaccountably, NBC-TV has switched to an old Edward G. Robinson gangster movie. It annoys him, and he tells his secretary to switch it off. Then the truth penetrates—and he hangs up. Later he cannot remember to whom he was talking on the phone.[6]

In Dallas the announcers are telling about the assailant, Jack Ruby. They say he runs a striptease joint, knows all the policemen, is arrested occasionally but seems drawn to them just the same, and likes to hang around headquarters giving out passes to his striptease parlor. Oswald, according to first reports, is badly injured.

At the time of the shooting, four out of ten television sets are on. An hour later the number has almost doubled. Telephone calls have risen to a peak.

6. Manchester, *The Death of a President*, pp. 606–7.

Coverage returns to Washington. The cortege has reached the Capitol. News of the shooting stirs through the crowd.

At 2:17 the world sees Mrs. Jacqueline Kennedy take daughter Caroline by the hand, and the two walk forward alone to the coffin. The mother walks with perfect self-control; the daughter looks to her for cues. The mother kneels and so does Caroline. The mother leans over to kiss the flag. The child slips her gloved hand under it and does the same. For millions it is the most unbearable moment in four days, the most unforgettable.

At about this time, at each network, teletype machines clack out the news that Oswald is dead. For once the urge to score a beat is restrained. Not until the Kennedy family has left the rotunda is the news relayed to the world.

The television spotlight is now fixed on solemnities. Millions are hypnotized by the river of faces passing the bier.

The television audience is aware that the new President is receiving visitors, but it has no notion what subjects are discussed. At 3 p.m., just after Jacqueline and Caroline have stepped back from the bier, the President receives John McCone, who has become head of the Central Intelligence Agency, and Henry Cabot Lodge, just in from Saigon. The subject is Vietnam, where things are going badly.[7] Behind the scenes the subject has high priority, but on the air it is not mentioned.

NBC decides to continue all night with coverage of the stream of humanity filing past the coffin. Late on Sunday word reaches the networks: at the cathedral ceremonies on Monday the technicians must wear formal morning clothes—striped pants and cutaways. The problem of renting clothes on a Sunday night is tackled by a volunteer; a list of sizes is compiled.

On Monday nine out of ten television sets are in use as the procession moves toward Arlington Cemetery. Some television people have slept as little as six hours in three nights. They go on, almost welcoming the absorption in the task at hand. Their dignity, intelligence, and judgment are extraordinary. Did the networks have any conception of the talent submerged in their news divisions, squeezed into daily one-minute and two-minute capsules?

In spite of the frantically assembled facilities, slips and fluffs are few. During the processions from the Capitol to the Cathedral, from the Ca-

7. *Ibid.*, p. 627.

thedral to Arlington, the action passes from camera to camera, newsman to newsman. The networks have only one-way communication to many of their men. William Small of the CBS Washington bureau speaks to one of his announcers via one-way phone: "Now Stan, it's moving a little closer to you now. You don't see it yet, you don't have to talk yet, just wait a few moments, then when you see the head of the cortege, then you can mention it. Now, do you have it? Don't answer me, because your mike is open, but that's what you are to do."

STAN: Okay, Bill.

The two words, spoken clearly in the midst of solemnity, are momentarily puzzling to millions.[8]

Throughout the four days, past and present mingle. Even as television looks at the flag-draped coffin in St. Matthew's Cathedral, we hear the voice of John Kennedy in his inaugural:

KENNEDY: . . . ask not what your country can do for you; ask what you can do for your country.

By Monday night John Kennedy has become a legend.

The networks carried no commercials from Friday through Monday. Some stations followed the example; others did not. Ten per cent of the stations, according to William Manchester, began slipping in commercials on Sunday morning. The networks estimated that the four days had cost them three or four million dollars each, including loss of revenue; but some of this would be recouped, because many sponsors agreed to a later rescheduling of commercials. To stations in their local operations, the cost may have amounted to an additional ten million.

On Tuesday schedules began to return to normal, but not entirely. Schedules for the following week were reviewed and changes made. The *Route 66* episode scheduled for Friday had the title "I'm Here To Kill a King"; another episode was substituted. The Ed Sullivan program had scheduled a Russian puppet theater but decided to postpone it.[9]

Praise for the achievement of the industry was widespread and warm. American television, it was felt, had helped the nation pull together. The world impact of its coverage, via satellite and film, was equally important. Rumors of vast conspiracies were heard in many countries but were mitigated and deflected by the coverage.

8. Small, *Interview*, p. 12.
9. New York *Times*, November 24, 1963.

The achievement had different meanings for different people. Leonard Goldenson, president of American Broadcasting-Paramount Theaters, felt that broadcasters had been able to make this contribution because they had had access to developing events. "Unfortunately," he said, "this is not always the case." Their exclusion from trials was a case in point. Broadcasters had now earned the right, Goldenson said, to be admitted to such events.[10]

To the American Civil Liberties Union the events—at least insofar as coverage at the Dallas police headquarters was concerned—had an opposite meaning. The ACLU was aghast that the processes of justice had taken on "the quality of a theatrical production for the benefit of the television cameras."[11] Bar associations made similar statements. They blamed Dallas officials for succumbing to television, but blamed television for creating the atmosphere and pressures to which the officials had succumbed.

Such views could be found also within the industry. William T. Bode of WCAU-TV, Philadelphia, wrote to *Variety:*

> I think radio and television helped kill Lee Harvey Oswald. I think that unless we in the industry own up to this, discuss the situation, and resolve it to the best of human ability, the chances remain that such a situation may occur again.[12]

Implications for the future of television were discussed in even harsher terms in other letters to newspapers. Some called for drastic changes.[13]

> To the Editor of the New York *Times:*
> The shooting of President Kennedy was the normal method of dealing with an opponent as taught by countless television programs. This tragedy is one of the results of the corruption of people's minds and hearts by the violence of commercial television. It must not continue.
> Gilbert E. Doan
> Nazareth, Pa.

> To the Radio-Television Editor:
> In view of recent tragic events, it seems to be highest time for the TV programs to be completely overhauled. The overwhelming number of "pistol-packing" programs are a disgrace and one would think that the time is here for a complete reappraisal of this art medium.
> Dorotheas H. Hecht
> New York

10. Goldenson, *Responsible Broadcasting,* pp. 3–5.
11. *Civil Liberties,* January 1964.
12. *Variety,* December 11, 1963.
13. New York *Times,* December 8, 9, 1963.

Concern over a climate of hate and violence came from many sources. A Dallas minister, Rev. William A. Holmes, received a visit from a fourth-grade schoolteacher who was distraught over the reaction of her class when she announced the assassination of President Kennedy. It had brought handclapping and cries of "Goodie!" The minister mentioned the incident in a sermon and in a CBS-TV network interview, saying it apparently reflected attitudes absorbed at home. In response he received telephone calls denouncing and threatening him; police decided to offer him armed protection and to move him and his family to another home.[14]

The sequence of murders jarred many Americans deeply. "My God, when will it end?" cried presidential counsel Theodore Sorensen when he learned of the killing of Oswald. Llewellyn Thompson, State Department expert on Russian affairs, considered it a diplomatic catastrophe.[15]

Martin Luther King was deeply affected. When he learned of the death of Kennedy, he told his wife, "This is a sick society." He felt that he also would meet his death violently.[16]

Talk about a "sick" society was heard on the air for a few days but soon subsided. If there was anything wrong in the climate of American life, the industry as a whole had no feeling of having contributed to it. It emerged from the four days with a glow of warmth and a sense of achievement. It had shown its maturity, scored a triumph, and now deserved a less regulated existence. The time had come, said numerous spokesmen, to end government harassments.

Easier times were expected. A few months before the assassination of President Kennedy, Newton Minow had left the FCC for private business, and the industry sighed with relief. His youthful successor, E. William Henry, was perhaps of similar disposition, but he was not as articulate and was not expected to have comparable influence. He was also a "Robert Kennedy man" and not considered likely to remain.

In December 1963 *Sponsor* magazine noted that several "Minow-inspired programs"—the children's series *Exploring* and *Discovery* were mentioned—had already fallen by the wayside.

The industry felt comfortable about President Johnson. He was the first President to come from the ranks of broadcasting management. At the time he took office the family broadcast holdings, which had started

14. *Ibid.*, November 28, 1963.
15. Manchester, *The Death of a President*, pp. 608–9.
16. *Life*, September 12, 1969.

in 1943 with a $17,500 investment in a radio station, were estimated by *Life* to be worth $8,000,000.[17] Johnson had always taken an active part in managing the properties, and it was felt he would understand industry problems. He was considered a friend of President Frank Stanton of CBS, who advised him on communication matters and seemed to have readier access to the White House than the chairman of the FCC.

The industry, in view of its expenses and losses in the assassination period, felt justified in getting back to business. There would soon be another presidential election—conventions, campaigns, election returns. Now it was time to recoup.

This swing came at a moment when new crises confronted America. One was in Vietnam, about which Americans knew almost nothing. Networks still had no bureaus there, although correspondents came and went on temporary assignments. The American stake in Vietnam seemed on the increase: the 800 advisers sent by Eisenhower had grown to 15,500 men under Kennedy. But now there was crisis. Three weeks before the assassination of Kennedy, there was action in Saigon. Ngo Dinh Diem and his brother, Ngo Dinh Nhu, were both assassinated, and turmoil followed. The United States seemed to face a choice of withdrawing from Vietnam or doubling its bets. All this faced Johnson the day he took office—as the problem of Cuba had faced Kennedy.

Most viewers did not yet know where Vietnam was, but they would learn.

17. *Life,* August 21, 1964. The stations were placed in temporary trusteeship during Johnson's presidency.

4 / JUGGERNAUT

Some day they'll give a war,
and nobody'll come.
CARL SANDBURG, *The People, Yes*

And as in uffish thought he stood,
The jabberwock, with eyes of flame,
Came whiffling through the tulgey wood,
And burbled as it came!
LEWIS CARROLL, *Jabberwocky*

For a while, little seemed to have changed. On the air, now and then, the liberal flame burned brightly. President Johnson kept it burning with a fervent televised plea for a civil rights law—and startled audiences by ending with the words of the song, "We shall overcome—some day!"[1] By thus aligning himself with the Negro rights movement as though he had been a lifelong marcher, and then applying his enormous persuasive talents to members of Congress, Johnson accomplished what Kennedy had been unable to accomplish. The bill became law in July 1964—shortly before the nominating conventions.

Throughout the early months of 1964 the civil rights drive seemed to gain momentum—with continuing effects on broadcasting. Young "freedom riders" from the North were going to Mississippi to help rebuild churches and homes of Negroes, to serve in "freedom schools," and above all, to show solidarity. In June the disappearance of two freedom riders, Michael Schwerner and Andrew Goodman, along with a local black youth, James E. Chaney, became a focus of news interest. When their bullet-ridden bodies were found, and Ku Klux Klansmen were indicted for the murder, the case won worldwide attention. Network cameramen turned up at the Chaney funeral service in the little First Union Baptist Church in Meridian, Mississippi, and the telecast brought many viewers close to the movement. In New York the young wife of Michael Schwerner remarked on television how tragic it was that it took the murder of some white boys to fasten national attention on what Negroes had long endured. The mother of Andrew Goodman said on television, "My thoughts are with Mrs. Chaney." Then Mrs. Chaney flew north from

1. March 15, 1964.

Meridian to be at the funeral of Schwerner and Goodman. She too wanted to show solidarity.

Along with freedom riders others went south. In March 1964 the communication division of the United Church of Christ, under the Reverend Everett C. Parker, began tape-recording the output of the two television stations in Jackson, Mississippi: WLBT, the NBC-TV and ABC-TV affiliate; and WJTV, the CBS-TV affiliate. On the basis of the recorded evidence, the United Church asked the FCC not to renew the station licenses. It pointed out that Negroes represented 45 per cent of the viewing area but were ignored by the stations. In addition, news about desegregation efforts, in the nation and locally, was ignored and even—in the case of network programs—eliminated by the stations of Jackson, with the result that its "entire population, Negro and white, receives a distorted picture of vital issues."

Each station was found to devote 16 per cent of its time to commercials. No black artists were used. Each station carried more than four hours of religious services—none involving Negroes. The United Church argued that the rigged program pattern, far from serving the "public interest," was a disservice to both white and black viewers.

So completely were the mores of the area taken for granted that such complaints seldom confronted the FCC formally. The commission was mildly embarrassed. By a 4 to 2 decision—the dissenters were Chairman E. William Henry and Commissioner Kenneth Cox—it renewed the licenses for one-year terms, pending good behavior. In so doing it refused public hearings demanded by the United Church and took the position that the group had in any event no rights in the case, since the "public interest" was represented by the FCC itself. The church group refused to accept this, and promptly challenged the decision in the U. S. Circuit Court of Appeals for the District of Columbia. The action foreshadowed a legal struggle with implications for many southern stations, although most—reassured by precedent—did not take the threat very seriously. Monitoring in other southern cities was begun by the United Church.[2]

This dispute helped stimulate a new activity among freedom riders. They organized radio workshops to develop plans for "freedom stations." One group proposed a station to be called Radio Free Southland.[3]

The agitation put pressure on northern as well as southern stations. In

2. New York *Times,* April 16, 1964; *Variety,* June 16, 1965.
3. New York *Times,* January 31, 1965.

July 1964 the pioneer station KDKA, Pittsburgh, approaching its forty-fourth anniversary, hired its first Negro performer.[4]

On television, programming that echoed the stirrings of the Kennedy years still turned up intermittently in 1964. On NBC-TV *That Was the Week That Was,* modeled after a British Broadcasting Corporation series, was offering lighthearted but sharp satire—a rare commodity on television. On CBS-TV *The Defenders,* produced by Herbert Brodkin, was showing special vitality, even digging into the industry's own problems—in January with a play by Ernest Kinoy, about television blacklisting; and in May with a play by Albert Ruben, about a correspondent deprived of his passport for "unauthorized" travel to China. The latter was a clear reminder of the William Worthy case. Both these scripts explored with integrity legal ramifications and nuances of their subjects; such programs were found to be winning for the series a loyal audience in the legal profession.

Equally unusual was the ABC-TV documentary *Cuba and Castro Today,* made by Harry Rasky, one of the emigrants from Toronto. A Canadian, he had no difficulty in going to Cuba. Dealing with Cuban representatives at the United Nations, he expressed interest in making a film there and was assured he would be welcome. "I told them very clearly, 'I'm not a communist and I'm not a communist sympathizer, but I hope I am objective,' and they said that was good enough for them."[5] He arranged to take as cameraman a Hungarian refugee living in Canada, Michael Lente. (Lente was afraid he would be seized and shipped back to Hungary, but this did not happen.) Having made arrangements, Rasky interested ABC-TV in backing the project. Lisa Howard, a relentlessly adventurous ABC reporter, begged the network to let her follow; she had once interviewed Castro and claimed to have secret phone numbers of intermediaries through whom he could be reached. The network agreed to let her go; its news division, now under former presidential press secretary James Hagerty, obtained State Department "authorization." All went to Havana via Mexico. At the Havana Libre—formerly the Havana Hilton—Lisa Howard used her phone numbers.

One evening they found a row of guards in the hotel corridor. Someone told them that was a sign—something would happen.

4. *Variety,* July 8, 1964.
5. Rasky, *Interview,* p. 25. The following is based on this interview; other sources as noted.

Hours later, long after midnight, Castro turned up. Sitting on the bed in Lisa Howard's room, he talked five hours as the cameras turned. It was the first of a number of filmed conversations with Castro. Meanwhile the film group moved about freely and shot interviews with people of diverse opinions.

> STUDENT AT CARLOS MARX SCHOOL: Well, this revolution to me is so great for me that I don't know how to express to you all those things that I have for this revolution. If I can die for this revolution, I die. If I can fight, I fight, because this is only my life.

> ANTI-CASTRO LADY: Well, I really don't think that the majority of the people like Mr. Castro. The only thing is we are living under a big panic.[6]

Dissenters spoke freely but most preferred to be photographed at angles or in lighting arrangements that made recognition difficult.

Traveling at high speed through farm areas in Castro's automobile, Lisa Howard asked:

> LISA: But your agricultural problems are more simple to deal with than your political problems?
>
> FIDEL: I don't know. I think it's easier political problems than technical problems.

They encountered a "cola" that was white, from a Czech formula. Farm workers were rewarded for extra achievement with weekends at the Havana Libre. The Rasky group photographed docks crowded with ships, most of them from communist countries but many from non-communist countries. China was providing rice and newsprint; farm machinery came from Czechoslovakia, Romania, Russia. Che Guevara, at that time minister of industries, spoke in one interview—filmed in his office at 2 a.m.—of contracts for the purchase of complete plants from France, Britain, Japan. He was asked if these represented a failure of United States anti-Castro pressures.

> CHE: Yes.
>
> LISA: A serious failure?
>
> CHE: That depends on the American self-esteem.

The group found Castro armed with information on many subjects, and he spoke fluent English. They were surprised that his conversational personality—quietly earnest—was so different from his arm-waving platform

6. Script quotations from *Cuba and Castro Today,* ABC-TV, August 19, 1964.

personality. Most Latins, observed Rasky, seem to have "a different face for public oratory than they do in personal oratory." Castro could be responsive; also very agile.

> LISA: Dr. Castro, many people we have spoken with on the island believe that Cuba is a police state. Do you think so?
>
> CASTRO: Many people? Many people? And you could speak with a hundred, a thousand people about this is a police state?

Lisa Howard accepted it as certain that Cuba had begun training guerrillas for use elsewhere. She reported having heard estimates as high as 1500 trainees.

> CASTRO: Why do you believe that a revolution can be made from another country? If it were possible, the United States government had made one hundred counterrevolutions in Cuba, and I don't understand how you think it's so easy to make a revolution from Cuba.

The program, quoting diverse views, maintained an objective stance. Toward the end it quoted Senator J. William Fulbright, chairman of the Senate foreign relations committee, to the effect that a re-evaluation of American policy toward Cuba might be overdue, even though it might "lead to distasteful conclusions."

To some elements of the industry the effort to be objective about Cuba seemed, in itself, suspect. The network found no sponsors and scheduled the program on Sunday afternoon. As a further precaution it preceded and followed the film with fund appeals for refugees from "Castro tyranny"[7]—a device that was becoming standard for troublesome programs.

Though some programming gave the feeling of a continuing Kennedy atmosphere, an opposite trend—encouraged by prosperity and a growing sense of ease about Washington—was increasingly evident.

ADVERSARY SYSTEM

By 1964 James Aubrey, who had said he could double the CBS net income—it had stood at $25 million when he took office—was well on his way to doing it. In January the evening television schedule for the following fall was already sold out, and even fringe periods looked profit-

7. New York *Times*, April 20, 1964.

able. Preventing even grander profits were programs like *CBS Reports,* which were sometimes sold to sponsors at less than cost. Fred Friendly, who in March 1964 became president of CBS News, found he would have to battle to keep it in the schedule.

On one occasion Aubrey, according to Friendly's recollection, told him that "in this adversary system" they would of course be at each other's throats. "They say to me, 'Take your soiled little hands, get the ratings, and make as much money as you can.' They say to you, 'Take your lily-white hands, do your best, go the high road and bring us prestige.'"

The "adversary" aspect seemed to stimulate Aubrey, and he tended to make annual budget meetings an arena for combat. Presenting CBS-TV financial achievements with color slides in which the cost of news programs was depicted in—to use Friendly's terms—"fire-engine red," he addressed the chair: "You can see, Mr. Chairman, how much higher our profits could have been this year if it had not been for the drain of news." Aubrey wanted *CBS Reports* offered at a profitable figure and, if not sold, dropped.[1]

Chairman Paley, a suave executive, disliked crude approaches of this sort. Yet Aubrey generally got his way, and the trade press carried rumors that *CBS Reports* might be scrapped.

A disdain for the "high road" trickled down through network hierarchies. When an Aubrey lieutenant engaged producer-writer William Froug, who had been producing *Twilight Zone,* for the Hollywood position of excutive producer in charge of drama, it was emphasized that Froug was not to produce *Medea* but mass entertainment. The executive summarized: "Your job is to produce shit."[2]

The view that affiliates had similar preferences was often expressed, and with some evidence. Edward W. Barrett, former broadcasting and publishing executive who had become dean of The School of Journalism at Columbia University, attended a 1964 broadcasting convention and heard a broadcaster talking about his station. Asked what network it was affiliated with, this man answered: "NBC, damn it."

"Why do you say it that way?"

"Well, they're ruining it with all this damned news and documentary stuff. . . . We lose audiences every time we have to put that stuff in place of entertainment. Take that nonsense of extending the Huntley-

1. Friendly, *Due to Circumstances Beyond Our Control . . .*, pp. xi–xii, 195–6.
2. Interview, William Froug.

Brinkley program from fifteen minutes to a full half hour. That alone costs us way over $100,000.[3]

If there was restiveness about news programs, it was because sponsorship money was available for other things. Even Sunday afternoon, long a "cultural ghetto," was turning into a gold mine, mainly because of the sensational rise of professional football. CBS paid $28 million for television rights for the National Football League games for 1964 and 1965 and instantly recouped the investment with two $14 million sponsorship contracts—one with Ford Motor Company, the other with Philip Morris and its subsidiary the American Safety Razor Company.[4] Additional spot announcements, to be sold to regional sponsors, would represent profit.

A factor in the rise of football as television fare was the "instant replay," so tellingly used in the Oswald murder. A few weeks later, on New Year's Eve, it was used in the Army-Navy football game, and in 1964 it became a standard sports technique. While one camera showed the overall action "live," other cameras followed key players in close-up, with each camera linked to a separate videotape machine. Within seconds after a play, its crucial action could be re-examined in close-up, or even unfolded in startling slow motion. This accomplished incredible transformations: brutal collisions became ballets, and end runs and forward passes became miracles of human coordination. Football, once an unfathomable jumble on the small screen, acquired fascination for widening audiences.

The mania produced new social phenomena. Football games were often withheld from television in cities where they were played. Such television blackouts caused weekend emigrations to motels within reach of other stations. New York games brought a boom to Connecticut motels.[5]

In 1964 CBS, flush with profits, purchased a baseball team—the New York Yankees—and acquired a new stake in sports promotion. Questions of a conflict of interest were raised in government committees, argued, and quickly forgotten. The growing involvement in sports was regarded by some executives as a useful hedge against rising talent costs. Any future strike by actors might be countered by an expanded sports schedule. Meanwhile sports were making weekend periods fabulously valuable.

Saturday mornings, once regarded as a time for do-good programs to

3. *Columbia Journalism Review*, Spring 1964.
4. New York *Times*, February 15, 1964. The 1962 and 1963 rights had gone for $9 million.
5. *Ibid.*, November 22, 1964.

please women's groups, were becoming profitable for a different reason. Toy manufacturers were adopting year-round rather than seasonal advertising schedules and were the main Saturday morning sponsors, backing a parade of animated films, largely violent.

Weekday mornings and afternoons were increasingly profitable. Daytime serials had at first seemed a failure on television. But when they were expanded from the 15-minute form inherited from radio to a 30-minute form, success followed. By 1964 daytime serials were an addiction comparable to the radio-serial addiction at its zenith, and were the mainstay of New York activity in television drama. They were especially profitable for CBS-TV.

The late evening hours were profitable far into the night. In 1962 the NBC-TV *Tonight* series became the domain of Johnny Carson, and turned into an even greater bonanza than it had been under Jack Paar. These hours were still feature-film time on CBS-TV, but a switch to programming on the *Tonight* style was considered inevitable. The early morning hours were also highly profitable, especially for NBC-TV, with its *Today* series. Almost all hours shared in the rising affluence.

The prosperity of television had various off-shoots. When the New York Giants baseball team decided to become the San Francisco Giants and the Brooklyn Dodgers to become the Los Angeles Dodgers, it was in large part because of a television scheme—Subscription TV, Inc., organized by Matthew Fox, who had become prosperous distributing feature films to television. He brought Pat Weaver back into the spotlight as president of Subscription TV.

The company planned to supply programs to homes in San Francisco and Los Angeles by television cable. The subscriber would pay an installation fee and then pay for programs actually selected via a telephone dial system. The subscriber was promised events that would not be available on commercial television: first-run movies, major sports events, opera. The Giants and Dodgers, arriving on the West Coast with fanfare, would eventually be seen *only* via Subscription Television, Inc., not by commercial television. Most major motion picture companies were spellbound by the prospect of a new kind of "box office." Figures projected by Matthew Fox and Pat Weaver suggested that both film industry and major league baseball would gross revenues even exceeding those earned through "free television." Motion picture theaters were up in arms at the prospect of a new form of competition. Broadcasters were likewise alarmed. If the system proved successful, major attractions would probably become unavailable to sponsored television—they would switch to Subscription TV and other such regional systems. "Free" television might become a ghost town of the unwanted. The year 1964 loomed as the year of decision as the sale of subscriptions and the wiring of homes got under way in the Los Angeles and San Francisco areas.

One aspect of Subscription TV, Inc., was of special interest to entrepreneurs. Because it planned to operate entirely within California, its cables would not cross state lines, and the plan would apparently avoid FCC jurisdiction.

Another scheme was developing unforeseen glitter in 1964: CATV. The letters had originally stood for "community antenna television," but the industry was switching to the term "cable television."

Community antenna systems had developed early in television; they even had a history that went back to the beginnings of radio.[6] During the television freeze period their function was to bring programs to towns that did not yet have stations of their own. After the freeze they remained important for small isolated communities or those with bad reception, such as towns surrounded by mountains. The subscriber paid an installation fee to have his set connected by cable to the community antenna system; then a continuing subscription fee, usually $5 a month. He needed no antenna of his own. The CATV system might bring him half a dozen to a dozen stations—in a few instances they included distant stations brought in by microwave relay. The cables to subscriber homes were usually strung on telephone poles under a contract with the local telephone company.

At first there was no thought of additional programming created by CATV systems themselves, but the possibility existed. Such programs might be provided to subscribers as a free service or at an extra charge.

As television spread during the 1950's, the function of CATV was at first regarded as diminishing, but it was only shifting. Many communities had only one or two stations and welcomed a system that brought in others. Special reception problems persisted and multiplied. Cities had ghost images bouncing from tall buildings, and other mysterious inter-

6. The first community antenna system was apparently launched in Dundee, Michigan, in 1923. Bringing in remote stations by means of a tall antenna tower, it offered their programs to local subscribers by wire at $1.50 per month. The system was described in the May 1923 issue of *Radio Broadcast* by University of Wisconsin student Grayson Kirk, later president of Columbia University.

ferences such as diathermy machines. A CATV system could solve such problems.

In the early 1960's the broadcasting industry woke up to the fact that fortunes were being earned by the cable systems. Were they a boon to television, or a threat?

Film distributors saw a threat. United Artists, having sold a film to a television station in Scranton, supposedly for a Scranton premiere, might find the station complaining that Scranton viewers had already seen it—thanks to a CATV system that had relayed it from a Pitsburgh telecast. Clearly control of copyrights was at stake. Broadcasters planning new stations on available UHF frequencies were likewise disturbed. Success would depend on inclusion in the CATV system; they would be at its mercy. They turned to the FCC for support.

Did the FCC have jurisdiction? The CATV systems (except the few that used microwave relays) did no broadcasting. They sent no waves into the ether. They only *received* waves—as did individual viewers. As of 1964 most FCC members felt it was not a legitimate concern of the commission.[7]

Meanwhile the CATV industry grew. In 1964 over a thousand systems were in operation. The majority of them were members of an aggressive National Community Television Association. Late that year Commissioner Frederick W. Ford of the FCC decided to leave the commission to accept the presidency of the NCTA. The industry had its own trade press; the monthly *TV & Communications,* launched in 1964, was devoted solely to CATV and was supplemented by weekly bulletins titled *Cable Television Review.*[8]

There was growing awareness that a nationwide linking of cable systems could turn into a coast-to-coast subscription television system. Broadcasters were hedging their bets by investing in cable systems. The three major networks and many group station owners became investors in cable systems.

The promoters of all these visions plunged ahead, luxuriating in a new laissez-faire era. But the center of the boom was sponsored television.

The number of commercials that could be crowded into a given time period amazed some observers, although most viewers, inured gradually,

7. It changed its mind two years later.
8. *TV & Communications* was later retitled *TV Communications.* The association was renamed the National Cable Television Association.

seemed to take it for granted. In 1964 a housewife who sat down for a morning television break to watch WCBS-TV, New York, could catch a Mike Wallace newscast at 10 a.m., followed by reruns of *I Love Lucy* and *The Real McCoys* and an episode of *Pete and Gladys*. The 120 minutes would include interruptions for 45 to 50 promotional messages for products and services. On a typical morning:

10:02	Beechnut Coffee ("turns lions into lambs")
10:09	Ajax ("knifes through waxed-in dirt")
10:14	Cheerios
10:15	Betty Crocker Cake Mix
10:20	Fashion Quick Home Permanent
10:20	Hudnut Shampoo With Egg and Cream Rinse
10:25	Playtex Nurser ("less spitting up and colic")
10:26	Playtex Living Stretch Bra ("lasting stretch that won't wash out")
10:28	Cotton Producers' Institute ("the fiber you can trust")
10:29	Jack Benny program promotion
10:29	Imperial Margarine ("sniff the wonderful aroma")
10:29	Sea Mist Window Cleaner ("washes dirt away")
10:30	Scott Tissue
10:37	Action Chlorine Bleach ("three active ingredients")
10:43	Fritos Corn Chips
10:44	Glass Wax ("gobbles up window dirt")
10:50	Duranel Cookware
10:57	Playtex Living Stretch Bra
10:58	Playtex New Living Girdle ("live a little")
10:58	Red Skelton program promotion
10:59	Shambleu Shampoo
10:59	La Rosa Pizza Pie Mix
11:00	Clorox
11:08	Endust ("cuts your dusting time in half")
11:09	Vanish Toilet Bowl Cleaner ("just brush and flush")
11:12	Westinghouse Iron
11:16	Playtex Cotton Bra
11:17	Playtex Panty Girdle ("try the fingertip test")
11:23	Cotton Producers' Institute
11:27	Ajax Laundry Detergent ("stronger than dirt")
11:28	Danny Kaye program promotion
11:29	Glass Wax ("the one right cleaner for city windows")
11:29	Sandran Vinyl Floors ("you can lie down on the job")
11:30	Rexall One-Cent Sale
11:36	O. J. Orange Juice
11:44	Playtex Living Stretch Bra

11:45 Playtex New Living Girdle
11:45 Pillsbury's Best Flour
11:45 Pillsbury's Mashed Potato Flakes
11:46 Sweet 10 Sweetener ("no bitter aftertaste")
11:51 Fritos Corn Chips
11:57 Right Guard Deodorant ("24-hour protection")
11:58 Pillsbury's Best Flour
11:59 *Petticoat Junction* program promotion
11:59 SOS ("loaded with soap")
11:59 *Eye on New York* program promotion[9]

News analyst Eric Sevareid observed in 1964:

SEVAREID: The biggest big business in America is not steel, automobiles or television. It is the manufacture, refinement and distribution of anxiety. . . .

Logically extended, this process can only terminate in a mass nervous breakdown or in a collective condition of resentment that will cause street corner Santa Clauses to be thrown down manholes, the suffering to be left to pain, and aid delegations from Ruanda-Urundi to be arrested on the White House steps.[10]

With Sevareid all problems were sublimated into fine prose. Other observers were more indignant and pressed for action, but the regulatory spirit was on the decline. When it appeared, it was quickly squelched. President LeRoy Collins, who was periodically disturbed about advertising excesses, suggested on one occasion that broadcasters should restrict cigarette advertising. The NAB soon afterwards fired Collins from his $75,000-a-year job.[11]

The experience of FCC chairman E. William Henry was even more enlightening, and also concerned the NAB. The NAB television and radio codes were constantly held up by industry spokesmen as shining examples of self-regulation, though regarded by many broadcasters as a charade. Most of its edicts had built-in escape hatches,[12] and the few clear rules—such as those dealing with the time to be devoted to commercials—were

9. WCBS-TV, New York, May 5, 1964. The period also included an "Every Litter Bit Hurts" message and a multiple sclerosis message.
10. Sevareid, *This Is Eric Sevareid*, pp. 71–2.
11. He became director of community relations under the 1964 civil rights law at $25,000; he explained, at the confirmation hearing, that he had left the NAB with a $60,000 severance payment.
12. Typical was a television code edict on *violence*, which was to be used "only as required by plot development or character delineation." The code writers seemed to pretend that plot and character were beyond the control of writer and producer. *NAB Television Code*, II, 2.

Martin Luther King, who had a dream.

Wide World

CBS

Bill Cosby narrates the documentary "Black History: Lost, Stolen, or Strayed," for the CBS series *Of Black America.*

Twentieth Century-Fox

Diahann Carroll and Marc Copage in *Julia.*

KBPL

In Portland, Oregon, a fourth grade class broadcasts over KBPS, radio station of the public school system.

NET

In Berkeley, California, a student edits a sequence for "Life Style," an *NET Journal* documentary made with the assistance of University of California students.

In San Francisco, KQED's *Newsroom* series probes regional issues.

KQED

American Samoa: Paradise Lost? —an NET survey of westernizing influences.

NET

Johnny Cash and Bob Dylan in a Nashville recording session.

Left and below: *Johnny Cash*—a *PBL* film by Arthur and Evelyn Barron, later seen as a theatrical feature.

Continental

Continental

Johnny Cash and a member of the Sioux tribe visit an Indian burial ground.

ON THE BALL

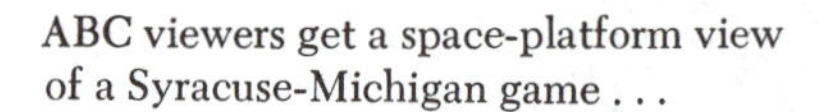

ABC viewers get a space-platform view of a Syracuse-Michigan game . . .

ABC photos

. . . a close-up in a Princeton-Rutgers game . . .

. . . a telescopic shot of the Green Bay Packers *v.* Chicago Bears.

widely ignored. An FCC sampling of stations in 1963 found that 40 per cent had advertising exceeding the code limits. The NAB television code had an enforcement machinery which was among its most absurd features. If a subscribing station was charged with violating the code and was found guilty by an NAB review board, the station (according to the rules) would lose the right to display on the screen the NAB "seal of good practice." Since the seal meant nothing to viewers and its absence would be virtually impossible to notice, the machinery meant nothing. In fact, no station had been deprived of the seal.

Chairman E. William Henry favored a modest approach to the problem of over-commercialization. Since the industry had defined its standards in its own codes and constantly spoke of them with respect and admiration, it seemed to Henry logical that the FCC should adopt those standards officially. At license renewal time the FCC could then inquire: has the station observed the industry's own proclaimed standards?

The industry reaction to this was one of horror and outrage; it stimulated instant countermoves in Congress, where Representative Walter E. Rogers of Texas introduced a bill forbidding the FCC to take any action to limit commercials. NAB memos marked "URGENT URGENT URGENT" were dispatched to all stations.

> Broadcasters should immediately urge their Congressmen by phone or wire *to vote for H. R. 8316*. . . . A vote for the bill is a vote of confidence in the broadcasters in his district. A vote against the bill would open the door to unlimited governmental control of broadcasting. . . .[13]

The bill passed, 317 to 43. The Senate ignored the affair, but the House vote served as due notice to the FCC. It dropped the notion of curbing commercials.

Chairman Henry, settling into frustration, was tempted into Minow rhetoric. That fall he referred to television schedules as an "electronic Appalachia." The words were shrugged off by the industry.

Back in the 1920's Herbert Hoover, as Secretary of Commerce, urged broadcasters to restrain their commercial impulses; he was sure that, for their own good, they would heed his advice. He said that if a presidential message ever became "the meat in a sandwich of two patent medicine advertisements," it would destroy broadcasting.[14] When Hoover died on

13. NAB memorandum, February 21, 1964.
14. *Radio Broadcast*, December 1924.

October 20, 1964, NBC broadcast a tribute which was at once followed at its key station by a beer commercial, a political commercial, and a cigarette commercial. The ex-President was triple-spotted into eternity.[15]

As he died, the broadcasting industry was in a state of almost delirious prosperity. A Pittsburgh station—WIIC—had just changed hands for $20,500,000, the highest price ever paid for a station. Even network radio was again profitable, and radio as a whole had earned a $55 million profit during the preceding year.[16] In mid-Manhattan a suave, impeccably dressed skyscraper designed by Eero Saarinen for CBS was rising as a monument to its role in the new era. Significantly it was almost wholly devoted to executive and sales functions, with most television program activity being banished to a rebuilt Borden bottling plant on less expensive real estate.

Against this background of prosperity, the industry was preparing for another presidential election.

GIRL WITH A DAISY

Many Americans, going about their business in an environment that—judged by television—was prosperous, vigorous, and forward-moving, first became aware of Vietnam through the 1964 presidential campaign as seen on the home screen. The awareness came not from things President Johnson told them—he scarcely mentioned the subject—but from issues raised by Senator Barry Goldwater of Arizona, who was nominated by the Republicans at their San Francisco convention. Goldwater saw Vietnam as a possible issue, and he needed issues.

Goldwater had been talking with some of the generals, and they told him that the bogged-down action in Vietnam—which in 1963 had come close to disaster—was a disgrace that could and should be resolved by firm measures. Goldwater seldom specified what measures should be taken, but in an ABC-TV interview with Howard K. Smith he mentioned some possibilities, one of which was to defoliate Vietnamese forests with a "low yield atomic device." The remark was widely discussed and gave the impression that Goldwater had a casual attitude toward dropping atomic "devices."

The Democratic Party in this campaign had the service of the Doyle

15. WNBC-TV, October 20, 1964.
16. *Television*, October 1964; *Variety*, October 14, 1964.

Dane Bernbach advertising agency—a young agency with a reputation for sophistication. President Kennedy had taken a fancy to its Volkswagen advertising and to its Avis campaign: "We try harder—we have to—we're only number two." It may have reminded Kennedy of his 1960 situation. In September 1963 he asked his brother-in-law Stephen Smith to talk to the agency about working on the 1964 presidential drive; later President Johnson decided to hold to the Kennedy choice.

Although the Democrats had in the past had trouble finding an agency willing to handle their "account," Doyle Dane Bernbach was at once interested. Perhaps rising prosperity had something to do with it.

The Doyle Dane Bernbach campaign work was novel in several respects. Determined not to disturb viewing habits, it concentrated on spots. It even avoided the 5-minute hitchhike programs featured in 1956, requiring partial pre-emptions. Some of the Doyle Dane Bernbach spots were less than a minute long and could be used at station-break time along with other spot commercials. The emphasis on spots was in harmony with current advertising trends.

For speeches the Democrats pre-empted only eight programs throughout the campaign—in contrast to the Republicans, who pre-empted thirteen.[1]

The Doyle Dane Bernbach spots, in addition to being the heart of the campaign, had another unusual feature. Earlier spots, such as those for Eisenhower and Stevenson, had been built around the candidate. The principal Doyle Dane Bernbach spots were not. This may have been partly a matter of necessity; the nomination of Johnson, though a foregone conclusion, did not take place until the end of August, more than a month after the Goldwater nomination. Meanwhile the Doyle Dane Bernbach spots dealt—without mentioning him—with Goldwater.

Its most celebrated spot showed a small girl picking petals from a daisy. This was accompanied on the soundtrack by a count-down. Then came an atomic explosion, darkness, and a brief statement.

> VOICE: These are the stakes. To make a world in which all God's children can live, or go into the dark.

In another spot a girl was seen eating an ice-cream cone. There was the ticking of a Geiger counter. A motherly voice was meanwhile explaining about Strontium 90, a radioactive fallout product found to con-

1. *Television Age*, November 23, 1964.

centrate itself in milk. Again a viewer was reminded of Goldwater's apparently casual attitude toward nuclear "devices" and perhaps his opposition to the test-ban treaty.

The Goldwater nomination represented a westward power shift. Goldwater backers included right-wing extremists who looked on some eastern members of the party—particularly those called "liberal"—as akin to communists. Goldwater, in a light mood, gave expression to this western chauvinism with the suggestion that someone should saw off the Atlantic seaboard and let it float out to sea. Doyle Dane Bernbach produced a spot in which someone did just that.

Though most Republicans, reversing opposition policies of earlier years, had long ago accepted Social Security, it was still anathema to some conservatives. In the New Hampshire primary Goldwater made a slighting reference to Social Security, causing speculation that he might abolish it. He denied this, but Doyle Dane Bernbach kept memories of the remark alive with a spot in which two hands were seen tearing up a Social Security card.

At the Republican convention in San Francisco, Governor Nelson Rockefeller of New York was loudly booed by Goldwaterites apparently determined to prevent him from delivering a scheduled speech. Several eastern leaders received similarly brusque treatment. Doyle Dane Bernbach offered a picturesque reminder of this and made a subtle appeal to their followers with a spot that showed campaign posters of the defeated leaders trampled in the dirt of a convention hall floor.

The Doyle Dane Bernbach strategy in all these spots was to panic Goldwater into over-reacting. They considered this strategy too subtle to explain to the Democratic National Committee and never did explain it to them. The agency dealt with the White House through Lloyd Wright and William Moyers—both Texans close to Johnson. But the strategy was apparently successful because Goldwater began fulminating against the spots. "The homes of America," he protested, "are horrified and the intelligence of Americans is insulted by weird television advertising by which this administration threatens the end of the world unless all-wise Lyndon is given the nation for his very own." Repeated assaults of this sort helped make the television spots the talk of the campaign. Astonishingly, the girl-with-a-daisy and ice-cream-cone spots were each used only once in paid time. The agency did not reschedule them, not because of orders to withdraw them—though some Democratic leaders considered them "unfortunate"—but because the attention won at first showing was so overwhelming that re-use seemed redundant. ABC News and CBS News carried news items about the daisy girl. *Time* did a cover story on her. The New York *Times* carried a news story about her. The hubbub helped reinforce the trigger-happy association pinned on Goldwater—and to imply that a vote for Johnson was a vote against escalation in Vietnam.[2]

All this carried a deep irony, for President Johnson had apparently already decided to escalate the war in Vietnam. Massive bombing of the north was planned for the new term. Preparations, in terms of supply, were no doubt contributing to the election-year prosperity.

Johnson hoped the bombing would dispose of the Vietnam problem; if not, he was prepared to go further. But during the campaign he said as little as possible. One statement, used repeatedly with variations, became his Vietnam campaign formula. On August 12:

> JOHNSON: Some others are eager to enlarge the conflict. They call upon us to supply American boys to do the job that Asian boys should do.

August 29:

> JOHNSON: . . . a war that I think ought to be fought by the boys of Asia.

September 21:

> JOHNSON: We don't want our American boys to do the fighting for Asian boys.

October 21:

> JOHNSON: We are not going to send American boys nine or ten thousand miles away to do what Asian boys ought to be doing for themselves.[3]

The oddity of this refrain caused little comment at the time. It seemed to imply that the killing of Asian boys—"the job"—was necessary and had to go on but "ought" to be done by their own kind. Americans, glad to note that they were exempted from whatever plans the President had, were ready enough to accept the statement.

While the Democrats relied heavily on spots, the Republicans relied

2. Robson, *Advertising and Politics*, pp. 13–39 and Appendix D.
3. Quoted, Schlesinger, *The Bitter Heritage*, pp. 44–5.

more on speeches but also on a half-hour television film—which, like the Democratic spots, became a storm center. Titled *Choice,* it was sponsored by Mothers for a Moral America and was scheduled to premiere on 145 NBC-TV stations on Thursday, October 22, at an afternoon time when the audience would be largely composed of women. Its theme was that the United States was undergoing a moral decay which the film attributed to the Democrats, including President Johnson. The decay was depicted through glimpses of topless dancers, pornographic magazines, marquees of nudist films—and rioting. The purpose, said one of its creators, was to rouse strong emotions, and to turn the anti-city feelings of rural people against the Democratic administration.[4]

The film associated sexual emancipation and the rise of nudism with Negro protest movements; all were considered aspects of the breakdown of "law and order." The phrase "law and order" was thus turning into a coded appeal to segregationists, and the film was considered by Democrats an attempt to woo a "white backlash" vote. Rioters in the film were mostly blacks but included white teenagers.

In an attempt to pin the moral decay on President Johnson, the film had intermittent shots of a speeding Lincoln Continental screeching across the screen, with beer cans being thrown from the window. These shots were a reminder of reports that the President had driven his Lincoln at high speed near his LBJ ranch, and had been seen drinking beer at the wheel. The producers acknowledged that the shots were staged. The President was not mentioned by name.

Two hundred prints of the film were distributed throughout the United States to be ready for saturation showings after the network premiere. But before it got on the air, protests began. Democratic Party chairman John M. Bailey had a preview and called it the "sickest" program in the history of television campaigning. He suggested that its "prurient" appeal mocked its moral pretensions. NBC was reported to have qualms and to be asking for cuts. The furor persuaded Senator Goldwater to have the telecast canceled, but local showings were held.

In San Francisco, Republican Party headquarters was reported showing the film in its front window. By this time the Democrats regarded the film as an asset to their own cause, and distributed handbills announcing the "sexiest show in town . . . free . . . see Goldwater family movies . . . see them shimmy, see them shake." They also used a barker on a sound

4. New York *Times,* October 21, 1964.

truck to send people to the Republican showings, and distributed leaflets with mock endorsements for Goldwater from "Daddies for Decent America."[5] After the campaign, according to the *Wall Street Journal,* the Democratic Club of Washington, D.C., used a print of the Republican "feelthy pictures" as a fund-raising attraction.[6]

The Republicans bought their television and radio time through Erwin, Wasey, Ruthrauff & Ryan. On four occasions, for campaign speeches, it pre-empted the network period occupied by the satiric series *That Was The Week That Was.* This series, returning to the air on November 10 after Johnson's landslide victory—the popular vote was 43,126,506 to 27,176,799, and the electoral vote 486 to 52—had a mild revenge. Film footage of Goldwater talking was so cut that he *seemed* to speak the message: "Due to circumstances beyond our control, the political address originally planned for this period will not be heard."[7]

The campaign saw the emergence of Lyndon B. Johnson as television impresario. Reporters were already familiar with his television obsession. At the White House he had three television sets lined up side by side so that he could watch all three networks simultaneously. When displeased he might phone a network president or an offending newsman. Early in the Democratic convention he phoned about camera work. He wanted the camera kept on the speaker at the rostrum and not switched to the strife-torn Mississippi delegates—a switch that seemed to play up disunity. During the convention he extracted maximum drama from the choice of a running mate. In the midst of the guessing game he arranged for Senator Thomas J. Dodd, the television-violence crusader, to pay a White House visit at the same time as Senator Hubert Humphrey. Johnson had no thought of Dodd as running mate, and the vice presidency was apparently not even mentioned between them. The ploy served merely to give Dodd a moment in the sun, and to build suspense. During the presidential nominating speeches Johnson flew to Atlantic City with a planeload of reporters, and while in flight followed the proceedings intently by television. Shoveling peanuts into his mouth, he watched Governor Connally of Texas nominating him. A reporter made a comment; the President was irked. "Shsh! he's nominatin' me!"[8]

Among television footnotes left by the campaign, one concerned Sub-

5. *The Reporter,* November 19, 1964.
6. *Wall Street Journal,* January 15, 1965.
7. *That Was The Week That Was,* NBC-TV, November 10, 1964.
8. Interview, Robert Goralski.

scription Television, Inc. On California ballots the voter had a chance to vote on Proposition 15, which declared "pay TV" to be against the public welfare. Avidly supported by theater owners, the proposition won a substantial majority of voters, who wished to keep television "free." The Subscription Television, Inc., backers planned a judicial appeal, but meanwhile the operation came to a halt and was dismantled.

Biographical campaign films played a prominent role at various stages. A curious example involved a meteoric primary drive for Henry Cabot Lodge, who roundly defeated all Republican rivals—including Goldwater—in the New Hampshire primary. At the time, Lodge was in Saigon as U. S. Ambassador. Since he could not campaign, his backers dusted off the almost unused 4:15-minute film of 1960 based on family photos, silent footage, and endorsements by Eisenhower and others. With slight editing it was brought up to date and telecast constantly—thirty-nine times in three weeks—in paid time on New Hampshire's only television station, WMUR-TV, Manchester. The film *was* the campaign.[9] Lodge's enormous victory, which seemed a demand for his candidacy, brought him home from Saigon to campaign in other primaries. If an absent Lodge could roll up such a vote, a campaigning Lodge was expected to sweep all before him—but somehow he failed to. The film makers, who had avoided shots of Lodge talking, were apparently the only ones not surprised at the outcome.

A biographical film played an important role in the transition of Robert F. Kennedy to elective politics. Entering the race for U. S. Senator from New York, he turned to film maker Charles Guggenheim, who was winning recognition as an extraordinarily skillful creator of campaign films. He had made films for various politicians including Senator George McGovern; the McGovern film brought him to Kennedy's attention. Although Guggenheim was based in St. Louis, his help was enlisted for the New York drive.[10]

Guggenheim studied campaign problems faced by Kennedy, who was regarded by some as a "carpet bagger" and as "ruthless." Guggenheim said, "I'll have to humanize you," and Kennedy agreed. Guggenheim persuaded many people to record reminiscences about Kennedy at various periods. They included: a school mate who used to walk to public school

9. Wyckoff, *The Image Candidates*, pp. 145–52.
10. Guggenheim subsequently transferred to Washington, D. C. Campaigns of Edmund G. ("Pat") Brown in California, Abraham A. Ribicoff in Connecticut, John J. Gilligan in Ohio, Milton Shapp in Pennsylvania, used Guggenheim films.

with him in Bronxville, N.Y. (subtly deflecting the carpet-bagger charge as well as the rich-boy aura); a governess who described him as the most considerate of the Kennedy boys; a high school coach who liked his team spirit; a politician; a diplomat; an author. Such material was combined with a parade of pictorial items including home movies, family-album photos, college yearbook items, newsreel footage.

One of the most crucial elements was a reminiscence by author Harry Golden. He was seen talking on the steps of the New York Public Library; his voice continued over a portion of the film dealing with civil rights. Kennedy had been expected by political pundits to lose the "Jewish intellectual vote" because of the ruthless image. Golden's few words were of utmost importance in this context. Speaking of the Alabama conflict with Governor Wallace, Golden said that Robert Kennedy had "smashed the caste system in the South . . . ended it." This, he felt, was of the essence of humanity. "*We* know that better than anyone."

The film had saturation scheduling on television stations throughout New York State, at a probable cost of several hundred thousand dollars.

Most books on campaigns have discussed and quoted speeches and press statements as though they *were* the campaign. Such materials have been readily available for quotation. Campaign films, on the other hand, have not been available in libraries, and in any case raise the question of what to quote. A narration may not be an index of what a film has "said." Television films and spots have thus eluded traditional scholarship procedures and have tended to be ignored. But candidates in 1964 clearly considered them of key importance.

Television itself became a campaign issue. Television newsmen underestimated Goldwater's strength before the Republican convention; the Goldwaterites, in turn, regarded the newsmen as hostile to their cause. At the convention they sported buttons with such legends as "Stamp out Huntley-Brinkley." Goldwater later revealed that his followers had plans ready to silence the networks if they proved troublesome. "We had every cable of every radio company marked up in the loft of the Cow Palace," he told reporters several years later. "If anybody got a little too obnoxious to our—us—they would always have cable trouble."[11] When ex-President Eisenhower in his convention speech referred casually to "sensation-seeking columnists and commentators," he was amazed to find it producing a

11. *That Was The Election That Was,* National Educational Television; quoted, New York *Times,* September 19, 1967.

five-minute ovation, while Goldwater followers shook their fists at the network booths. The Goldwater forces, controlling the convention, severely restricted the movements of network newsmen, who in other conventions had freely roamed the aisles. NBC newsman John Chancellor was arrested on camera, fading out with, "This is John Chancellor, somewhere in custody."[12]

Reporters were taken aback by the virulent attacks. The Goldwater forces seemed to regard network newsmen, along with writers of the New York *Times* and other eastern publications, as part of a conspiracy allied with "the communists." All this reflected a paranoia that also pervaded the military leadership, which readily blamed its troubles on civilian interference engineered by communists. Its frustrations in Vietnam, like the communist victory in China, seemed to require scapegoats in American government and news media. Not since the bitter conflicts of the Depression period, wrote James Reston, had correspondents been so deluged with "vulgar and personal abuse."[13]

Paranoia has been a persistent element in American politics. In 1964 it was gathering intensity. But this was not confined to speeches, tracts, and political in-fighting. It pervaded also that other and more significant political weapon—the entertainment telefilm.

PARANOID PICTURES PRESENTS

Sociologists Paul F. Lazarsfeld and Robert K. Merton have pointed out that even in their evasions, mass media are political. Their influence has stemmed, say these writers, "not only from what is said, but more significantly from what is not said. For these media not only continue to affirm the *status quo* but, in the same measure, they fail to raise essential questions about the structure of society."[1]

In this sense the overwhelming absorption of tens of millions of mid-twentieth-century Americans in football games and struggles against cattle rustlers was a political achievement, in a class with the imperial Roman policy of bread and circuses.

But telefilms have been political in other ways than through evasions. Public acceptance of an Eisenhower-Dulles foreign policy based on good

12. *Variety*, July 22, 1964.
13. New York *Times*, July 8, 1964.
1. Lazarsfeld and Merton, "Mass Communication, Popular Taste and Organized Social Action," in Bryson, *The Communication of Ideas*, p. 107.

guy/bad guy premises may owe something to a telefilm mythology of similar obsessions. When Eisenhower described the world as "forces of good and evil arrayed as never before," he was offering a picture viewers could recognize. We and they were lined up across an international arroyo.

In the years 1964–66 telefilms turned more specifically to international struggles, with emphasis on clandestine warfare. The timing was not an accident. References to CIA action in various coups and upheavals were beginning to appear in print. Besides Cuba and Guatemala there was mention of Laos, Malaysia, Indonesia. In 1963 the drift of many rumors was confirmed by Allen Dulles himself in his book *The Craft of Intelligence.* Written in retirement, it was partly a memoir and partly a prideful justification of CIA machinations. The 1964 publication of *The Bay of Pigs* by Haynes Johnson, and *The Invisible Government* by David Wise and Thomas B. Ross, gave the public—to the consternation of the CIA and the State Department—a more detailed picture of CIA work.

To many Americans, accustomed to a national image of clean uprightness—the cowboy—the revelations were disturbing and called for some adjusting. They seemed to require either indignation or rationalization. For most people, rationalization was the easier solution. If our government had really developed a "department of dirty tricks" to organize putsches, unseat rulers, and murder when necessary, all masked by elaborate fictions, it must have been brought on by dire necessities. A deluge of spy fiction, latching onto a timely topic, provided the rationale, and got Americans used to the idea. On television it was *The Man From U.N.C.L.E., The Girl From U.N.C.L.E., Get Smart, I Spy, The Man Who Never Was, Mission: Impossible,* and others—some amiable, some witty, some melodramatic. So successful was this spy cycle that its subject matter took over numerous other series. Amos Burke, millionaire cop of *Burke's Law* who rode around in a Rolls Royce, became Amos Burke the millionaire secret agent. *The FBI* became concerned chiefly with communist agents. *77 Sunset Strip* acquired a spy-story obsession. Even comedy series like *I Dream of Jeanie, Mr. Ed,* and *The Lucy Show* took up spy themes.

Like earlier action telefilms, the new cycle concerned struggles against evil men who had to be wiped out. Bosomy girls fitted easily into the picture. But in one respect the new wave departed from precedent. Older action heroes, especially the cowboy, had maintained a code of honor and fought fairly. This tradition was rapidly vanishing. When the U. S. Navy in a *77 Sunset Strip* episode ("The Navy Caper") hired a private eye to try to steal one of its top-secret gadgets—to test its own security arrangements against enemy powers pursuing the same objective—the hero's instructions were: "You can lie, steal, cheat—whatever the enemy might do." What followed was an epic adventure in deception and counterdeception. At the start of each *Mission: Impossible* episode an agent —member of the Impossible Missions Force—received instructions via a tape recording. The instructions included a standard passage:

> VOICE: As always, should you or any of your IMF be caught or killed, the Secretary will disavow any knowledge of your actions. This tape will self-destruct in five seconds. . . . Good luck. . . .

The official lie was thus enshrined.[2]

Justification for deceptions was provided by the vastness and monstrousness of schemes devised by enemy forces, which were "arrayed as never before." An *Amos Burke* episode depicted agents of SEKOR—the enemy—as ready to paralyze Washington, D.C., by introducing a special ingredient into gasoline supplies entering the city. This ingredient, sprayed into the atmosphere through automobile exhausts, would stupify the inhabitants and allow the enemy to destroy the White House, the Pentagon, the State Department, and even—perhaps most heinous—the spy organization where Burke worked. The paralyzing fumes would be spread through the city unwittingly by an anti-war group that was about to demonstrate. Thus the program neatly pictured peace demonstrators as dupes of an international anti-American conspiracy. In an *I Spy* episode ("Weight of the World") the Red Chinese had a scheme for dropping bubonic plague bacteria into a big-city water supply. In a *The Man From U.N.C.L.E.* episode ("Her Master's Voice Affair") the Red Chinese had subverted the headmistress of a fashionable Long Island girls' school, and with her help put the girls—all daughters of prominent fathers—into a state of mass hypnosis. The daughters became obedient zombies ready to annihilate their fathers whenever they got the verbal cue over the school's public-address system. In a *The Girl From U.N.C.L.E.* episode an Arab ruler was killed by enemy agents by means of poisoned popcorn, and the fate of the free world hung on the marriage of his successor to a certain popular princess—who, however, had been kidnapped by the

2. *Writing "Mission: Impossible,"* p. 7.

enemy. The girl agent from U.N.C.L.E. temporarily impersonated her and went through with the wedding—she had trouble staving off the passionate sheik—until other U.N.C.L.E. agents arrived to put things right.

Each series depicted huge enemy networks for undercover warfare. Some had names like KAOS (*Get Smart*), or THRUSH (the *U.N.C.L.E.* series), or SEKOR (*Amos Burke*). Viewers, asked to identify these, readily said "the communists." Some series used terms like "communists," "reds," "iron curtain countries," "people's republics," or simply "the enemy," while generally avoiding names of specific countries. *Mission: Impossible* writers were encouraged to use specific names in early drafts; nonspecific designations could be put in later. In one episode the Impossible Missions Force masqueraded as a traveling carnival troupe that had the task of rescuing a cardinal from a maximum-security prison behind the iron curtain. The action was clearly located in the first draft but was later placed "in the Balkans." All this avoided the problem of coping with embassy protests from countries with whom the United States had diplomatic relations. This sort of problem did not apply to China; any kind of villainy could be pinned on China. The same applied to East Germany; the hero of *The Man Who Never Was* assumed the identity of a dead industrial leader specifically to thwart East German machinations. Cuba could also be mentioned with little inhibition.

There was frequent interest in Latin America. In an *I Spy* episode ("Return to Glory") the hero agents had been assigned by the State Department to visit a deposed Latin American dictator to find out—"unofficially"—what help he would need to return to power. The dictator did not turn up at the rendezvous; his disappearance, the sinister work of enemy forces, became the problem faced by the agents. In a *Mission: Impossible* episode the team had the task of re-rigging, in a Latin American country, a voting machine which had been fixed by a local despot to give him victory over the "Libertad" party.

In many ways the spy cycle showed Hollywood talents in brilliant form. Each of the leading spy series had special qualities. *The Man From U.N.C.L.E.* had charm and sophistication. Against melodramatic action the heroes maintained a casual air that punctured spy-drama stereotypes of earlier years without undermining tension. Climaxes involved spectacularly graceful karate or jujitsu activity. *I Spy* adopted a similarly cool manner and scored an added coup: the spies were a white man and a black man—played with a warm, improvisational quality by Robert Culp and Bill Cosby. Their "cover" roles were a touring tennis professional and his trainer. When Cosby was offered the role, he was wary: would this be another second-banana part, a faithful aide like the Lone Ranger's Tonto or the Green Hornet's Kato or Jack Benny's Rochester? Assured that in this spy role he would have special skills, equal prominence, and romantic interest, and aware that Culp was equally determined that a feeling of equality and warm friendship be maintained, Cosby accepted. Virtuoso performances resulted. *Get Smart,* similarly debonair, added a running romance and a unique zany quality evolved with precision by Don Adams and Barbara Feldon. Adams's bumbling agent Smart established a catchphrase—"oops, sorry about that"—that swept the world. *Mission: Impossible* added an element of technical expertise; its international "capers" generally involved sophisticated gadgetry, always technically plausible. This aspect exerted such fascination that science teachers suggested devices for future episodes. This series, like *I Spy,* was in the vanguard of racial shifts in television drama. It adopted a device common to a number of series—a *group* of heroes, one a Negro—Barney Collier, played by Greg Morris. Conforming to a necessity of the moment, he was a specialist, an electrical engineering genius. Only Barney could defuse a time bomb. The group also included a woman, Cinnamon, who —according to the instructions to writers—"does magnificently what only half the world can do at all—be female."

Most of these spy series were not violent in the primitive fashion of westerns and gangster films. The means were more discreet and arcane, and seldom noisy. *Mission: Impossible* arranged, whenever possible, for one foreign agent to murder another by mistake. This was regarded as a salutary policy *vis-à-vis* congressional committees.

The novel surface features of these programs gained most of the critical attention. But more significant was the fact that all shared a common premise, which had to be accepted for the programs to have meaning: that Americans lived, at home and abroad, amid unscrupulous conspiracies that required response in kind. "The villains are so black," *Mission: Impossible* told its writers, "and so clever that the intricate means used to defeat them are necessary."[3]

In its depiction of communist villains, the spy telefilm of the 1964–66 period departed from hallowed precedent. Reds had once been boorish, with a dirty look. A few years of intercultural visits may have made this

3. *Ibid.*, p. 1.

less credible. On the *U.N.C.L.E.* programs as well as *I Spy* and *Mission: Impossible,* the enemy might be as svelte and handsome as "us." Female operatives were as cool and antiseptic as ours and looked as if they used the same hair sprays, skin creams, and deodorants. Our agents occasionally had passing affairs with them—a custom popularized in the James Bond books and films and adopted more circumspectly and ambiguously on television. The standards on both sides were identical. Each saw the whole world as its arena. The moral basis which had once been regarded as a *sine qua non* of film conflict, on or off the air, was vanishing. Conflicts were drained of principle and reduced to a question of "our side" against the other side.

Among the few critics who took note of this trend, Robert Lewis Shayon of the *Saturday Review* wrote forcefully:

> My candidate for this season's most harmful television program is *Mission: Impossible*. . . . The heroes of *Mission: Impossible,* for pay and at government instigation, interfere directly in the affairs of foreign nations with whom we are at peace and from whom no direct threat to our safety emanates. They break the laws of these nations. . . . In the United States the program series tends to legitimatize unilateral force for solving international problems at a time when our nation recognizes, or at least verbalizes, the desperate urgency of collaborative efforts among nations for world order. It pretends that individual Americans are morally impeccable when they break the laws of a foreign nation under the shield of our ideology . . . in emergent nations the viewer may say: "The Americans are telling us, in these programs, that this is the way to run a society. . . ."[4]

But such criticism was not common. The spy cycle moved high in rating charts, and its ingredients were emulated far and wide. On children's series especially, international conspiracy dominated. *The Lone Ranger,* now animated, depicted struggles with a mad scientist able to destroy mankind through control of the weather. On *Mr. Terrific* the government's Bureau of Special Projects turned to the hero to locate and thwart a defecting scientist who had invented a power paralyzer. *Tarzan* became the champion of emergent nations against communist conspiracies. On *Superman* a pet parakeet, which a scientist had taught his devastating explosives formula, was stolen by a foreign agent. On the cartoon series *Gigantor,* Mr. Ugablob had a plan to conquer the world through a freeze ray.

4. *Saturday Review,* November 19, 1966.

A group of outer-space series was saturated with a similar atmosphere. On *Star Trek* ("Balance of Terror") the crew of the Enterprise faced a decision that might trigger an intergalactic war: should they engage the powerful alien spacecraft that had already destroyed several earth outposts? On the cartoon series *Astroboy,* an agent from the planet Xenon was recruiting brilliant earth children for some special mysterious tutoring. On *The Outer Limits* "invisibles" from space planned to take over the earth. *The Invaders* depicted each week the arrival of space creatures who could assume human form; their task was to prepare for a total takeover. When shot, they shriveled and vanished. This made for a special-effects triumph and also had plot and propaganda value. Because each vanquished invader left no trace of his existence, the dupes of the world remained unconvinced that anyone was planning the conquest of the world and the extinction of our civilization. But each week brought new landings of the mysterious infiltrators.

During this period the involvement in Vietnam skyrocketed as 15,500 "advisers" grew to an armed force approaching 500,000. U. S. spokesmen explained the need largely in rhetorical terms, speaking constantly of communist plans to "take over the world," and the need to halt aggression. Both President Lyndon Johnson and Secretary of State Dean Rusk constantly tried to recapture the consensus of World War II by depicting the Vietnam expedition as a continuation of older struggles against tyranny.

World War II had inspired a number of popular radio drama series, notably *The Man Behind the Gun.* The Vietnam war inspired no television drama series about Vietnam—producers and writers shied away from the subject. But the period did bring a television eruption of military drama about other United States wars: *McHale's Navy, Combat, Rat Patrol, Gomer Pyle U.S.M.C., Twelve O'Clock High, Jericho, Hogan's Heroes, Wackiest Ship in the Army, Mr. Roberts, F Troop, Wild Wild West.* Most dealt with World War II; a few—*F Troop* and *Wild Wild West*—with earlier wars. Some—*Jericho* and *Wild Wild West*—featured the undercover atmosphere of spy programs. In most, military life was richly amusing as well as heroic.

It was not the conscious intention of producers to buttress administration arguments linking Vietnam with World War II. But the rash of heroic and amusing World War II series, in conjunction with the flood of enemy-conspiracy drama, probably did just that.

On television the wars merged in curious ways. *Jericho* on CBS-TV dealt with an Allied team of World War II agents, usually operating behind Nazi lines in Europe or Africa and engaging in sabotage, intelligence, ambush or rescue missions. Such names as de Gaulle and Hitler cropped up in early scripts. Producer Stanley Niss received a call from a CBS executive who wondered, "Will kids know who de Gaulle is?" Investigation tended to show that most youngsters did not know de Gaulle; in fact, many had never heard of Hitler.[5] It seemed best to avoid these confusing names; also, to avoid constant references to Germany and Italy as enemies, since they had meanwhile become NATO allies and purchasers of telefilms. So terms like "the enemy" won preference. It became a timeless, symbolic drama: Americans once more involved in heroic action for freedom. No doubt many viewers unconsciously identified "the enemy" as communists.

For younger viewers, toy manufacturers added their contribution to the atmosphere. The Saturday morning "children's series," heavily supported by toy manufacturers, were saturated with violence on earth and in space. In addition, many commercials featured violent gadgetry, including war toys. Christmas 1966 was a war Christmas.

Few adults watched Saturday morning television. With children watching in rapt attention, most adults felt they could go about their own business. A typical Saturday morning at WNBC-TV in New York began quietly with *Crusader Rabbit* (7:30–8), built in noise level with *Colonel Bleep* (8–8:30), whose interplanetary space command was engaged in unrelieved hostility; *Dodo, the Kid From Outer Space* (8:30–9), of similar content; *Super 6* (9–9:30), whose superheroes smashed evil-doers throughout the universe; *Atom Ant* (9:30–10), the fighting, "up-and-at-'em" ant; *Secret Squirrel* (10–10:30), a spy series; *Space Kidettes* (10:30–11), whose struggles against the "meanest pirate in the universe" required "plutonium disintegrators" and "space agitator ray guns"; *Cool McCool* (11:30–12 noon), whose business was danger; *The Jetsons* (11:30–12 noon); and *Top Cat* (12–12:30). But a special quality of the morning came from commercials and public service announcements.

For boys there were things like Mattel's Fighting Men with machine guns and tanks—"everything real fighting men use"—and the G. I. Joe army toys, that included a "ten-inch bazooka that really works" and gas masks "to add real dimension to your play battles."

5. Interviews, Stanley Niss, David Victor.

There were things for girls too: Mattel's Cheerful Tearful Dolls that could "cry real tears" as well as smile; and Kenner's automatic knitting machine—"a fast fun way to knit."

A piquant addition was provided by public service announcements which included advice to "keep America beautiful," read books ("open your mind—read"), support the Young Women's Christian Association (to "help you become more interesting"), visit the United Nations, and buy U. S. Savings Bonds ("underwrite your country's might").[6]

A visitor from another planet watching United States television for a week during the Vietnam escalation period might have concluded that viewers were being brainwashed by a cunning conspiracy determined to harness the nation—with special attention to its young—for war. Of course there was no conspiracy. Manufacturers were making things for which they saw a market, promoting them through advertising agents, producers, and broadcasters who believed in serving the client. In so doing, all avoided anything that might seem to undermine current government policy—and thereby gravitated toward its support.

Almost every segment of American business was feeling the financial injections of the 1964–66 Vietnam buildup, which created—according to U. S. Labor Department estimates—more than a million jobs during that period, with indirect effects elsewhere throughout the economy.[7] More significantly, over half the federal tax revenue was going into the war, and more than half of that was going directly to American corporations that were the chief suppliers. On the list of the hundred largest government suppliers, over half were companies heavily involved in broadcasting—as licensees, manufacturers, or sponsors. Restraints from these interdependencies operated all along the line. There was no conspiracy—there were merely innumerable parallel incentives. They tended to make television entertainment an integral part of the escalation machine.

Producers of programs like *The Man From U.N.C.L.E.* and *Mission: Impossible* were positive they were only providing "entertainment." Norman Felton, the former *Studio One* producer who made MGM a television success with *Dr. Kildare, The Man From U.N.C.L.E.*, and other series, insisted that children "can see and hear a news program which concerns Vietnam and know the bullets are real. They can go to a James Bond movie and know the bullets are not real." He felt sure this was true of

6. WNBC-TV, New York, December 10, 1966.
7. Associated Press, September 13, 1967.

The Man From U.N.C.L.E., but the evidence was ambiguous. The United Nations received many letters from people who thought U.N.C.L.E. was a United Nations agency and who wanted to work for it. Most inquiries came from young people, but not all. A United Nations official received a business visitor from abroad, who asked to see the second basement. The puzzled official asked why. The visitor said: "Isn't that where U.N.C.L.E. has its headquarters?"[8]

The Man From U.N.C.L.E., Get Smart, I Spy, Mission: Impossible all received wide foreign distribution. *Mission: Impossible* reached virtually all countries with which the United States had friendly relations. While viewers abroad, like those at home, accepted the programs as "entertainment," the political impact of such material did not go unnoticed. In December 1965 Hollywood writers were notified by Radio Free Europe that it was anxious to buy radio drama scripts in two categories to beam behind the iron curtain: (1) western, and (2) "James Bond type stories dealing with espionage, intrigue, counterespionage. Address: Gordon Davis, Radio Free Europe, Munich 22."[9]

The producer of *Mission: Impossible*, Bruce Geller, was as certain as Norman Felton was that his program was just "entertainment" and not an influence. Speaking of violence on television, he said: "TV is a mirror. Breaking the mirror won't do anything."[10]

By 1966 television had been for four years—according to Roper reports—the chief news medium for most Americans. For young people especially, it was the main source of information and impressions about the world. It was a question whether the entertainment telefilm or the news program was bearing the brunt of this journalistic function.

News programs had minimal audiences among children, most of whom were telefilm addicts. The authors of *Television in the Lives of Our Children* revealed a curious pattern in youthful viewing. Most children were heavy television viewers, watching mainly entertainment telefilms, into the high school years. Between the sixth and tenth grades a split occurred. Some remained heavy viewers, keeping about the same viewing habits; the authors considered them fantasy-oriented. Others became very light television viewers, who might turn to educational television, but largely to print; the authors considered them reality-oriented.[11]

8. New York *Times*, December 26, 1965.
9. *WGA-w News Letter*, December 1965.
10. Interview, Bruce Geller.
11. Schramm, Lyle, and Parker, *Television in the Lives of Our Children*, pp. 24–74.

The findings suggest that the telefilm could be accepted as a satisfactory representation of the world until—and unless—the child gained a foothold in the real world. Then the world of the telefilm fell apart and had to be rejected.

The telefilm appeared to have become the chief journalistic medium for a large part of the population, while newscasts and newspapers performed this function for another part.

The power of the telefilm as an instrument of journalism obviously stemmed from the opportunities it offered for emotional identification. The action in the international telefilms offered roles with which the young viewer could unconsciously merge. Attention was not based on curiosity for information but on deeply enlisted drives for power. Under these circumstances any hint of information or quasi-information was strongly retained. *Fortune* magazine, discussing the hero-villain pattern of most television drama, complained: "The young viewer expects the news to fall into the same dramatic pattern."[12] To some extent producers of newscasts *made* it do so, but the news often resisted the process.

Network news telecasts, besides being saddled with off-peak hours, labored under the handicap of dealing with tidbits of information. Information sticks in the mind to the extent that it fits into a framework of attitudes and emotions. The telefilm provided such a framework; the news program, in most cases, did not. Much of the information in news programs tended to slip down the drain.

The journalistic potency of "entertainment" and the relative impotence of fact-based news reports would seem to explain several public-opinion mysteries.

During the 1960's the historian Henry Steele Commager often protested—and expressed amazement at—the double standard in American opinion on various topics. He saw the United States intervening in a struggle that had been primarily a Vietnamese civil war. Sending a half million men to Vietnam, it also kept submarines stationed in the Pacific ready to rain atomic destruction on major cities of China, to deter its entry into the war. Throughout this period China sent no soldiers to Vietnam, although it sent equipment and was certainly interested in the outcome. It sent no planes, although the United States had thousands of planes in action in the area, which sometimes strayed over Chinese territory. Throughout this time the United States fulminated against China

12. *Fortune*, October 1969.

for its aggressiveness, and refused to deal with China diplomatically because of its alleged aggressive nature. It was somewhat as though China had placed a half million men in Mexico to control the outcome of a civil war, bombed some of its cities round the clock, placed submarines in the Atlantic to threaten New York, Philadelphia, and Washington with atomic attack, and meanwhile fulminated against the United States for aggressive behavior.

Most Americans, living in a near-vacuum of information about China, with little to go on except items from Hong Kong and State Department handouts, seemed to accept the picture of Chinese intransigence. Their readiness to do so could trouble and amaze an historian like Commager, but he did not watch telefilms, where it was open season on the Red Chinese.

John Quincy Adams said of the United States that "she goes not abroad in search of monsters to destroy." But the telefilm told its young audience it was indeed its destiny to search out monsters—search and destroy.

REVOLT IN THE NEWSROOM

Even before the escalation there was tension in Saigon between reporters and American officialdom—military and civilian. "The brass wants you to get on the team," observed Peter Kalischer of CBS, a visitor from his post in Djakarta; "but my job is to find out what the score is." Charles Arnot of ABC charged that he had been fed lies, half-truths, and misleading information by government spokesmen. Early in 1965 Edward P. Morgan, ABC commentator, found the atmosphere "one of the most rancid I have ever seen in thirty years of reporting."[1]

The tensions involved problems of international justification. The "advisers" arrangement was meant to maintain a show of adherence to the Geneva accords, which the United States had pledged to honor. But in 1963–64 correspondents already saw Americans flying jet bombers in Vietnam—"we aren't supposed to mention that," reporters were told.[2]

The drastic escalation plan involved a decision to drop the adviser masquerade and shift to new justification. One rationale was that North Vietnam was guilty of "aggression" against South Vietnam and that Amer-

1. Quotations from *Columbia Journalism Review*, Fall 1965.
2. Interview, Bernard Birnbaum.

icans were helping to repel the aggression. The slipperiness of the charge did not become clear until later.

At the start of 1965, when the massive United States troop buildup in Vietnam got under way, the Pentagon estimated the "insurgent" forces in South Vietnam at 140,000. Among these, according to the Pentagon's own estimate, were only 400 North Vietnamese—a fact revealed only much later by Senator Mike Mansfield, a member of the Senate foreign relations committee. The North Vietnamese, like the United States, apparently began by supplying training.[3]

A United States White Paper issued early in 1965 also charged aggression on the basis of supplies from the north. Reporters, aware that most weapons captured from guerillas were American weapons, were dubious about this charge from the start, but could not document their doubts. Again statistics from the Pentagon itself, released much later, revealed that the charge was largely a hoax. During the two-year period before escalation, 22,200 American weapons had been lost to the guerillas (8500 in 1963, 13,700 in 1964). During those same two years 10,400 weapons were captured from guerillas, almost all American. An eighteen-month listing (June 1962–January 1964) included only 179 weapons from communist countries—mainly the Soviet Union, China, Czechoslovakia. Items identified as of North Vietnamese origin consisted of two machine guns, sixteen helmets, and assorted sweaters, socks, and belts—scarcely an impressive stockpile of aggression.[4] But such information was withheld from the American public during the buildup.

Meanwhile the Johnson administration had found new justification. In August 1964 the navy reported an "unprovoked" North Vietnamese attack on two United States destroyers, the *Maddox* and the *Turner Joy,* which were described as on "routine patrol" off North Vietnam. President Johnson immediately ordered retaliatory bombing of North Vietnam and obtained from Congress the so-called Tonkin Gulf resolution, which authorized "all necessary measures" to repel armed attack and prevent "further aggression." The resolution was used to justify all later escalation. Details of the alleged incident were, at the time of the resolution, extremely vague, but until later congressional inquiries the original navy version of the incident won acceptance.

Underpinning all rationales was a repeated emphasis on a worldwide

3. New York *Times,* November 15, 1966.
4. Gettleman (ed.), *Vietnam,* pp. 317–18; Schlesinger, *The Bitter Heritage,* p. 35.

conspiracy and the argument that "they" had to be stopped in Vietnam or "they" would sweep the world. This was a state of mind that received more reinforcement from telefilms than from newscasts.

Most correspondents, while chafing over official concealments and distortions, at first accepted the necessity of the American presence. They accepted the ways of bureaucracy as a fact of life. Toward information officers they maintained, to some extent, a club feeling. "Joe is doing his job, I'm doing mine. In his shoes, I too would be stuck with official lies."

Correspondents seem to have been largely unconscious of the extent to which they prejudiced their own coverage by adopting the vocabulary of the military handouts. A Vietcong attack, even on a military post, was always a "terrorist attack"—a term not used for American operations, not even for massive bombings, though their purpose was to "break the will of the enemy."

The year 1965 brought, along with the troop buildup, the establishment of network bureaus in Saigon and a swelling migration of American television newsmen. Many, like young John Laurence of CBS, arrived with a strong belief in the American cause; he had asked CBS for Vietnam duty. Disenchantment came slowly. The sense of involvement in a dubious enterprise was strongly resisted by newsmen as by combatants. To some extent the nature of the war made this easy. Reporters seldom saw "the war" or "the enemy."

Harry Rasky was sent by ABC-TV in 1965 to produce the film *Operation Sea War Vietnam* in the Gulf of Tonkin. Arriving in Saigon, he engaged a Vietnamese film technician to assist him, then flew to the aircraft carrier *Kitty Hawk*, which was to be the center of their activity. From there, around the clock, pilots were going on bombing missions. But they looked with intense curiosity at Rasky's assistant—the first Vietnamese most had seen. Their lives were confined to the *Kitty Hawk*. For rest periods they went to Hong Kong.

On the *Kitty Hawk* an Alice-in-Wonderland feeling enveloped a correspondent. There were 5300 people on board—a small American town afloat halfway around the world. Everyone was public-relations minded; people could not have been nicer. A ship publicity brochure revealed that the *Kitty Hawk* had three soda fountains, a barber shop with thirteen chairs, a fine drug store, and a splendid hospital including an operating room. At night there were movies. Producer Rasky was invited to sit with the admiral as they watched *Who's That Sleeping in My Bed?* While

Dean Martin chased Elizabeth Montgomery (of the *Bewitched* television series) around the bed, bombs were being loaded on the *Kitty Hawk*'s planes. Soft music played over the ship's own radio station. A western was on the ship's television station.

A Vietnamese New Year truce gave occasion for a kind of water ballet by the entire strike force 77 of the Seventh Fleet. Twenty ships—carriers, cruisers, destroyers—were arranged for a picture portrait while jets formed a giant 77 in the sky overhead. That shot, Rasky kept thinking, must have cost the taxpayer millions. One admiral sent a message to another: "From the Commodore of the Tonkin Gulf Yacht Club to the Chairman of the Board, a salute." That day's program included Elizabeth Taylor in *Cleopatra.*

Rasky and his assistant went aboard a destroyer that had the task of shelling the coast. Over the hills a spotter plane called out locations and directed the guns. On the ship orders were given, and the guns rattled and shook the ship. Over the hills puffs of smoke drifted gracefully into the sky. "VC in the open," the spotter reported. The ship's guns blasted again and again: 138 rounds were fired. At one point the camera jammed and Rasky asked the captain to hold up a moment; he readily complied. The spotter, circling over the hills, asked: "What program is that you're filming down there?"

"It's for ABC."

"You know what night it will be on? . . . I'll be home soon."

"February, March, I guess. . . ."

"I'll have to see the summer reruns."

The Vietnamese technician kept saying, "What are they shooting at?"

"VC, they say."

"How do they know?" The Vietnamese later became sick and did not eat lunch.

Rasky was not allowed to write the narration. This was done in New York, where the film was also edited. ABC engaged Glenn Ford as narrator. The final film had spectacular shots of fleet maneuvers, but said almost nothing that Rasky had felt impelled to communicate. He therefore poured his feelings into a series of articles in the form of letters to his newborn daughter, which were syndicated in Canadian newspapers as *Letters to Holly*. "They conveyed more about what I felt about what I saw than my film did, and that distressed me."[5]

5. Rasky, *Interview*, pp. 15–18.

This was a common experience. All along the line, the message was processed into something palatable. Things omitted, words added—these could temper meaning and even reverse it.

In the field newsmen received overwhelming cooperation. When Malcolm W. Browne, AP bureau chief in Saigon who turned ABC correspondent, was ready to record a network report, he was surprised at the readiness of a field commander to provide a background effect of 105-millimeter howitzers firing. They fired, thought Browne, about a thousand dollars' worth of ammunition—all wasted, since the item was not used by ABC.[6]

There was little direct censorship. Network cameramen were not permitted on North Vietnam bombing runs, and they never accompanied the CIA-trained assassination teams maintained by army and navy; but correspondents could otherwise go where they liked, with a considerable sense of freedom.

Yet subtle pressures were at work. Vietnam newscast items were videotaped back home by the Defense Department and flown to Saigon for the command to see. This electronic clipping service enabled them to know exactly what each newsman was filming and reporting. If they did not like what they saw, the newsman soon learned about it.

The most startling item to reach the air in 1965 was a report by Morley Safer. A village was said to have aided the Vietcong and had to be punished. Marines moved in. The Safer report showed a Marine flicking his cigarette lighter to set its huts afire; some 120 huts were burned. This scene ignited in some viewers anger against the war; in others, anger against television. The Defense Department made it clear to CBS that Safer was *persona non grata.* CBS upheld Safer and those who had put his film on the air, but top CBS officials let Fred Friendly, president of CBS News, know that they felt uneasy about Safer. Such crises inevitably encouraged self-censorship.[7]

A film made for WABC-TV, New York, involved curious processing. An American unit in Vietnam received intelligence reports that the Vietcong had entered a town during the night. At dawn American troops moved in and demolished the town. Camera crews then came in, and filmed the smoldering debris. It became a WABC-TV documentary, which the

6. New York *Times,* November 6, 1966.
7. *Variety,* December 22, 1965; Friendly, *Due to Circumstances Beyond Our Control . . . ,* p. 214.

station ingeniously titled *Dong Xoai: the Town the Vietcong Couldn't Kill.*[8]

American troops were almost never out of reach of AFRTS radio programs, and carried them everywhere via transistor radios; in March 1966 AFRTS brought television to Saigon and later extended it via relay transmitters. At the same time, American soldiers felt a compulsion to listen to Hanoi Hannah; one reason was that she broadcast American casualty lists. It was chilling to hear from North Vietnam that a friend from Wisconsin had been killed near Danang—but difficult to refrain from listening. The lists were apparently obtained from Pentagon casualty lists relayed back to North Vietnam through diplomatic channels. Hannah also went in for inviting, sexy talk. She began with "Greetings, dirty imperialists."

By the end of 1965 the United States role in Vietnam was being denounced in a number of American magazines, and was the subject of protest meetings and marches. Criticism erupted in newspapers columns, notably those of Walter Lippmann and James Reston, and in letters to the editor. It entered newscasts via footage of protest demonstrations, including many in foreign capitals. Some church groups, especially the Quakers, were condemning the war. Law professors were questioning its legality. A few generals criticized it on military grounds. Television newsrooms, too, were the scene of debates—but during 1965 they seldom reached the air, particularly in prime time. Discussion was usually relegated to fringe periods. In newscasts a protest march was likely to be covered with a shot or two of a bearded youth, as though to categorize it as a "hippie" event.

The magazine *Commentary* had as its television critic Neil Compton, a Canadian, who covered American television from north of the border, sometimes comparing it with Canadian television. In the October 1965 issue he mentioned two overriding impressions:

> . . . first, the great networks seem to express a massive political consensus; second, they are commercial to a degree which even an outsider used to television finds overwhelming. The two phenomena are not, of course, unrelated.

Compton felt that Vietnam issues were "fiercely and frankly" discussed in American magazines and newspapers.

8. Interview, Arthur Alpert.

> But anyone who relied on NBC, ABC, and CBS television would have been far less well informed than his Canadian counterpart even though the total time devoted to Vietnam on American networks was undoubtedly greater.

He found events reported as though "in a vacuum." Comments were generally "variations on the official line." Official propaganda was often on a juvenile level. Compton mentioned the brigadier general who on CBS-TV called the Vietcong cowardly for fighting and running by night. There seemed to be constant effort to insulate the viewer from world opinion through assurance that demonstrations in London and elsewhere were "leftist." Compton mentioned one program which seemed to him a notable exception—an August 1965 ABC-TV special titled *The Agony of Vietnam,* written by Stephen Fleischman and narrated by Edward P. Morgan. It stated, "We cannot afford to see Vietnam only through American eyes," and resolutely surveyed world opinion, pro and con.

Compton also reviewed that other form of television journalism, the telefilm, and noted that two archetypal American figures, the cowboy and the gangster, had merged in the secret agent.[9]

The consensus observed by Compton was, within network news divisions, under attack by reporters but resolutely upheld at executive levels. Especially at CBS, anything that might prove unduly disturbing or inflammatory was regarded with disapproval. Demanding "objective" reporting, the policy seemed to insist on information drained of meaning.

Facts that might embarrass the administration were unwelcome. CBS president Frank Stanton and NBC president Robert Kintner were in frequent touch with President Johnson and seemed to identify his policies with the "national interest." When a Murray Fromson broadcast in January 1966 mentioned the United States air bases in Thailand—springboard, since February 1965, of incessant bombing of the North Vietnamese[10]—Stanton protested to CBS News president Fred Friendly, saying all correspondents had agreed to embargo this information. Sam Zelman, CBS bureau chief in Saigon, denied this. Friendly suggested to Stanton that the North Vietnamese undoubtedly knew they were being bombed, and that only the American public would be gulled by suppression of the Thailand involvement. But Stanton said the United States government might be embarrassed—which was certainly true.

9. Compton, "Consensus Television," *Commentary,* October 1965; "Camping in the Wasteland," *Commentary,* January 1966.
10. Lomax, *Thailand,* pp. 82–5.

In 1966 the surface consensus began to show signs of cracks and fissures. Two prominent members of the Senate foreign relations committee, Chairman J. William Fulbright and Senator Wayne Morse, were attacking the Johnson Vietnam policy. In view of the importance of these two men they could not be ignored, and CBS News arranged a half-hour program, *Fulbright: Advice and Dissent,* in which the Senator was interviewed by Eric Sevareid and Martin Agronsky. Fulbright expressed his growing disenchantment over the Tonkin Gulf resolution and the use made of it. According to Friendly, Stanton was upset about the broadcast, saying: "What a dirty trick that was to play on the President of the United States. . . ."[11]

The Senate attacks persisted and necessarily brought further interviews and discussions by news divisions. Both at networks and affiliate stations, scheduling conflicts often intervened. At CBS-TV they produced a major crisis.

The Senate foreign relations committee scheduled hearings which loomed as the first full-scale senatorial debate on Vietnam policy. Administration supporters and critics would be heard. Fred Friendly, urging live CBS-TV coverage, wrote an intra-network memorandum:

> Broadcast journalism has, once or twice every decade, an opportunity to prove itself. Such an opportunity were the events leading up to World War II; such was the McCarthy period. The Vietnam war—its coverage in Asia and in Congress—is another such challenge.

At CBS John A. Schneider had just become president of the television network, succeeding James Aubrey. Aubrey had suddenly departed amid a flurry of rumors concerning his personal life and charges by stockholders of conflicts of interest in his program decisions.[12] On all of this the network refused comment, but it emphasized its confidence in Schneider. He had risen in network ranks from its sales division and was considered solid and reliable. Coverage of the Vietnam hearings would be his first major decision.

As the hearings began in February 1966, CBS-TV and NBC-TV were on hand. The first testimony was on behalf of the administration and was by David Bell, the director of foreign aid programs. He was followed by retired Lieutenant General James Gavin, who opposed current strategy on military grounds.

11. This and the following are based on Friendly, *Due to Circumstances Beyond Our Control . . .* , pp. 219–65; other sources as noted.
12. *Life,* September 10, 1965.

The following testimony was to be given by former Ambassador George Kennan, considered one of the most effective critics of Vietnam policy. His objections were political rather than military. NBC-TV planned to continue coverage; Schneider reserved his decision.

As Kennan began testifying at a morning session, CBS cameras were on hand. CBS News president Fred Friendly was in his office at CBS, facing two screens—one the CBS program; the other, the NBC program. At 10 a.m. Kennan appeared on the NBC-TV screen; on the other appeared a rerun (the fifth) of *I Love Lucy,* followed by a rerun (the eighth) of a *The Real McCoys* episode. John Schneider had decided—or been instructed—not to carry the testimony.

He later explained his reasoning. Very few "decision makers" were at home in the daytime, he said. Also, the hearings would confuse the issue for many people. He had saved the company a great deal of money but said this had not been a factor in his decision. Stanton, by way of supporting argument, pointed out that Kennan no longer held an official position.[13]

Fred Friendly, sending in his resignation, helped to make the case a front-page issue. The bitter tensions over Vietnam were causing many resignations—at the White House, the State Department, the U. S. Information Agency, and elsewhere. "Health," "personal reasons," "family reasons" were generally cited. Administration aides seemed to feel they should withdraw in a way that would not embarrass the President. Discussing this long afterwards, James Reston wrote that they "gave to the President the loyalty they owed to the country."[14]

Friendly used no evasion. In his letter, released to the press, he wrote:

> I am resigning because CBS News did not carry the Senate foreign relations hearings last Thursday. . . . I am resigning because the decision not to carry the hearings makes a mockery of the Paley-Stanton Columbia News division crusade of many years that demands broadest access to congressional debate. . . . We cannot, in our public utterances, demand such access and then, in one of the crucial debates of our time, abdicate that responsibility. . . . I now leave CBS News after 16 years, believing that the finest broadcast journalists anywhere will yet have the kind of leadership they deserve. . . .

13. New York *Post,* February 11; New York *Times,* February 16, 1966. Cancellation of each half-hour daytime program and its commercials was estimated to cost the network $5,000 to $30,000 in revenue, depending on the number of commercials.
14. New York *Times,* March 9, 1969.

Variety, referring to Friendly's exit as "cataclysmic," called it the end of the Murrow era at CBS. Murrow himself had died the year before, after surgery and lingering illness.

Although NBC-TV carried most of the hearings, and CBS-TV, stung by criticism, resumed its coverage, many affiliates of both networks failed to carry the broadcasts, or carried only portions. The *Columbia Journalism Review* later found that seventy-one cities ("markets") had not had access to the full hearings.[15]

Obstructions of this sort constantly muted or sidetracked television discussion about Vietnam. Under the title *ABC Scope 1966*, ABC produced a series of reports and discussions, which it offered to stations during evening time. But many affiliates failed to carry it while others—including WABC-TV, New York—videotaped the program to carry it at a less favorable time. A West Coast affiliate ran it before breakfast. A February 1966 United Nations Security Council debate on Vietnam was not carried live by any network. CBS News did prepare a half-hour summary, edited and narrated by Richard C. Hottelet and scheduled at 4:30 p.m., but WCBS-TV, New York—the network's "flagship" station—declined to carry it because it would have interfered with the movie scheduled on *The Early Show*. Documentarian Albert Wasserman, admired for his work on the *NBC White Paper* series, was sent by NBC to film a South Vietnam election, which was being hailed by the administration as proof of democratization in South Vietnam, although the government there considered talk of peace to be treasonable. Wasserman's film, *Vietnam: War of the Ballot*, reflected the conflicting currents in Saigon. It was eventually scheduled to follow a major golf tournament, but when it became clear that the tournament would run beyond its allotted television time, NBC-TV postponed *Vietnam: War of the Ballot* to 11:15 at night. In November 1965 CBS commentator Eric Sevareid scored a startling scoop with an article in *Look:* he revealed that Adlai Stevenson, hours before his death, had confided to him that he was on the point of resigning in protest from his position as U. S. Ambassador to the United Nations. Stevenson's reason was that the Johnson administration had twice sabotaged peace negotiations conducted by United Nations Secretary General U Thant. Observers wondered why Sevareid revealed this sensational news in a magazine rather than on a television news program; both network and Sevareid declined to comment. Across the country most broad-

15. *Columbia Journalism Review*, Spring 1966.

casting stations, along with most newspapers, ignored the Sevareid revelations—according to a *Variety* article on the "Era of No-Guts Journalism."[16]

Until this time Sevareid himself, in his short "think pieces" on CBS news programs, had tended to avoid the subject of Vietnam. He was troubled by doubts but felt President Johnson should be given the benefit of any doubt. He told himself, "Sevareid, you don't know anything *about* this." Then he began to feel that he owed it to his viewers to find out and to speak on the subject. He began to read everything he could lay his hands on, and in the spring of 1966 he journeyed to Vietnam. Here he was soon overwhelmed by a sense of the hopelessness of the entire Vietnam intervention. He found an endless mosaic; one might pick up a tiny piece but never glimpse how it might fit into the total. No one—at any level—could see the shape or meaning of the total. He felt the same hopelessness about the machinery available for reporting the war. The bits of film appearing nightly on television screens could not possibly represent the story. "The facts didn't yield to the equipment." Returning, he was allowed a half-hour to express his dismay; it was broadcast on June 21, 1966, over CBS-TV, and entered in the *Congressional Record.*[17]

Thus, slowly, with constant pressure, opposition to the American involvement moved forward in television and radio. Circulating on the edges, it pushed toward the center. But counterpressures were also mighty and unrelenting, and were led by the President himself.

IMAGE CONTROL

No President ever worked harder to cajole, control, and neutralize the news media than Lyndon Johnson. The weight of his efforts was usually directed toward television, as though he largely discounted the influence of other media. In the end his hopes seemed to hang entirely on keeping television in line.

In the first year of his regime, riding on extraordinary legislative successes—especially in civil rights—he lived on good terms with the various media. In 1964, at the suggestion of a television newsman, he gave a gala party for White House correspondents and their wives and children. It was a smash hit; many of the children had their pictures taken with the

16. *Variety,* December 29, 1965.
17. Interview, Eric Sevareid; *Congressional Record,* v. 112, pp. 14125–6.

President's famous beagles, Him and Her. After the President's triumphant re-election, top network officials were invited to celebrate with a swim in the White House swimming pool. Later he became more edgy in relations with the media, particularly when his dispatch of troops to the Dominican Republic and his Vietnam escalation drew criticism. Hostile questions in regularly scheduled press conferences caused him to concentrate on a strategy of impromptu background sessions, to which only selected reporters were invited. He would relax and favor the group with pungent language and candid opinions about people and events. According to newsman Robert MacNeil, Lady Bird sometimes sat in, "modestly not hearing the President's riper remarks."[1]

Correspondents already regarded President Johnson as a strange phenomenon. He had a deep fund of knowledge about the workings of government, but his approaches to newsmen—sometimes folksy, sometimes wheedling, sometimes angrily demanding—were often so blatant as to seem naïve. He often engaged in long, rambling soliloquies in which extraordinary vanity rose to the surface. He sometimes seemed bent on merciless self-exposure.

A "not for broadcast" background memorandum in a network news division file describes a plane ride with President Johnson during which he talked continuously while newsmen listened—silently. The President drank a glass of milk and ate chocolate cookies.

> JOHNSON: You guys can be big men in your profession if you treat me well. . . . But, damn it, you never learn. I used to think wire service copy was the Holy Writ. It ain't so. Hell, you guys distort all the time. . . . Why do you guys give George [Reedy] such a hard time? . . . Reedy's got a good brain. He knows what he's talking about. If he doesn't know, he comes and asks me. . . . And I'll tell you who knows more about what's happening in the White House than anyone else. Jack Valenti. He gets in there at six in the morning. Doesn't leave until one in the morning. And all day he's snooping around, reading over people's shoulders, reading what's on their desks. . . .
>
> And all this talk about Jack Kennedy knowing more about foreign affairs than I do. Why, I appointed him to the foreign relations committee. And when he became President he used to call five or six men on a desk at the State Department. Hell, I call Rusk. If he doesn't know, I'll get me another one.

1. MacNeil, *The People Machine*, p. 301.

Without a word from reporters, he talked on for almost an hour. Meanwhile he ate four cookies, commenting: "Lady Bird would kill me if she saw me." He was still talking as the plane landed. Brushing crumbs from his lap, he got up. "This has been damned interesting."[2]

During 1965–66 press references to a "credibility gap" in the Johnson administration multiplied, and relations with newsmen became increasingly prickly. If he considered a question hostile, he could vent full fury or scorn on the questioner. Turning on one reporter, he said: "Why do you come and ask me, the leader of the western world, a chicken-shit question like that?"[3] Television newsmen became familiar with his consuming interest in what they broadcast about him. His three-set lineup of receivers seemed to keep him aware of every word. At CBS Walter Cronkite, leaving the air after his nightly *CBS Evening News With Walter Cronkite,* often found his secretary holding the telephone aloft. "White House on the line." It might be a presidential aide, or it might be the President himself, demurring at a detail. White House correspondent Dan Rather got frequent angry phone calls at home, often laced with strong language. A protest often culminated in an inquiry about sources. "Where did you get that?" Sometimes the President categorically denied reports which were later confirmed. When Robert Goralski of NBC broadcast a report that General Maxwell Taylor had resigned, Johnson called NBC president Robert Kintner to deny it and demand a retraction, which was promptly broadcast. A few weeks later the White House announced Taylor's resignation, named his successor, and released the text of Taylor's resignation—the date of which confirmed Goralski's original report. Some correspondents acquired the impression that Johnson had a congenital urge to mislead. Others explained it as a dread of losing the initiative. He wanted to keep his "options" open and therefore insisted on controlling the timing of every move and announcement.

His relations with the media reached a bizarre climax in the summer of 1965. Professor Eric F. Goldman of Princeton, who had succeeded Arthur Schlesinger, Jr., as intellectual-in-residence at the White House, suggested that relations with intellectuals could be cemented with a White House Festival of the Arts. It would be a day of readings by leading writers and performances by celebrated musicians, while paintings and sculptures would be on view in the White House and its gardens. Television cov-

2. Background memorandum, April 29, 1964.
3. Davie, *LBJ,* p. 15.

erage would make the occasion a national event in celebration of American culture, and place the White House first in the recognition of talent. The President, according to Professor Goldman's "scenario"—the term remained standard at the White House—would appear at a reception in the afternoon and speak briefly.

The festival, suggested in February 1965, was scheduled for June 14. The interim had brought the Dominican intervention and the start of continuous bombing of North Vietnam.

At first, the festival invitations to writers and artists brought a rush of acceptances. But then the poet Robert Lowell withdrew his acceptance. He released to the press a letter he had written to the President. Noting a "chauvinistic" trend in American policy, he expressed fear that the nation was drifting toward catastrophe—perhaps nuclear catastrophe. "At this anguished, delicate, and perhaps determining moment, I feel I am serving you and our country best by not taking part in the White House Festival of the Arts."

If this angered the President, the novelist John Hersey infuriated him even more. Hersey told reporters he would *not* withdraw because he felt he "could make a stronger point by standing in the White House—I would hope in the presence of the President—and reading from a work of mine entitled *Hiroshima*."

The White House Festival of the Arts was planned as a television special, and a network was preparing to film the events for this purpose. The Hersey maneuver threatened to turn it into a coast-to-coast attack on Johnson policies. Adding fuel to the blazing anger in the White House, Hersey sent Goldman the text of what he proposed to read—including preliminary remarks.

> HERSEY: I read these passages on behalf of a great number of citizens who have become alarmed in recent weeks by the sight of fire begetting fire. . . . We cannot for a moment forget the truly terminal dangers, in these times, of miscalculation, of arrogance, of accident, reliance not on moral strength but on mere military power. Wars have a way of getting out of hand.

Professor Goldman was summoned to lunch with Mrs. Johnson. "The President and I," she told Goldman, "do not want this man to come here and read these passages in the White House." She made clear that Professor Goldman was to handle the matter with a phone call, but he declined, saying it would harm rather than help the President. "I am sorry. . . . I must refuse to call Hersey."

The President had by this time had enough of "these people" and ordered the festival blacked out—no television cameras, no spectacular. The festival took place on schedule; Duke Ellington played, the Joffrey ballet performed, and Hersey read—but not on television. The President appeared momentarily, his face as "hard as caliche soil." The wall between him and "these people," thought Goldman, had become as impassable as the Berlin Wall.[4]

The President did not cease for a moment his labors to control the public image of his war. In October 1966 he scheduled a trip to Manila, South Vietnam, and Thailand, which revealed, wrote columnist Pete Hamill, the extent to which the mass media had become prisoners of their own techniques.

> At every stop so far, Johnson and his people have set up situations that lead the TV men into their traps. They are assured of good film, lively quotes, and pictures—above all, pictures—which give the illusion of action.

The President's bubble-top limousine was flown halfway around the world for this purpose. A *Life* photographer lay on its floor as it moved through Manila crowds. From a pickup truck moving ahead of the limousine, aides frantically tossed paper flags to the crowd so that people could be seen waving them in television footage. At Camranh Bay in South Vietnam, stage management got "almost a bit too thick," thought the New York *Times,* when a young man in full field pack with a grenade launcher on his back was sent to eat with the President at a mess hall table, for the benefit of the cameras. It apparently had to be documented that way because the President had already taped a radio message about his Vietnam visit to be broadcast later to the American people.

> JOHNSON: I went there to visit our men at our base in Camranh Bay. Many of them only recently had come from the battlefield. Some were in field dress, carrying their packs and rifles. . . . I thought of all the battlefields in this century where Americans have fought: Belleau Wood and the Argonne, the Solomons and Bastogne, the Pusan perimeter and the 38th parallel in Korea.

4. Goldman, "The White House and the Intellectuals," *Harper's Magazine,* January 1969.

The men who had died in those places, said the President, had died "for the same cause that brought the men I saw at Camranh Bay to a place called South Vietnam." The Vietnam visit lasted 144 minutes, and the junket was off for Thailand, where the President took part in royal pageantry and in the presence of King Bhumibol Aduldet signed an International Education Act with eight pens, while cameras whirred.[5]

Back home Johnson was constantly on the air. He tried various techniques and gadgets: contact lenses, a new type of prompter, a slow delivery, a fast delivery. He called critics "nervous Nellies" and cast doubt on their loyalty. Dean Rusk followed suit. Castigating reporters, Rusk said: "I know what side I am on—that of the United States. All your news publications and networks won't be worth a damn unless we succeed."[6]

Rusk, like Dulles, pursued a passport policy designed to prevent travel to mainland China, North Vietnam, North Korea, Albania, Cuba. But this policy was challenged with increasing frequency by individuals—writers, teachers, churchmen—ready to risk a court test on constitutional grounds. One ground was that the travel ban infringed freedom of the press. Another was that the right to travel was reserved "to the people" under the Tenth Amendment.[7] Another was that the government could not invoke "war powers" because Congress had never declared war on North Vietnam. This argument challenged the legality of the war itself.

Among those who traveled to North Vietnam in 1965 was Christopher Koch of the Pacifica radio stations—WBAI, New York; KPFA, Berkeley; KPFK, Los Angeles. This group of noncommercial stations subsisted on listener subscriptions and gifts, and welcomed views from diverse sources including extremists of left and right. In so doing, the stations had won increasing attention precisely because of what *Variety* called the "no-guts journalism" of the dominant media. But in winning a cult following, the group of stations had also won indignant opposition—a fact which the FCC, under E. William Henry, had faced squarely at a 1964 license renewal:

> We recognize that as shown by the complaints here, such provocative programming as here involved may offend some listeners. But this does not mean that those offended have the right, through the

5. New York *Post,* October 28; New York *Times,* October 30, 1966.
6. *Variety,* February 21, 1968.
7. "The powers not delegated to the United States by the Constitution, nor prohibited by it to the States, are reserved to the States respectively, or to the people." Article X, Bill of Rights, Constitution of the United States.

commission's licensing power, to rule such programming off the air waves.[8]

That this seemed dangerous doctrine to many commercial broadcasters was a reflection of a schism on the American scene. For the rise of Pacifica was symptomatic of a widening movement—a subculture developing across the country, especially in cities and on college campuses. Its keynote was dissent.

SUBCULTURE

It was perhaps the most fateful development of the Vietnam war. Fateful for the nation, it had crucial implications for broadcasting.

Throughout the rise of television the worldwide thrust of American business, allied with military interests, had dominated the medium—its themes, myths, taboos, business practices. It had permitted discussion and dissent, but industry leaders tended to look on these as defense mechanisms, a safety valve. Ad lib discussion programs and documentaries on controversial topics were thought of in this light and generally neutralized by scheduling, style, and low budgets. With proper management, they could be ornaments of the system. On the whole the industry reflected dominant power groups in its definition of what was good and admirable.

Harold A. Innis in his *Empire and Communications,* a study of past empires and their media, showed that shifts from one kind of empire to another have always involved the development of new media and of monopolies controlling them.

But monopoly control has always, by its restrictions, encouraged the development of other media on the edges of society. Beginning as organs of dissent, they sometimes laid the basis of new monopolies.

The very completeness of military-industrial control over television was a chink in its armor. Throughout the rise of the Vietnam war and the military atmosphere it involved, many Americans were turning from commercial television and responding to new media—or old ones in new form: underground film, off-Broadway theater, cafe, cabaret, folk song, poster, newsletter, demonstration, march, rock festival, sit-in, teach-in, love-in. All offered potent messages shunned by prime-time television; this was their strength. To some extent noncommercial television, along with segments of radio, became a part of this movement of dissent.

8. FCC statement, January 22, 1964.

The babel of voices represented diverse groups, but dissent united them, at least temporarily. Moved by fury against a repulsive war, many began to question hypocrisies and self-deceptions that made it possible. The mechanisms of social control became an obsessive interest of many young people. On college campuses activist students were ferreting out the relationship of glossy new buildings and research programs with funding agencies in government or big business. Was it the function of colleges to develop the discerning mind or, in the words of a student in the CBS documentary *The Berkeley Rebels,* to "turn out shiny new parts for General Motors?"[1]

Commercial television was a similar target. Virtually a symbol of the "establishment," it was lampooned in hundreds of underground films and magazines. Its routine frauds were a favorite subject—along with Vietnam—of underground cabarets.

Television—the rejection of it—seemed to determine the life-style of the new subculture. For a generation television had showed people how to dress, talk, behave. Junior high school students had been carbon copies of heroes and heroines of commercials and telefilms. The underground reversed the process. It despised the clean-shaven hero, the office haircut, the lacquered hair, the necktie, the suburban home, the Detroit car. To many a hard-working suburban father who had for decades followed the rules and obeyed the commercials, the anarchy of style seemed to threaten social foundations. But in truth the uprising was also a return to an earlier America. The bearded Walt Whitman had said that he found the sweat of his armpits "more sweet than the perfumes of Araby." The young activist might have worded it, "than the leading deodorant." Thoreau had said that to cooperate with a government that was breaking the law was to condone its crimes. Any honest man, he said, had to "withdraw from this co-partnership." This the young activists were doing.

To the television industry the subculture seemed at first a negligible phenomenon. Broadcasters reassured themselves—and their viewers—with such terms as "hippie," "yippie," "teenie-bopper," "peacenik." Hippie activities often provided newscasts with a final one-minute light touch. But amid the condescension the subculture made inroads.

The phonograph-record field was producing anti-establishment successes—in folk-song, rock 'n' roll, and folk-rock styles—that were difficult to ignore. Stations vacillated between permitting and forbidding the some-

1. *The Berkeley Rebels,* produced by Arthur Barron for CBS News, June 14, 1965.

times cryptic, sometimes explicit messages of the subculture. Thus WMCA, New York, which called itself "the home of the Good Guys"—these were its disk-jockeys—at first permitted the scheduling of "Home of the Brave."

> The school board says he can't come to school no more
> Unless he wears his hair like he wore it before.
> The PTA and all the mothers
> Say he ought to look like the others.
> Home of the Brave, Land of the Free,
> Why don't you let him be what he wants to be?[2]

But the station, apparently grown fearful of encouraging student dissidence, suddenly reversed itself. Many stations developed similar anxieties over "Eve of Destruction" by P. F. Sloan, with its haunting refrain, "You don't believe we're on the Eve of Destruction."

> Don't you understand what I'm try'n' to say?
> Can't you feel the fear that I'm feelin' today?
> If the button is pushed there's no running away.
> There'll be no one to save with the world in a grave.
> Take a look around you boy,
> It's bound to scare you, boy,
> Ah, you don't believe we're on the Eve of Destruction.[3]

But the song got enough radio exposure to help make it a major hit.

"The Universal Soldier," by the Cree Indian folk singer Buffy Sainte-Marie, precipitated similar terror and on-and-off policies, as did "Waist Deep in the Big Muddy," by the singer Pete Seeger. In 1967 CBS, having decided to end its seventeen-year blacklisting of Seeger as a performer, permitted him to be booked on *The Smothers Brothers Comedy Hour*. But the network felt new tremors when he decided to sing "Waist Deep in the Big Muddy." The song was based on an actual training-camp episode, in which some Marine trainees had died by drowning. The lyric began:

> The captain told us to ford a river,
> And that's how it all begun—
> We were waist deep in the Big Muddy
> But the big fool says push on.

2. "Home of the Brave," by Barry Mann and Cynthia Weil. Copyright © 1965 by Screen Gems-Columbia Music, Inc. By permission.
3. "Eve of Destruction," by P. F. Sloan. Copyright © 1965 by Trousdale Music Publishers, Inc., 1330 Avenue of Americas, N.Y., N.Y., 10019. All Rights Reserved. Used by permission.

The sergeant protests but the captain tells him not to be a "nervous Nellie" and they push on—with the captain himself finally drowning. The song ends:

> . . . every time I read the papers
> That old feeling comes on—
> We're waist deep in the Big Muddy
> And the big fool says to push on.[4]

During rehearsal CBS asked if the last lines could be left out, but Seeger considered them essential, as did the Smothers brothers. Later, before the telecast, CBS cut the whole song from the videotape. But the resulting hubbub gave the song a role of honor, and led to a shift. In a return engagement on the Smothers series, Seeger was allowed to sing "The Big Muddy"—in full.[5] Thus the subculture inched forward.

While hatred of the Vietnam war was the leading underground impulse, it tended to merge with other drives. "Make love, not war," was a slogan displayed in every peace march, and neatly fused the peace drive with the revolt against sexual taboos. The Negro rights movement fused with both. It was in 1967 that Martin Luther King began persistently to identify the Negro rights crusade with the peace movement. He did so partly on economic grounds, since he saw war costs scuttling anti-poverty programs. He did so also on broader grounds: the readiness to slaughter Asians seemed to him of a piece with racial callousness at home. The anti-war and Negro rights movements became closely linked in underground films, plays, publications.

After 1965 the underground film movement began a rapid expansion and ceased to be "underground." The collapse of state film censorship aided this development.[6] Film-making was bursting out on every college campus, with alienation and dissent as dominant themes. An underground newsreel was organized, dealing with such topics as draft resistance, demonstrations, police riot-control methods. Its films were welcomed by campus groups and a sprinkling of theaters. The groups also began circulating newsreels and documentaries from Vietcong and North Vietnamese

4. "Waist Deep in the Big Muddy" ("The Big Muddy"), words and music by Pete Seeger. TRO—copyright © 1967 by Melody Trails, Inc., N.Y., N.Y. Used by permission.
5. New York *Times*, October 8, 1967; February 15, 1968.
6. N. Y. Court of Appeals, in reviewing a censorship case, declared the New York State film censor board unconstitutional on June 10, 1965. On July 27, 1966, the Kansas Supreme Court took similar action in regard to the Kansas film censor board. With these two actions, state film censorship virtually ceased to exist.

sources, as well as Vietnam war footage obtained from Japan, France, Poland, East Germany. During 1966–67 the networks began intermittent use of such material, usually superimposing a warning that it was "propaganda from communist sources." The administration regarded the use of the material as little short of treason.

One of the most notable films on North Vietnam to reach the television screen was actually commissioned by CBS, but was seen on National Educational Television. It was the work of Felix Greene, a British citizen who had been American representative of the British Broadcasting Corporation before World War II. He settled in the United States, but retained his British citizenship. This enabled him to travel to such places as mainland China and North Vietnam. His 1964 book *A Curtain of Ignorance,* the result of three sojourns in China during the 1950's and 1960's, documented the tragic results of the Chinese-American information gap. Troubled by Chinese misinformation about the United States, he was even more appalled by United States misinformation about China. In 1967 the San Francisco *Chronicle* decided to send him to North Vietnam, and CBS commissioned him to make a color-film record of his observations. The result was an 85-minute documentary, *Inside North Vietnam.* Press previews caused fervent discussion. Cleveland Amory, writing in the *Saturday Review,* found the film "superb" from start to finish.

> It is objective and, if anything, understated—but it is so moving it will make you first ashamed, then angry, and finally utterly determined to make everybody you know see it.[7]

At a preview Felix Greene fielded a barrage of questions from foreign correspondents—some friendly and some hostile. He made it clear—as did the film—that he had not been given access to military installations. Then why did he call the film "uncensored?" "Because once I had shot something my footage was never inspected by North Vietnamese authorities. And I could go anywhere I pleased so long as it wasn't military." Why had he not also gone to South Vietnam? "I applied to go to South Vietnam and Saigon wouldn't let me." Why? "Because I'd been to China."

He had photographed only what he found happening. He said he had recreated nothing. There had been a bombing at Phat Diem in which seventy-two people were killed in a church. The bombers, apparently realizing their error, later returned to drop candies, transistor radios, and

7. *Saturday Review,* February 3, 1968.

other gifts. Greene was told that these items had been put in a circle and, to the accompaniment of patriotic songs, burned. Such an occasion could have been recreated for the camera, but nothing of this sort was done.[8]

The film, by showing evidence of wide use of anti-personnel bombs, belied Defense Department claims of being concerned only with military targets. It also undermined the State Department version of the war by providing a close look at an "enemy" described by Rusk as living in a reign of terror and fighting at the behest of madmen. Instead the audience saw, amid bomb craters, a people smiling and laughing. Girls picked up rifles to shoot at bombers, then sat down to be demure again. Their lives were extraordinarily disrupted and their work ceaseless, but strangely joyful. This reinforced reports of various travelers, but the impression was more tellingly conveyed by camera images. To Amory, the film's most notable achievement was to show "what kind of people we are fighting—and why their record against us is bound to go down in history alongside Thermopylae, Stalingrad—or, for that matter, Valley Forge."[9]

CBS, which had an option on the film by virtue of having provided 20,000 feet of raw film and having paid laboratory costs and an advance, had meanwhile decided not to use it, but only to insert fragments in *CBS Evening News With Walter Cronkite*. As a result, *Inside North Vietnam* became available to National Educational Television, which planned to use a 49-minute segment followed by a panel discussion.

The announcement caused extraordinary rumbles. Former U. S. Representative Walter Judd, who had not seen the film, wrote to congressmen.

> Dear Friend,
>
> I hope you will read and sign the accompanying letter protesting the proposed showing on the Educational TV network of Felix Greene's film on North Vietnam. . . . When American youth are giving their lives in a war against a ruthless enemy, surely we have an obligation to protect their families and the public against anything that strengthens that enemy.

The enclosed letter of protest, addressed to John White as president of National Educational Television and describing the film as "nothing more nor less than communist propaganda," was signed by thirty-three congressmen—not one of whom had seen the film. The letter stated that if

8. *Variety*, January 17, 1968.
9. *Saturday Review*, February 3, 1968.

this was NET's idea of public service, a change of management was obviously needed.

Perhaps CBS was happy to have avoided the congressional attack, but NET went ahead with its plan. The discussion following the film was between television newsman David Schoenbrun and political scientist Robert Scalapino—one a critic, the other a defender, of United States policy in Vietnam. *New Yorker* reviewer Michael Arlen found the film unmistakably partial but very moving. Referring to "all those scenes of bombed towns and villages, of leveled huts, and craters, and silent children," he commented:

> . . . even if Saint Peter himself, and all the other admirals, should one day explain and make meaningful these scenes, these facts of life, it seems that they are indeed facts of life and that it is better to glimpse them now, even through prejudiced eyes, than not at all.[10]

The *Inside North Vietnam* case exemplified—and no doubt furthered—the slippage of commercial television and the shift elsewhere. A 1967 Louis Harris poll noted "a growing television boycott" among the college-educated.[11] The industry itself was torn by doubts. Men who had been pillars of commerical television, like Pat Weaver, were suddenly propounding the need for noncommercial television as an alternative voice. Fred Friendly, ex-president of CBS News and now a Ford Foundation consultant, had become a leading crusader for noncommercial television. Noncommercial television stations had—in 1965—finally reached a hundred in number, and more were being organized. The system had a significant coverage and, in spite of pinched budgets, was building a loyal following. In February 1967 a commission established by a Carnegie Corporation grant urged establishment of a Corporation for Public Broadcasting, aided by a tax on television sets, to promote noncommercial television. It mentioned the need for sums of at least $100 million a year to finance a meaningful network. NET had seldom had a tenth of that sum; during its first decade it had survived on Ford Foundation allotments of $3 to $6 million a year.

In the spring of 1967 the Ford Foundation stimulated the hubbub by earmarking $10 million for a Public Broadcast Laboratory. It was to produce a major Sunday evening series, *PBL,* to be available to the 100-odd

10. Arlen, *Living Room War,* pp. 161–2.
11. New York *Times,* February 2, 1967.

educational stations, most of which would be linked by AT&T cables—at a cost of several million. Av Weston of CBS News was engaged as executive producer. The series was to explore the possibilities, in theme and treatment, of an adequately financed system released from commercial restraints.

PBL, debuting in November 1967, at once reflected a world in sharp contrast to that of prime-time television. Mindful of its function as an alternative voice, it dipped into work of fringe theaters, cabarets, and underground films, and inevitably reflected the angry subculture. The thrust of the message was anti-war, anti-racist, anti-establishment; the techniques, sometimes drawing on the absurdist theater, were strange and seemed outrageous to many television viewers. *PBL* gave a platform to diverse ghetto spokesmen, confronting them with representatives of other groups—including gentle liberals—in sessions that sometimes turned into vituperative shouting matches. To many people who favored "discussion" but felt it should observe the decorum of the old *University of Chicago Round Table*, *PBL* was often horrifying. But it gave a sense of the rage sweeping through the American subculture, and the fever of ideas it was thrusting into circulation.

The very first *PBL* program presented a bizarre off-Broadway success by Douglas Turner Ward, titled *Day of Absence*, about a southern town from which all blacks have suddenly vanished. Finding itself unable to function, the town begs them to return. To many viewers the really jolting aspect of the production was that whites were played by Negroes in "whiteface," with makeup applied in the blatant style of the blackface comics of another era. The enactments were likewise in a brutal cartoon style. Members of the *PBL* supervisory board, headed by Dean Edward W. Barrett of the Columbia School of Journalism, were deeply disturbed. A number of southern stations, informed by advance rumor, did not carry the program.

The crisis made it clear that noncommercial television, working under boards often selected for fund-raising potential and political influence, could be as tightly tethered as commercial stations. Yet noncommercial television was to some extent giving expression to the ferment of the subculture—not only through *PBL* but also through *NET Journal*, *Black Journal*, *NET Playhouse*, *The Creative Person*, and various local series.

Many observers were puzzled when President Johnson, obsessed with maintaining a Vietnam consensus—or the appearance of a consensus—

boarded the noncommercial television bandwagon. He espoused the idea of a Corporation for Public Broadcasting and urged its adoption. With astonishing smoothness the idea moved through the legislative process and became law on November 7, 1967—two days after *PBL*'s shock debut. The meaning of the President's sudden interest in noncommercial television was not clear but observers soon felt they saw straws in the wind. He recommended an appropriation of only $4,500,000 for the first year, and followed this with an appointment that stunned noncommercial television enthusiasts. As chairman of the new corporation he chose Frank Pace, Jr., a former Secretary of the Army and a former chief executive officer of General Dynamics—in other words, an embodiment of the military-industrial alliance. Chairman Pace at once expressed his enthusiasm for his new post and said he had already commissioned research on an important idea—how public television might be used for riot control. The President's support had created vast expectations among supporters of noncommercial television. Now they wondered if it was being hugged to death.

In 1967 every subject tended to become Vietnam. Networks, with a haunted intensity, looked for safe documentary topics that might lure a sponsor, and during the year they came up with *Venice, Nurses, The Pursuit of Pleasure* and other subjects. But some seemingly remote topics turned out to be, in the context of the day, Vietnam. On ABC-TV B. F. Goodrich sponsored a magnificent film shot in Westminister Abbey, entitled *Hall of Kings* and starring James Mason, Lynn Redgrave, and Siobhan McKenna. The film was the work of the Canadian Harry Rasky. Turning to history, he found he could convey feelings he had not been able to express in *Operation Sea War Vietnam.* The men and women whose words rolled out so magnificently were concerned with empire—its rise and decline. Many were words of warning. From the gloom of Westminster Abbey emerged William Pitt, Earl of Chatham, crippled and near death, but rising to demand an end to a war in America:

> PITT: If I were an American, as I am an Englishman, while a foreign troop was landed in my country I never would lay down my arms. Never! Never! Never!

And there was Kipling, who had once urged Englishmen to take up the "white man's burden" to subdue and lead "the lesser breeds," but now pleaded—in the words of "Recessional"—for a humble and a contrite heart. And there, like an early voice in the cause of "love, not war," was Elizabeth Barrett, with verses she had sent to the poet Robert Browning:

> BARRETT: The face of all the world is changed, I think,
> Since first I heard the footsteps of thy soul
> Move still, oh, still beside me . . .[12]

Two months later NBC-TV broadcast a "documentary drama" that likewise made no reference to Vietnam but—to many viewers—meant Vietnam. This was *The Investigation*, by Peter Weiss, adapted from a Broadway production by its director, Ulu Grosbard. All dialogue was from nazi war crimes trials relating to Auschwitz extermination operations.

Only a few years earlier millions of Americans had been able to watch the television drama *Judgment at Nuremberg* with a comforting sense of moral superiority.[13] It was about Germans—nazi Germans. *The Investigation* was also about them but about a great deal more. Again and again the world of the defendants offered familiar echoes.

> I only did my duty. . . . I believe in my country. . . . Even now, corporal punishment would stop a good deal of delinquent behavior. . . . They were to blame for everything. . . . Personally, I have always behaved decently. . . . We were dealing with the annihilation of an ideology. . . . It was my duty as a soldier. . . . I still don't see how else I could have acted.[14]

The charter that had established the nazi trials was very much on the minds of young Americans. It stated: "The fact that the defendant acted pursuant to the orders of his government or of a superior shall not free him from responsibility."[15] This was one of the bases for defiance of the draft and was widely discussed in the subculture. The NBC-TV special of *The Investigation*—not surprisingly, it was unsponsored—contributed to the discussion.

In 1967 Americans were becoming aware of atrocities that "sophisticated" technology had made possible on their behalf. A Vietnam "war crimes tribunal" in Stockholm, in which such figures as Jean-Paul Sartre and Bertrand Russell were taking part, was airing details. In May 1967 it heard testimony from a former member of the French General Staff, Jean-Pierre Vigier, who had visited North Vietnam and discussed the in-

12. *Hall of Kings*, ABC-TV, February 14, 1967.
13. A *Playhouse 90* special on CBS-TV, April 16, 1959.
14. *The Investigation*, NBC-TV, April 14, 1967.
15. Nuremberg Charter, Article 8.

tensive American use since 1966 of the "guava" or cluster bomb. It consisted of a "mother bomb" which in falling ejected hundreds of "guavas," or secondary bombs, each of which on contact spewed out hundreds of fine steel pellets—of little effect on steel or concrete but devastating to human bodies. He considered it exclusively an anti-personnel weapon and one of the most monstrous ever invented. Such testimony was noted only briefly on the air but discussed extensively in newsletters, magazines, paperbacks.

One of the most devastating paperbacks of 1967 was one which, oddly enough, was written in defense of American policy by an expert favored by the Pentagon. *Air War—Vietnam,* by Frank Harvey, was infused with the conviction that a tough job needed to be done and that Americans should learn to face the toughness of it. His revealing, authoritative book became a treasured resource for anti-war groups. Fully confirming reports about guava bombs, napalm, and other weapons, he added a grisly dimension with a vocabulary rooted in suburban affluence, and apparently common among airmen. With guava bombs, said Harvey, a pilot could "lawnmower" a large area, killing or maiming everyone along a path several hundred feet wide. A large bomb was a "swimming pool maker." And a defoliation operation had a slogan derived from television public-service spots: "ONLY YOU CAN PREVENT FORESTS."

Harvey explained that pilots in Vietnam had to go through a hardening process which was best done in southern areas where a beginner was not threatened with powerful ack-ack.

> He learns how it feels to drop bombs on human beings and watch huts go up in a boil of orange flame when his aluminum napalm tanks tumble into them. He gets hardened to pressing the firing button and cutting people down like little cloth dummies, as they sprint frantically under him. He gets his sword bloodied for the rougher things to come.

Napalm, Harvey explained, was especially good against people hiding in caves and tunnels "since it suddenly pulls all the oxygen out of the tunnel by its enormous gulp of combustion, and suffocating anyone inside." But a tendency had developed to use it routinely against huts along canals and rice paddies because an improved napalm could send its rolling fire over water as well as land. The author had no doubts about the necessity of all this. "We are a Have nation, and we intend to continue to be a

Have nation." He described the stakes as "money, influence, land, power, trade routes, resources, control."[16]

In startling contrast to *Air War—Vietnam* was a 1967 CBS documentary titled *Vietnam Perspective: Air War in the North.* Apparently oblivious to the horrors discussed by Harvey, it was full of administration rationales, punctuated with marvelous shots of hardware in action and the rhapsody of roars that went with it. It referred to civilian deaths, but "civilians are always killed in war," said the narrator. One might have reached the conclusion, wrote Michael Arlen in his *New Yorker* review, "that CBS is another branch of the government, or of the military, or of both." He was astonished that a network dared to present so "childish and unaware and chicken a piece of journalism."[17]

But a sense of the horror of the war, ceaselessly agitated by fringe media, inevitably spread further—as in a very different 1967 CBS-TV special, *Morley Safer's Vietnam.* A complex and eloquent work, it stuck to CBS rules of objectivity, while providing rich food for thought. Giving considerable attention to the official façade, it showed General Westmoreland visiting troops in the field.

> WESTMORELAND: How's your morale?
> GI: Pretty good, sir.
> WESTMORELAND: How's your food?
> GI: Real good, sir.
> WESTMORELAND: Son, what state are you from?
> GI: Texas, sir.
> WESTMORELAND: What part?
> GI: Southwest. Shullerville. . . .
> WESTMORELAND: How old are you, son?
> GI: Twenty years old, sir.
> WESTMORELAND: Twenty years. Where did you get your basic training?
> GI: Fort Polk, Louisiana, sir.

Safer later showed Westmoreland addressing the group:

> WESTMORELAND: It's a matter of great pride to me to see the high morale that has obtained with the troops—well, for the last year here. I attribute this to many things. First, they believe that they are performing an important mission. They take pride in doing a good job. They find this a very exciting experience. The food is good. The mail service is excellent, although from time to time there are delays, but these are exceptional.

16. Harvey, *Air War—Vietnam*, pp. 2, 55, 183–4.
17. Arlen, *Living Room War*, p. 45.

Safer showed a helicopter crew just back from a mission, drinking beer. He asked how it felt to "make a kill like that." Their answers appeared to confirm General Westmoreland.

CAPTAIN: I feel real good when we do it. It's kind of a feeling of accomplishment. It's the only way you're going to win, I guess, is to kill 'em.

PILOT: I just feel like it's just another target. You know, like in the states you shot at dummies, over here you shoot at Vietnamese. Vietnamese Cong.

ANOTHER (*off, interrupting*): Cong. You shoot at Cong. You don't shoot at Vietnamese.

PILOT (*laughing*): All right. You shoot at Cong. Anyway, when you come out on the run and then you see them, and they come into your sights, it's just like a wooden dummy or something there, you just thumb off a couple pair of rockets. Like they weren't people at all.[18]

In January 1968, shortly after General Westmoreland forecast an early victory, the guerrillas began lobbing mortars and rockets into major South Vietnamese cities, including Saigon and Hue, and followed by seizing footholds in them. To eradicate the infiltrators, American airpower began a block-by-block destruction of the very cities that had been the basis of the American position in Vietnam. The problem of refugees, already numbering in the millions, became almost catastrophic. Westmoreland asked for 200,000 more troops. President Johnson, already hard-pressed by opposition at home, now faced military peril abroad. His ordeal became overwhelming.

The way in which the agony of Lyndon Johnson was conveyed to the public epitomized American broadcasting at the height of affluence. Worry was not television material because it was not visual, but the President's troubles were ably described by Dan Rather in an early-morning report on CBS Radio, in the series *First Line Report.* These brief reports, usually from Washington, were less than five minutes long but had to be broken into segments to accommodate commercials.

On February 1 Rather opened his broadcast by saying that Lyndon Johnson was always a fretful sleeper, inclined to wake at 5 a.m. or earlier and to phone his military aides at once for latest war news. The President might then, said Rather, doze off again but even if news was good, he seldom slept more than five or six hours. The teaser ended with:

18. *Morley Safer's Vietnam*, CBS-TV, April 4, 1967.

RATHER: This week he has hardly slept at all. In a moment, the details.

At this point singer Petula Clark crashed in with the song: "And the beat goes on, and the beat goes on!" It was a musical commercial for Plymouth, promoting a "win-you-over sale." After it, Rather began to explain why the President, beset by worldwide problems, military uncertainties, political attacks, was hardly sleeping. Then Rather promised a comment—"after this." Another Plymouth commercial. Finally came Rather's comment on the anguish of a President.[19]

In February the Senate foreign relations committee scheduled hearings on the background of the Tonkin Gulf resolution. In the heightening crisis, CBS-TV and NBC-TV canceled daytime programs for direct pickups. Viewers learned that the ship which had reported blips on its radar, interpreted as attacking gunboats, had a defective radar. Some committee members doubted there had been an attack. Also, the patrol along the North Vietnamese coast, which had been described as "routine," seemed in the light of additional information to have been provocation designed to activate and reveal the location of North Vietnamese radar, as well as to draw North Vietnamese defenders away from coordinated action planned further south by the South Vietnamese. To some the Tonkin Gulf "incident," used to justify massive destruction and slaughter, began to seem a trumped-up pretext for attacks long planned.[20]

It was election year. The renomination of Lyndon Johnson had been a foregone conclusion, but Senator Eugene McCarthy made plans to enter the New Hampshire primary, contesting Johnson on the basis of the Vietnam war. Senator Robert Kennedy was moving toward opposition to the President.

Mounting bitterness marked the home front. Ghetto areas and college campuses were torn with riots and demonstrations. Attacks on ROTC offices, recruiting offices, draft boards were routine newscast items. Recruiters for Dow, makers of napalm—among many other products—were blocked from campuses by demonstrators. A group in Hartford calling itself the Americong announced that it would mark Memorial Day by napalming dogs of the community, to show what the United States was doing to people in Vietnam; the idea was lurid enough to win attention in newscasts. In New York an unlicensed Radio Free Harlem erupted

19. *First Line Report,* CBS Radio, February 1, 1968.
20. New York *Times,* February 4, 22, 25, 29, 1968; October 5, 1969.

intermittently on the air. Countless young men were resolving to go to jail rather than into uniform. Thousands had moved to Canada to avoid the draft. A colony of deserters was forming in Sweden. Use of drugs was on the rise at home and in Vietnam. Colonies of runaway teenagers congregated on New York's lower East Side, in San Francisco's Haight-Ashbury section, and in slum areas in other cities. The girls painted flowers on their foreheads, meaning "make love, not war."

But some of the young were not ready to flee or "freak out." The unhysterical, reasoned anti-war candidacy of Senator Eugene McCarthy drew many to his banner. They became McCarthy volunteers and painted flowers on the family car.

According to Gallup polls taken in the early months of 1968, supporters and opponents of the war had become evenly balanced. On television the early-evening network newscasts seemed to assume the task of keeping America on an even keel. Chet Huntley on NBC relayed official statements and body-counts of killed Vietcong without hint of approval or disapproval. David Brinkley sometimes twitched an eyebrow; he had said in *TV Guide* that he opposed the war and felt the bombing should halt,[21] but he did not say it on the air. On *CBS Evening News With Walter Cronkite,* anchorman Cronkite gave all news items the same weight, seeming to avoid intonations that might imply degrees of significance. To many there was something reassuring about this unwavering delivery; Cronkite himself suspected that it was the key to his popularity. He radiated steadiness and honesty.

President Johnson watched these programs constantly. He had written off the book world; it hardly mattered. As to columnists, it was the administration view—conveyed in background comments—that Walter Lippmann was senile. But television was different.

The organizations behind these network news programs had grown enormous, and the years of crisis had given them a growing following. In the late 1950's they had found sponsors elusive, but now they were well sponsored. By expanding to half-hour length in 1963 they had taken a small step toward peak time. The shift to color film in 1965–66 had affected some topics, including war. Mud and blood were indistinguishable in black and white; in color, blood was blood. In color, misty Vietnamese landscapes hung with indescribable beauty behind gory actions.

Some television executives, like James Hagerty, were convinced that

21. *TV Guide,* July 1, 1967.

television had brought the face of war home to the American people. Many others felt with syndicated columnist John Horn that television had trivialized the war, making it "of no more consequence than a movie star's latest marriage, the arrival of the Beatles, a Senator's pronouncement, a three-alarm fire"—all links in a chain of unevaluated events used to sell mouthwash, pills, and cigarettes.[22]

In January 1968 Walter Cronkite decided he had to see for himself. He had been to Vietnam before—in 1965—but resolved to go again. "And this time," he told CBS executives, "let's say what I think about it." When he returned he appeared on television saying that the United States might have to accept a stalemate in Vietnam. Two weeks later, on an NBC telecast, Frank McGee said that the United States was losing the war in Vietnam. To destroy Vietnam in order to "save" it, he implied, might not be a policy of wisdom.[23]

These moments marked a divide for television and Lyndon Johnson. According to some observers, the defection of Cronkite especially shook the President.

On March 12 he was shaken again. In the first presidential primary of the year, New Hampshire voters showed massive support for Senator Eugene McCarthy; a McCarthy triumph was also predicted in the Wisconsin primary.

About this time the President requested television time to address the nation, and his aides began writing the speech. The first drafts were like all the other speeches: justification, denunciation of Hanoi, scorn for critics, pleas for unity, and touches of the "Abraham Lincoln syndrome" that was overtaking Johnson. He seemed constantly to see his ordeal—or to try to see it—in terms of the vigils of other wartime leaders, especially Lincoln. But all of this had been said, and some advisers felt it would not do. These included Clark Clifford, who on March 1 had replaced Robert McNamara as Secretary of Defense. The tinkering led to more and more drastic revisions. Finally the President made his decision. Characteristically, he "kept his options open" until the last moment. As the speech went on the air on the evening of March 31, it gave no hint of its bombshell ending. An advance text had not included the ending. All this contributed to its impact.

The words near the end had been written by Johnson himself. He ex-

22. Quoted, *Variety*, February 1, 1967.
23. NBC-TV, March 10, 1968.

pressed confidence that the United States would be a "land of greater opportunity and fulfillment" because of what his presidency had achieved. Those gains were important, he said, and must not now be jeopardized by "suspicion and distrust and selfishness and politics." Therefore—

> JOHNSON: I have concluded that I should not permit the presidency to become involved in the partisan divisions that are developing in this political year. Accordingly, I shall not seek, and I will not accept, the nomination of my party for another term as your President.

The speech brought a stunned reaction throughout the world. Morley Safer, from London, reported the British consensus to be: "Nothing became Mr. Johnson so much in his presidency as his decision to leave it." The Associated Press reported incredulity in Vietnam, where many heard or watched the speech. One soldier said, "I can't believe it." Another, "It's not true." A civilian doctor said, "Man, that's wild. It's a whole new ball game."[24]

Lyndon Johnson had removed himself from the eye of the storm. He had gained time. But the storm was not over.

The stage was set for bitter election struggles. The politicians and broadcasters got ready.

MEANWHILE BACK IN PRIME TIME . . .

Most Americans during the 1966–68 upheavals were not demonstrating, marching, or rioting. Most were doing their jobs and relaxing over television. In homes with children the set was likely to be on sixty hours a week. The breadwinner watched after dinner. He might catch the early-evening news, or part of it, but the rest of the evening was the main thing.

There was also some radio listening by all members of the family, usually as background to other activity.

These media were for many a psychological refuge, a fortress. Their pattern of viewing or listening could easily make it so.

Radio, which had been the start of the broadcasting empire, had gone back to its beginnings, relying mainly on phonograph records. But most stations now aimed at unity of mood, to hold specific audience groups. There were frantic-music stations for the young, "easy listening" for the

24. Associated Press, April 1, 1968.

tired, "soul" for the black, "country" for plain folk, "good" for others. News came in small capsules, often with big titles like *Total Information News*. The music-and-news station, backed everywhere by American advertisers, was a worldwide phenomenon; disk-jockeys calling themselves The Good Guys erupted even on "pirate" stations operating from ships around the British Isles—almost all financed by American capital. Along Madison Avenue in New York girls in pirate costume drummed up business for this novel form of international freebooting, which for a time earned small fortunes. Britain put them out of business in 1967, but they had meanwhile persuaded BBC to adopt their style of programming.[1]

At home the alternatives to disk-jockey programming were marathon talkers, including telephone talkers, who seemed to take radio back to its amateur pre-history; a handful of ambitious all-news stations; and noncommercial stations on starvation budgets.

The specialized, fragmented nature of radio programming made it easy for the average listener to get push-button confirmation of his tastes and prejudices. Once he had "his" station, he was in "his" world.

Television offered similar asylum—especially if one avoided news, which network scheduling made it easy to do. But even news programming, always strong in human-interest emphasis, tended toward an air of reassurance.

During peak hours almost everything implied a reassuring view of American society. The only exceptions were occasional documentaries, which were often so at odds with the prime-time view of the world that they aroused indignation. According to government statistics 30 per cent of Americans were living at a poverty level, but such information reached few people, and when it did, it tended to remain an abstraction, seldom visible in human form. When networks scheduled documentaries on the subject, some people invariably attacked them as propaganda—sometimes "communist" propaganda. Martin Carr, who directed the eloquent documentary *Hunger in America* for CBS News, discovered soon after the telecast that the FBI was investigating him.[2]

Except for the intrusive and sometimes notable documentaries, evening television confirmed the average man's view of the world. It presented the America he wanted and believed in and had labored to be a part of.

1. Turner, *American Influence in British Broadcasting Since World War II*, pp. 64–79.
2. Interview, Martin Carr.

It was alive with handsome men and women, and symbols of the good life. It invited and drew him into its charmed circle. If the circle was threatened, it was surely not by flaws within itself but by outside evil-doers.

Hollywood, where most of this was made, was also a kind of refuge and a fortress—although a troubled one. It had its dissident spirits who made independent features that seemed to belong to the subculture. But much of the solid part of Hollywood was busy with telefilms. The activity was now firmly established at Metro-Goldwyn-Mayer, Paramount, Twentieth Century-Fox, Warner and the others—the aristocracy of the big-studio days. They had gone through upheavals but a sense of continuity with the past was even more notable.

Visitors could take guided tours at $3 each through Universal City, home of *McHale's Navy, The Virginian, The Name of the Game,* Bob Hope programs, and scores of others. Here on a chicken farm Carl Laemmle had begun making films in 1915; now it was a television fort dominated by the dark executive tower of MCA, owner of Universal since 1959. Tourists rode in candy-striped Glamor-Trams, guided by nubile girls and bronzed young men in quest of stardom but momentarily detoured. The young guides performed memorized commentaries studded with jokes and information supplied by promotion department writers, often updated to build current attractions. The patter also suggested that nothing had really changed; the whole Hollywood legend was reinforced. "This spiral staircase was used in the original *Phantom of the Opera* with Lon Chaney." A dressing room was "Lana Turner's dressing room—note the white velvet couch." Familiar humor linked present and distant past. "This used to be a chicken farm; now we only raise hams." The tour wound past a Bank of America branch near the MCA tower. "Jack Benny is said to have a special tunnel from his dressing room to the bank." At a special-effects building tourists were told—and no one smiled—"This is where we add the realism to your movies." The tours, begun in 1964, drew a million visitors a year during the following years—the Vietnam escalation years. The tours stopped for convenient pauses at profitable souvenir stands; television actors were sometimes pressed by management to be on view at strategic spots. The tours became a favorite topic of Hollywood humor. Bob Hope quipped that he had a three-year contract at Universal—"one in pictures, two in the souvenir stand."[3] But the tours

3. *Variety,* April 2, 1969.

also caused explosions. While watching rushes of *The Virginian*, its production staff on one occasion was astonished to see a candy-striped tour car enter the distant landscape.[4]

MCA-Universal and the other giants were making theatrical features as well as telefilms, but these lines of enterprise were merging. Television, devouring both, had been satisfied with old features but was exhausting the backlog. "The biggest crisis faced by the TV networks," said CBS-TV vice president Michael Dann in 1966—the year of the credibility gap—"is the fact that we're running out of movies."[5] In 1967 CBS decided to face the crisis by going into feature-film production, buying the old Republic lot and turning it into the CBS Studio Center. ABC had similar ideas, with plans for production abroad. NBC made an agreement with Universal under which NBC would provide financial backing for Universal features; those selected by NBC would have their United States premieres on television—often under the title *World Premiere*—rather than in theaters.

The line between television and theater enterprise was becoming blurred in other ways. Some *The Man From U.N.C.L.E.* episodes, premiered on television in the United States, were expanded into features by MGM and premiered in theaters in other parts of the world. Disney used various procedures of this sort. The fusing of the two worlds was bringing television executives into top studio positions.

From television, huge sums were pouring into the studios; program budgets were reaching a scale not dreamed of in earlier television years. By 1968 the 90-minute *The Virginian* and *The Name of the Game* were each budgeted at $275,000 per episode. The 60-minute programs *Bonanza, Mission: Impossible, Land of the Giants, Star Trek,* were budgeted at $180,000 or more per episode. The 30-minute series *Bewitched, Hogan's Heroes, Green Acres, Get Smart,* were budgeted at $80,000 or more per episode.[6] It was a land of milk and honey but also of tensions. Everything

4. Interview, Norman Macdonnell.
5. *Saturday Review*, June 4, 1966.
6. Expenditures on a typical one-hour program may be suggested by outlays on one 1968 *Star Trek* episode: script (writer and consultants) $8783, cast $22,650, director $3825, camera $3066, production facilities $23,550, production staff $2983, set design $1832, set construction $3153, set operations $3634, wardrobe $3198, makeup and hair stylists $3311, electrical expenses $4082, rushes $9221, editing $7351, optical effects $15,030, music $3250, lab processing $7946, payroll benefits $7448, producer $10,427, and additional overhead allocations. "Wink of an Eye," *Star Trek*, Paramount Television.

depended on decisions of networks and, beyond them, of advertisers and their agencies. On the basis of ratings, fortunes rose and fell, and heads rolled with unnerving suddenness. Contempt for the needs of "the market" was often heard, especially from writers, directors, and actors, although the rhetoric of denunciation seldom changed anything. The machine had its momentum; the stakes were huge. Escape into "independent production" for theaters was a favorite fantasy; a few achieved it.

Bitterness erupted in many forms. In 1966 the Screen Actors Guild, in a brief addressed to the FCC, charged: "As television networks have been increasing and tightening their monopolistic control of program sources and production, the quality of programs has sharply dropped." Its members, said SAG, felt that torrents of stereotyped shows were "dissipating" their talents.[7] In 1967 Hollywood members of the Writers Guild of America debated the question, "Should the Guild concern itself with the quality of TV?" Then they voted Yes, 452; No, 58. Were writers to be only "tailors cutting someone else's cloth to measure?" asked WGA–w vice president Richard M. Powell, and the consensus was that they should not, but no one knew what to do about it.[8] The frustration erupted into bitter humor at annual WGA award banquets. The 1966 banquet featured a series of vignettes of writers at work. One showed a writer sharpening several pencils with utmost care. Having brought them to a fine point, he contemplated them for several moments, then stabbed himself with them as the lights blacked out. Another vignette showed a writer at the telephone:

> AL: Sure, sure, I understand. Listen, Marty, if your maid doesn't understand what I wrote, I understand. Sure—that's the audience.[9]

In the 1950's the informal relation between writer and producer had enabled many a writer to turn out dozens of scripts a year. Now this had vanished, overwhelmed by protocol. The network as underwriter of a series generally had the right to review each episode at several stages: synopsis, teleplay, revised pages, and screenings of roughly edited workprint and final print. Advertising agency representatives and sponsors might also see copies of the teleplay; a CBS policy statement permitted them to "participate in the creative process." All this encouraged memowriting. Directors and writers were inclined to feel that the Xerox ma-

7. *Variety*, March 6, 1966.
8. *WGA-w Newsletter*, November 1967.
9. *Ibid.*, April 1966.

chine had complicated their lives. A producer issuing policy instructions to a director could easily send "information copies" to a dozen agency representatives and sometimes seemed to have written the memo for this purpose—or perhaps for any government agency that might care to subpoena it. A surging tide of paper flowed back and forth.

Scripts were in constant revision. Each time a page was revised, it was remimeographed on paper of a different color. Some pages of the script might remain white in each version of the script while others became blue, pink, yellow, red, green—each indicating a new revision. The final script was always multi-hued.

The telefilm exhibited a number of persistent themes. Heroes with special magic powers remained numerous, and seemed to observers a metaphor for the national obsession with secret weapons. The atom bomb had been thought by many to be a key to world control, and perhaps the dream persisted in the super-heroes. They represented, according to Beverly Hills psychiatrist Murray Korengold, "the American ethic of hegemony and supremacy."[10] Women were included in the magic powers. In *I Dream of Jeannie* the heroine was a genie in a bottle found by an astronaut after splashdown near a tropical island. She becomes his constant companion, often frustrating foreign agents trying to obtain American space secrets. *Bewitched* and *The Flying Nun*—from the same studio, Screen Gems—also featured women with special powers. The trend coincided with a wave of magic characters in commercials, dramatizing the occult powers of detergents and cleansers, and made drama and commercial highly compatible. Animal heroes satisfied similar power obsessions and, like all supermen, inevitably became involved in international struggles. In a 1968 *Flipper* episode the dolphin-hero helps to prevent a spy from delivering an aborted rocket's instrument package to a hostile power. Television animals had sound instincts about enemies.

In contrast to the paranoia and hostility of the power programs, family series were uncompromisingly wholesome and offered a reassuringly warm view of the American home. Two monster-family series, *The Addams Family* and *The Munsters*, perhaps represented a countertrend, but even they were wholesome and lovable. A curious thread ran through the wholesome-family tradition. In many of the most successful series, a father or mother was missing. In *My Three Sons*, Fred McMurray had the problem of raising his sons "without the woman's touch." In *A Family Affair*—by the same producer, Edmund Hartmann, also for Don Fedder-

10. New York *Times* Service, July 23, 1968.

son Productions—a bachelor and his English valet were raising a family. *Here's Lucy* presented Lucille Ball as widow and mother. *Julia,* launched in 1968, applied the same formula to a black family group. *Bonanza* with Lorne Greene and *The Big Valley* with Barbara Stanwyck had the same aspect in western framework. *Lassie* built its original following in a similar situation. Various explanations were offered for this persistent thread: that the producers didn't really believe in marriage; that the series offered a satisfying escape-fantasy to both mates and children; that the series reflected the prevalence of the broken home; that the unattached character offered more plot possibilities; that the gap in the family provided a niche into which the viewer could fit himself and participate in the warm family life. Whatever the reason, the truncated family remained a television standard.

Imitation of success, especially one's own, was a persistent trend, and seemed to be an inevitable result of the huge stakes involved. The multiplicity of sponsors was also credited with a homogenizing effect. By 1968 the single-sponsor series had almost vanished; many had a dozen sponsors; ABC-TV's *The Big Valley* had twenty-six.[11]

If all went well, investment funds flowed back in profusion—from American and foreign use. By 1968 there were 140 million television sets abroad—almost twice as many as in the United States. More than a hundred countries had become markets for American telefilms.[12] A successful one-hour telefilm series could expect up to $7000 per episode from the United Kingdom, $6500 from Canada, $6000 from West Germany, $6000 from Japan, $4400 from France, $4250 from Australia, $4000 from Brazil —and, at the other end of the scale, $180 per episode from Kuwait, $150 from Hong Kong, $120 from Saudi Arabia, $110 from Guatemala, $90 from Taiwan, $60 from Nicaragua.

United States foreign earnings from telefilms were climbing:[13]

1958	$15 million		
1959	25	1964	70
1960	30	1965	76
1961	45	1966	70
1962	55	1967	78
1963	66	1968	80

11. *Variety,* September 25, 1968.
12. In May 1958 NBC International listed 102 countries as customers.
13. Motion Picture Export Association. The only pause in the upward climb resulted from a 1966 dispute with Australia over rates, and a temporary boycott resulting from it.

Some sales were being made in communist countries: Bulgaria, East Germany, Hungary, Poland, Romania, Yugoslavia. Their motives were often a subject for speculation. Did Poland buy *The Untouchables* out of sheer enthusiasm, or because it confirmed a Polish conception of American life?

A similar question was raised when Sweden, strongly anti-American in respect to war policies, promptly bought *Mission: Impossible.*

In numerous countries United States programming took the prime-time spotlight. This was made fairly inevitable by the dearth of rival products. In 1968 the Motion Picture Export Association was pleased to inform its members that in Italy "only a few films are being produced locally for television, since American-made TV films are available at far below what it would cost RAI to produce similar films or series."[14]

If this was true of Italy, a major film-producing country, it was all the more true of scores of other countries. In Nigeria, peak viewing time was 8 to 9 p.m. Programs occupying this time-slot in 1968 were the secret-agent series *Mission: Impossible* (Sunday), the western *The Big Valley* (Monday), the private-eye series *Mannix* (Wednesday), the westerns *Bronco* (Thursday) and *Bonanza* (Friday), and the espionage series *I Spy* (Saturday). On only one night was another source used: the BBC series *The Human Jungle* occupied the Tuesday period. Other evening hours were occupied by *Daktari, Star Trek, The Beverly Hillbillies, The Addams Family, Buffalo Bill, Jr.*[15]

In 1968 *Bonanza* was on the television screens of over eighty countries; *The FBI, Mission: Impossible,* and *The Fugitive* were not far behind.

In two non-communist countries, Britain and Japan, native products were the dominant prime-time force.

The American telefilms moving into foreign television were supplemented by other American items, including feature films from the same distributors; public-relations films supplied free by American business firms; and free items from the U. S. Information Agency. The last-surviving American newsreel, the Hearst-MGM reel, expired in January 1968, but USIA continued to supply newsclips in many countries. They were free, and the local television system could decide whether or not to identify the source. Most preferred not to.

USIA also provided some free television drama. In 1968 it was offering Latin American stations a Spanish-language series titled *El Periodista!,*

14. *International Television Manual,* 1968. RAI is Radiotelevisione Italiana.
15. *Programme Schedule,* Nigerian Broadcasting Company, channel 10, Lagos.

which could be sponsored commercially. Again, the source of the drama did not have to be revealed; the commercial sponsor could take the spotlight. USIA took up sponsored drama—offered as entertainment—for the sake of access to prime time.

El Periodista!, following a hallowed telefilm formula, concerned a crusading newspaper editor, Emilio. In one episode his newspaper is strike-bound. Settlement is prevented by the impossible demands of a group whom Emilio suspects of being communist. He feels they want to keep "the truth" from reaching the people.

At a forthcoming meeting of the journalists' guild Emilio plans to urge acceptance of a compromise, but the opposition group, led by one Arturo Latta, keeps him from the meeting by having thugs attack and beat him. However, he is found by two guild members and gives one of them his speech to read. So it happens that a cub reporter rushes to the meeting and reads Emilio's words:

> PACO: Let me speak bluntly. I accuse Arturo Latta of wishing to prolong this strike in order to aid and abet the purposes of the Communist Party! A party directed from foreign soil outside this hemisphere. He is in truth serving not our cause but theirs! His leadership must be rejected.

The speech denounces Latta's "duplicity," as Emilio, his forehead bandaged and shirt bloody, is helped in. His compromise proposal is approved. In a closing narration Emilio says no union can afford the Arturo Lattas and their foreign influence.[16]

Along with warnings against foreign influence, the U. S.-subsidized *El Periodista!* sometimes cautiously promoted the Alliance for Progress, as in an episode about a schoolteacher, Alma Gonzales. Her school is so poor that sheets of paper must be torn in bits and pencils halved to provide all children with writing materials. She herself does not receive a living wage, so she supplements her earnings with evening work in the newspaper file room. This also enables her at night to take a few pieces of paper, pencils, and paper clips to use in her class next morning, but a night watchman catches her at this and she is arrested, to be tried as a thief.

Emilio, the crusading editor, resolves to defend her. He pleads her case with the Minister of Education, who speaks sadly of the "impossible

16. "The Strikebreaker," *El Periodista!,* U. S. Information Agency.

odds" facing schools because of the poverty of his country. But Emilio tells him:

> EMILIO: There is a glimmer of light on the horizon. We now have a national plan which is a result of the Alliance for Progress.

It will help pay for milk "as well as pencils and paper," Emilio tells the Minister of Education, who apparently knew nothing about all this.[17]

Each *El Periodista!* drama had two or three breaks for commercials. Like all USIA productions, the series could not by law be shown in the United States. Many American taxpayers would probably have been appalled at the use made of their money.

Belief in private enterprise was a principal theme of United States propaganda abroad. Ironically, a project like *El Periodista!* tended to suppress private enterprise among our allies. Although the station manager welcomed the free offering, which enabled him to serve a sponsor—perhaps an American affiliate—at a good profit, the native film producer found himself pushed to the wall by competition of this sort, added to the formidable competition of the telefilm. The native voice was effectively shunted aside—perhaps into commercials, perhaps into fringe media.

Not surprisingly, the American Motion Picture Export Association could tell its members: "Very little TV material is filmed in Argentina." "There is no TV film production in Chile except news flashes and TV commercials."[18]

Prime time, in scores of countries, was becoming a psychological bond uniting like-minded people. It was a state of mind, a refuge, a fortress against change.

American prime-time devotees, like those abroad, accepted the idea that what they saw and heard during the evening hours was mostly "entertainment." It was so accepted because it confirmed their basic beliefs and view of the world. Its message, implicit or explicit, was almost always a confirming one. This was true of most drama and most comedy.

Occasionally a comedian might twit the President. Jack Paar observed that President Johnson had a unique speaking style. "He always seems to be speaking under our heads." This shocked some viewers, yet brought no issue into play. *Rowan & Martin's Laugh-In,* which exploded into popularity early in 1968 with a kaleidoscope of momentary jokes and black-

17. "Alma Gonzales," *El Periodista!*, U. S. Information Agency.
18. *International Television Manual,* 1968.

outs, went an inch further, but the jokes were gone before their meaning —if there was meaning—could be assimilated. A *Laugh-In* mock broadcast proclaimed: "LBJ REVIEWS U. S. TROOPS—IN CANADA." The Smothers Brothers went still further, jabbing at American militarism—as in their scheduling of "Waist Deep in the Big Muddy." Their sallies, if they survived network surveillance, brought cheers but also furious mail.

The comedians who dominated television year after year did none of this. They were court jesters who knew the line.

Bob Hope, a great man in radio and television for thirty years, played golf with Presidents and was a perennial toastmaster at award ceremonies. His one-liners, enlivened by an ever-present leer, had earned him fantastic wealth. A 1968 *Fortune* survey of millionaires put him in the $150 to $200 million group.[19] He had begun entertaining troops in World War II, and had never stopped. During the Vietnam war his appearances were filmed and became the substance of prime-time telecasts at home and a potent element of support for the American involvement. In December 1967, as domestic discord over the war was approaching a climax of indignation, Hope was on one of his Vietnam visits, bringing roars from 12,000 troops at Danang. His jokes—war after war—seemed to proclaim that all were the same war. "I hope your grandfather heard me at Appomattox. I was great." He also said: "I'd rather be a hawk than a pigeon." On his 1967 visit he was accompanied by singer Phil Crosby, son of Bing Crosby, who had been Hope's comedy teammate during World War II on numerous films and *Command Performance* broadcasts. In the Hope monologues, hammered out by half a dozen writers—one of whom had been with him since the 1930's—some jokes were reassuringly everlasting. Words that had gotten laughs for decades, critic Vincent Canby noted, were *freeway, smog, girdle, Crosby.* Occasional jokes reached into the headlines.

At Danang Bob Hope and Phil Crosby came out wearing long-haired wigs and carrying anti-war signs, and got roars of laughter. "Don't worry about the riots in the States," Hope told the troops. "You'll be sent to survival school before you go back there."

There was no note of bitterness in this, but it left no doubt where he stood. The long-haired people, with signs and so on, were fools or dupes. They were outsiders, not part of the prime-time world.[20]

19. *Fortune,* May 1968.
20. New York *Times,* December 19, 1967; April 23, 1968.

On April 1, 1968, as America was agog over the withdrawal of Lyndon Johnson, Jack Benny was on television with Lucille Ball. She had been queen of comediennes for almost two decades. The Desilu she had built with Desi Arnaz had been sold to Gulf & Western—the new owners of Paramount—for $17 million. Jack Benny had been a ratings leader even longer than Hope. For thirty-five years he had refined the same jokes. One concerned his famous $100 Stradivarius—"it's one of the few *ever* made in Japan." Even more celebrated was his miserliness. In the program with Lucy she worked in a bank which wanted a celebrity depositor, and her job was to persuade Benny to give up the cavernous underground vault beneath his house where he guarded his accumulated hoard of millions. For this program the Paramount scenic crews constructed an underground defense labyrinth that would have stymied the heroes of *Mission: Impossible*.

All of this—fortunes, success, glittering decor, beautiful people, and the jokes that kept them all in the public mind—were a continuing celebration of the American way. Viewers who stuck to prime time could dwell in that splendid world. Outside its orbit—pushing into newcasts but seldom beyond—were shouters, protesters, attackers. The very sight of them aroused deepening fury. With the withdrawal of Johnson, many viewers already knew what they wanted. A man who had stood up to Khrushchev could take care of the hippies.

SHOOT-OUT

At the start of 1968 Richard Nixon already seemed certain of the Republican nomination, and this was a miracle. In 1962—two years after defeat by John Kennedy—he had tried to reactivate his career by running for election to the governorship of California but had been soundly defeated by the incumbent Democratic governor, Edmund ("Pat") Brown. It was perhaps the most ignominious event in Nixon's political life, for he considered Brown a second-rater. When the outcome became clear at Nixon headquarters, his press aide Herbert Klein read a concession statement to reporters, and said the candidate would not be available for questions. But as he spoke there was commotion in the corridor and Nixon pushed in, surrounded by red-eyed followers. Quivering, he vented his frustration. "As I leave you, I want you to know, just think how much you'll be missing. You won't have Nixon to kick around any more. . . ." Leaving,

Batman

Twentieth Century-Fox

"Go to it, Batman . . .!"

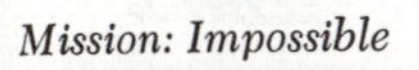

Mission: Impossible

Paramount

". . . will self-destruct in five seconds . . ."

Lost in Space

Twentieth Century-Fox

Land of the Giants

Twentieth Century-Fox

A hostile universe . . .

1968: VIOLENT DENOUEMENTS

Wide World Photos

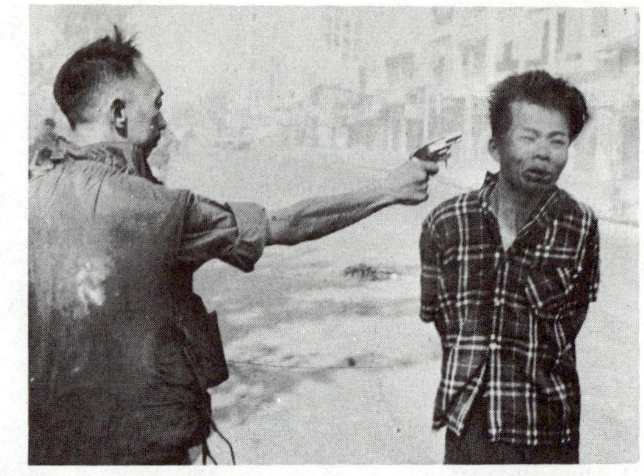

Saigon police chief Nguyen Ngoc Loan executes a Vietcong suspect, then tells photographer: "Buddha will understand." Some Congressmen considered television use of photo in bad taste.

Funeral of murdered Martin Luther King, seen by viewers throughout the world. At right, among the mourners, Robert Kennedy, Eugene McCarthy, Mrs. King, daughter Bernice.

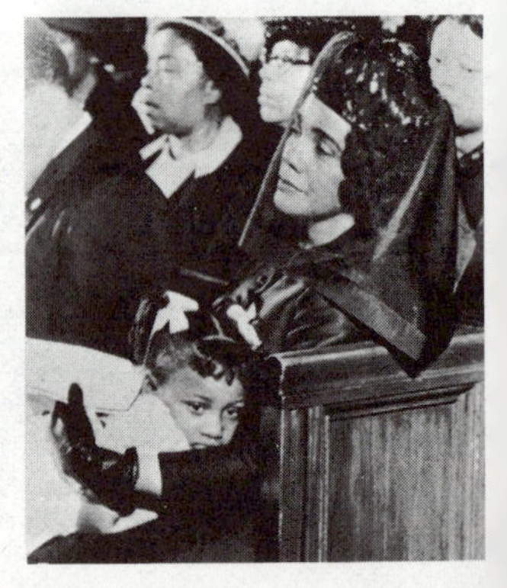

Wide World

As tribute to murdered Robert Kennedy, Democratic convention sees memorial film, adapted by Charles Guggenheim from television campaign film.

NBC

United Press International

NBC

NBC

Tapes, films, photographs arriving at network convention centers depict scenes of violence in Chicago streets.

1965: Early Bird communications satellite, owned by INTELSAT, managed by COMSAT.

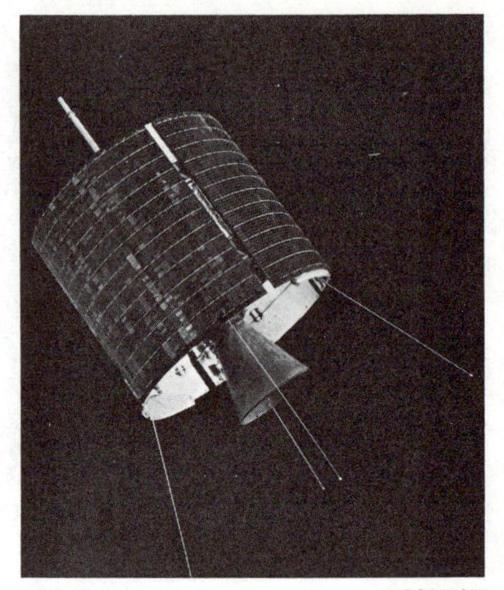

COMSAT

COMSAT

1968: Earth station for satellite communication at Etam, West Virginia.

Wide World

1968: A spot announcement from outer space. Donn F. Eisele, left, and Walter M. Schirra, Jr., right, aboard Apollo 7.

he muttered to aides, "It had to be said, it had to be said."[1] It was considered his farewell to politics.

He went into private law practice. In 1964 he said on *Today* over NBC-TV that he was not a candidate for the presidency but might accept a vice presidential nomination if his party wished him to.[2] Then he went on a world tour, described as private. He was accompanied by a Pepsi Cola executive and was serving as an international envoy for the company; he is said to have ironed out some ticklish Pepsi Cola problems in Taiwan. But it was also a chance to visit world leaders and was regarded by some observers as a try at re-establishing himself as a major figure. The wire services did not feel it worthwhile to send a reporter with him, nor did newspapers. Only NBC News felt the tour called for coverage and sent correspondent Herbert Kaplow and a camera crew with Nixon and the Pepsi Cola entourage. Occasional filmed items and statements appeared in NBC newscasts, generated some press attention, and helped Nixon to maintain a continued visibility. During his tour he called for vigorous military action against North Vietnam and rebels in Laos. On his return he backed Goldwater, which may have been crucial to his fortunes. Republican leaders who shunned Goldwater, as Governor Nelson Rockefeller did, aroused much party resentment. The defeat suffered by Goldwater brought party stalwarts back to Nixon.

Echoes of old days stirred briefly when Checkers, the cocker spaniel made famous by a 1952 Nixon telecast, died and was buried at the Bide-a-Wee pet cemetery with the headstone, "CHECKERS, 1952–1964."

While fury over the Vietnam war raged around Johnson, Nixon could afford to say little or nothing on the subject; it was not his war. He spoke widely at Republican fund-raising rallies.

Meanwhile the Democrats were tearing each other apart. In the face of apparent Johnson immobility, attacks on his war policies grew constantly angrier.

Among the angry men was Martin Luther King. His nonviolent crusade had seemed to move within sight of the promised land, but now he saw his cohorts breaking. Wartime inflation was widening the gap between rich and poor. Among black youth, said King, unemployment ran as high as 40 per cent. On newscasts, voices for "black power" outshouted the pleas for nonviolence. None of the causes of riots, warned King, were

1. Washington *Evening Star*, November 8, 1962.
2. *Today*, NBC-TV, March 6, 1964.

being solved. "Why have we substituted the arrogant undertaking of policing the whole world for the high risk of putting our own house in order?" He saw the United States as the chief generator of violence around the world. He planned a Poor People's March on Washington, for white and black. Against growing odds, he said it must be nonviolent.

On April 4, 1968, he walked out onto the balcony of a Memphis motel and was felled by a sniper. A bullet ripped into him, throwing him against a wall; blood spurted from his neck as he went down. Riots and looting quickly broke out in all major cities. Men and women wept—over the loss of a loved leader and over the tidal wave of violence it had let loose. In South Vietnam servicemen were described as stunned, although one said: "We have three hundred people dying here each week." In Saigon the Armed Forces Radio and Television Service canceled rock 'n' roll programs and substituted music by Morton Gould, Montavani, and Kostelanetz—quieter, but scarcely a tribute to King or his people. On the hour AFRTS brought bulletins, telling of troop movements from various locations to Washington, Chicago, and other danger areas. American embassies and libraries in many world capitals were confronted by anti-American demonstrations. In Berlin protesters rioted outside the U. S.-controlled radio station RIAS, until they were driven off by police clubs and "water cannon."[3]

Once more television audiences throughout the world participated in an extraordinary funeral. In the course of a seven-hour telecast, sent across the ocean by satellite, they joined mourners in an Atlanta ghetto church, and among them glimpsed faces they knew—Jacqueline Kennedy, Robert Kennedy, Ted Kennedy, Richard Nixon, Hubert Humphrey, Eugene McCarthy, and many of the entertainment world. Then they watched spellbound as King's coffin, on a farm wagon drawn by two Georgia mules, moved toward the cemetery followed by crowds on foot—crowds estimated at two hundred thousand people.

The presidential campaign resumed—on television, a thread of frenetic counterpoint to "regular" programming. Robert Kennedy had decided to run. The Eugene McCarthy showing in the New Hampshire primary had clinched his decision—to the dismay and anger of McCarthy supporters. But Kennedy felt that the quiet-spoken, sometimes mystical McCarthy could not be elected. The need of the hour, as Kennedy saw it, was to unite the disaffected—the young, the black, the poor—and Kennedy felt

3. New York *Times*, April 8, 13, 1968.

he could. During the next two months he was constantly in motion. Day by day television newscasts showed glimpses of him—hair wild and allowed to grow longer—as he pushed through frenzied crowds of people screaming, jumping, wanting to touch him. He responded with an ecstasy of his own. In his entourage now were people like Jack Newfield, a youth twice jailed in civil rights demonstrations, who had once picketed Attorney General Robert Kennedy but now felt that Bobby had discovered his true character and destiny. Clearly Kennedy, who had a capacity to arouse hate, also had a capacity to stir extraordinary hopes. Behind the growing euphoria was hard-driving organization. Once more the cameras of the talented film producer Charles Guggenheim were at work: delirious, touching campaign moments were made into television spots. The "humanizing" film of 1964 was updated, and hundreds of thousands of dollars were spent on television time. An adviser told Kennedy he could afford to spend even more on television if he weren't so rich. "You're too rich not to get out there and mix." His aide Richard Goodwin told him: "You have to go out there and do it all and you have to show that you don't have contempt for them, that you value who they are." Some felt that Robert Kennedy, in the process of doing this, was reborn.[4]

On the night of June 5, in a back corridor of the Ambassador Hotel in Los Angeles, surrounded by jubilant supporters celebrating his victory in the California primary, Robert Kennedy was shot down. A guest screamed: "Not *again!*" As celebrities crowded around the sprawled figure, a member of the ABC-TV production staff, William Weisel, found he too was wounded, bleeding. Within minutes the scene of pandemonium and consternation was on television, as the unconscious Kennedy was rushed to a hospital. Later a solemn President Johnson appeared on the home screen.

> JOHNSON: Let us put an end to violence. . . . Let us begin in the aftermath of this great tragedy to find a way to reverence life, to protect it. . . .

He said he had appointed a commission headed by ex-President Eisenhower's brother, Milton Eisenhower, to study the phenomenon of rising violence and find out "what in the nature of our people and the environment of our society makes possible such murder and such violence."

The following day Arthur Schlesinger, Jr., commented: "We are today

4. Halberstam, *The Unfinished Odyssey of Robert F. Kennedy*, p. 29; interview, Charles Guggenheim.

the most frightening people on the planet." He considered the Vietnam war the main factor in the trend to violence but noted that "the atrocities we commit trouble so little our official self-righteousness, our invincible conviction of our moral superiority." Some attacked the indiscriminate sale of guns; others, the role of television as merchandiser of violence. The attorney Morris L. Ernst, interviewed on television on *The Merv Griffin Program*, said: "We're being murdered by TV, not by the guns." Networks began a frantic pruning and juggling of programs. Scheduled segments of *Bonanza, Mannix, The Avengers, Big Valley* and *The Guns of Will Sonnett* were canceled. The producer of *Get Smart* blue-penciled a number of scripts ready for production. Scripts of *The Mod Squad* and *It Takes a Thief*—in production for the fall—were likewise revised. Network spokesmen said again that there was no demonstrated causal connection between entertainment and crime, but many in the industry no longer felt reassured by such statements. Several hundred Hollywood actors and writers signed a published statement in *Variety* and other trade papers: "We will no longer lend our talents in any way to add to the creation of a climate for murder. . . ."[5]

Once more, hundreds of millions of television viewers witnessed a ritual beyond belief. They saw the center of New York City paralyzed as millions passed the bier in St. Patrick's Cathedral. Then came a fantastic migration to the interment in Washington. A camera poised at a Pennsylvania station escalator in New York watched celebrity faces, in close-up, drift by in dream-like motion—political leaders, social leaders, media leaders. Among familiar show-business faces were Leonard Bernstein, Kirk Douglas, Shirley MacLaine, Kim Novak, Jason Robards, Jr. Crowds of watchers were seen lining the train route. In Washington, cameras showed the cortege pausing at the floodlit Lincoln Memorial while a marine band played and a group of the Poor People's March—now led by the Reverend Ralph Abernathy, King's deputy —watched quietly. It was close to midnight when the funeral ended at Arlington Cemetery.

Then programs, commercials, and campaigns resumed. Richard Nixon was duly nominated at a convention in Miami. In his acceptance speech viewers found an uncanny resemblance to the oratory of Martin Luther King. Where King had said, "I have a dream," Nixon said, "I see a day," and he saw a day eight times. The same kinds of visions were used—an attempt, apparently, to offset the conservative-Southern aura of men

5. New York *Times*, June 6, 17; *Wall Street Journal*, June 17; *Variety*, June 12, 1968.

around Nixon, as exemplified by the selection of Spiro Agnew as vice presidential candidate.

But greater tension surrounded Chicago, where the Democratic convention would be held weeks later—in August. City authorities anticipated trouble and prepared for it relentlessly. Police received special riot-control drills. Fences and barbed wire were installed around the convention area. As television engineers began their wiring preparations, they felt they were in an armed camp.

For many Americans, especially the young, the democratic process itself was facing a test. To many it seemed that issues were being settled by a shoot-out. However, there was still Eugene McCarthy.

He confounded observers. His campaigning remained easy-going—some said indolent. The delirium that had surrounded Kennedy campaigns was missing. Followers who hoped for anti-Johnson invective were disappointed. McCarthy seemed almost to try to dampen audiences with broad philosophic arguments. On a California telecast he said that America during the 1950's and 1960's had built up for itself a mission "to judge the political systems of other nations—nearly all the other nations of the world—accepting that we had the right to interfere with all those systems if we found them wanting."[6] He said all this with such calm detachment —without teleprompter—that it was difficult to realize he was setting aside assumptions that had governed American foreign policy for two decades.

The absence of theatrics, though troublesome to politicos, was the delight of others. Students came to him by thousands to stuff envelopes, lick stamps, campaign from door to door. Polls showed him favored over Nixon or Humphrey. A New York poll indicated he could carry the state over any Republican candidate, but that any other candidate would lose to Nixon. The vast support of the young—a children's crusade, some called it—seemed to be pushing the McCarthy drive beyond hopes or expectations.

But the delegates heading for Chicago included many politicians who had backed Johnson war policies and were not eager to repudiate their own records. Although Kennedy and McCarthy—the anti-war candidates —had polled 80 per cent of the Democratic votes in the primaries and had far out-polled war supporters, it was clear long before Chicago that many delegates leaned to Hubert Humphrey; he, too, had gone down the line with Johnson.

6. Quoted, New York *Times*, August 25, 1968.

It looked as if voters would have a choice between a Nixon who spoke for law and order; a Humphrey who spoke for law and order; and a third-party candidate, Alabama ex-Governor George C. Wallace, who was also for law and order—but more avidly and violently than any other candidate. When an ABC-TV cameraman happened to film Wallace shaking hands with Ku Klux Klan imperial wizard Robert Shelton, Wallace demanded the film. When it was refused, he told his bodyguards, "Take it!" They closed in and took it. ABC News director Elmer W. Lower wired a protest, but the incident was soon forgotten.[7]

Throughout the early months of 1968 the nation seemed to be heading for a climax of violence. The accumulating tension was perhaps most vividly reflected in the Sunday evening *PBL* series. Again and again it arranged confrontations in which youth activists, angry black leaders, and other dissidents had their say. One black leader said:

> HADEN: If you won't listen to me when I make an appeal for the Negroes . . . listen to me when I make an appeal for America. You claim you love America. Well, we love America. But you're driving us back and you're making a Samson out of us, and we are going to pull down the pillars.
>
> Should you try to pretend that I'm crazy because I want America to be saved?

The same *PBL* program showed suburban housewives of Dearborn, Michigan—mothers and grandmothers—getting pistol instructions from a city instructor. It showed him presenting them with pistol manuals. "Keep them with your cookbooks. They make a pretty good reference if you ever need anything like this." He supervises their target practice, explaining the targets are the same as those used by police and FBI. "Study the target, and you'll see what they consider to be the more vital areas." Some of the ladies feel strange about it all. "But if it's necessary—" says one lady. "I mean, you know, you do what is necessary." The gun, the *PBL* narrator tells us, is suburbia's tranquillizer.[8]

In June the hundred thousand people in the Poor People's March were building a shanty city—Resurrection City—in Washington, D. C. Its leaders visited government agencies and made statements on television newscasts but gained nothing.

In August a few of them straggled to Chicago, converging with other

7. New York *Post*, June 28, 1968.
8. *PBL*, NET, March 3, 1968.

groups of dissidents. These apparently came with diverse motives. Some no doubt hoped that a massive public demonstration would stop Humphrey and turn the tide toward Eugene McCarthy—or perhaps Ted Kennedy—and a repudiation of the Vietnam war. Some had no illusions and came to disrupt. The notion that a good cause needed "dirty tricks" was not their invention. Seriousness, hatred, and wild comedy bubbled into bizarre mixtures in the Chicago migration. In a North Side drinking spot a young man posted a notice asking for female volunteers to march in a "Bare Breasts for Peace Parade" to the Democratic convention hall. The notice was a prank but its writer received calls from more than thirty women offering to doff their brassieres to help stop the war.[9] A few young people came to Chicago because it seemed, suddenly, the place to be. The migrants numbered less than ten thousand, but seemed likely to make news—if only because of the huge police power mobilized to confront them.

The hundreds of television people who arrived with several hundred tons of television equipment—each network brought dozens of color cameras plus control equipment, mobile units, videotape machines, microwave transmission systems, receivers, paging equipment—faced extraordinary problems. The arrangements made by the city under the direction of Mayor Richard J. Daley, partly for "security" reasons and partly because of union difficulties, virtually confined live coverage to the hall itself. Demonstrators were to be kept in areas remote from the convention hall, such as Grant Park and Lincoln Park, where they were apparently to demonstrate for their own satisfaction. Television coverage of any demonstration would have to be done mainly by videotape mobile units or by cameramen on foot, with film and tape being rushed by courier to the convention hall facilities. Use of film would involve additional time-lags for development. The regulations posed an additional obstacle by forbidding mobile units to park on streets adjoining downtown hotels, the headquarters for various delegate groups.

Broadcasters became convinced that the regulations were a pro-Humphrey maneuver, designed to ensure that delegates would not be stampeded by mass demonstrations for a peace candidate. Under the Daley arrangements the convention would be completely insulated from demonstrators. The television viewer would be aware of them only if networks were able to film and tape their activity and intercut it with

9. New York *Times*, August 18, 1968.

convention hall proceedings. *Variety* predicted that the arrangements would clinch Humphrey's nomination "but contrariwise would cast a pall over his election chances . . . his victory at Chicago is bound to look like a political steamroller."[10]

The arrangements caused broadcasters and convention management to have an edgy relationship from the start. This was aggravated when NBC newsmen were found to have planted an electronic eavesdropping device in a secret platform committee meeting. Their action was at once repudiated by network executives, but it may have helped them learn that the platform, in its first draft, contained a clause denouncing television violence. Network pressure finally eliminated the clause.[11]

Network newsmen encountered even more hostility when they moved their mobile units toward police lines facing multifarious protesters near Grant Park. The police clearly considered the cameras an unwelcome presence, but the protesters cheered them. Taunting the police, they shouted, "The whole world is watching! The whole world is watching!" As a long, sleek NBC mobile unit, with camera and men perched on its roof, pushed toward the area of confrontation, the protesters continued to goad the phalanxes of policemen. "Sieg heil! Sieg heil!" Demonstrators were resolved on a protest march to the convention; police, who outnumbered them and were armed with nightsticks, guns, tear-gas, and mace, and reinforced by National Guard units, were resolved to stop them. Groups began to form marches and to attempt to bypass police lines at various points to enter the Loop area.

The film and tape arriving at the television control rooms suddenly began to look like slaughter. Networks threw it on the air as fast as they could. Viewers saw a dizzying kaleidoscope: nominating speech, headcracking, speech, tear gas, shouting crowd, wounded, paddy wagons, balloting, ambulances, speeches. Viewers sensed that the delegates were unaware of the bloody events; then, suddenly, that they were becoming aware. Rumor swept through the hall. Some got the news via phone calls from downtown hotels. Some listened on transistor radios or left to look at monitors. The convention managers plowed ahead with proceedings. Some delegates wanted to recess, but protests and questions were ruled out of order. The hall became a wild rumble. Senator Abraham Ribicoff of Connecticut, in the midst of a nominating speech, digressed to speak

10. *Variety*, July 31, 1968.
11. New York *Times*, September 2, 7, 1968.

of "turmoil and violence competing with this great convention for the attention of the American people." As he spoke bitterly of "Gestapo tactics in the streets of Chicago," television cameras focused on a furious Mayor Richard Daley, in close-up.

Among the hundreds injured in the melees were twenty-one reporters and photographers, clubbed by police while trying to cover the events. They included CBS-TV cameraman Delos Hall, who said that he was filming police action when he was clubbed from behind, and that he was then attacked by several policemen while he was lying on the ground; NBC-TV cameraman James Strickland, who was struck in the face when he tried to photograph the action involving Hall; and NBC's Aline Saarinen, who was apparently felled by mace. Numerous victims reported that uniformed policemen, contrary to regulations, had removed their badges and name-plates before going into action.[12]

Violence spread to the hall itself; several scuffles broke out. Television viewers saw CBS newsman Dan Rather assaulted by security police. The flare-up was rapidly suppressed, but not before anchorman Walter Cronkite had expressed fury.

As the convention rumbled on, bitter charge and countercharge overshadowed the proceedings. Frank Sullivan, information officer for the Chicago police, spoke angrily of the demonstrators as "allies of the men who are killing American soldiers." Dismissing the indignation against Mayor Daley, he said: "The intellectuals of America hate Richard J. Daley because he was elected by the people—unlike Walter Cronkite."[13]

The world saw Hubert Humphrey nominated in what appeared to be a fortified stockade, and guessed he had very little chance of election. Choice of the highly respected Senator Edmund S. Muskie as his running mate improved his chances.

Richard Nixon seemed to his campaign advisers to be sure of election. But they felt that his campaign had to be managed with care. In a strategy session, Philadelphia television producer Roger Ailes was reported as saying:

> AILES: Now you put him on television, you've got a problem right away. He's a funny looking guy. He looks like somebody hung him in a closet overnight and he jumps out in the morning with

12. New York *Times*, August 28, 1968; Associated Press, August 29, 1968.
13. New York *Times*, August 30, 1968.

> his suit all bunched up and starts running around saying, "I want to be President." I mean this is how he strikes some people. That's why these shows are important. To make people forget all that.[14]

Debates were ruled out. In various cities the campaign managers arranged telecasts in which Nixon faced a panel of questioners in an arena-like setting, with an audience seated in circular tiers. An important element in the staging was that Nixon had no lectern; he seemed totally unprotected. Alone before large numbers, he seemed to face enormous risks—but the questioners and audience members had been carefully chosen. The audience cheered the answers, and knew it was to mob Nixon enthusiastically at the close of the program. Nixon also made numerous pre-taped radio and television talks and spots. Huge issues were compressed into pellets of confidence.

> NIXON: The administration has struck out on keeping the peace abroad, on keeping the peace at home, on providing prosperity without inflation. I say three strikes are enough. Let's get a new batter up there!

On foreign affairs:

> NIXON: The American flag is not going to be a doormat for anybody at home or abroad in the next administration!

On poverty:

> NIXON: Instead of more millions on welfare rolls, let's have more millions on payrolls!

On Vietnam:

> NIXON: We need new leadership that will not only end the war in Vietnam but keep the nation out of other wars for eight years![15]

Nixon appeared so certain of victory, said court jester Bob Hope, that Whittier, California—Nixon's home town—had already started building a log cabin for him.

It was a grim and weary campaign. After the shoot-out period of the primaries and the Chicago convention, the campaign was anti-climax. The young people had gone home. The Republicans spent over $25 million on television and radio; the Democrats, more than $15 million.

14. McGinniss, *The Selling of the President 1968*, p. 103.
15. Quoted, New York *Times*, October 17, 1968.

Broadcast expenditures had almost doubled since 1964, tripled since 1960, quadrupled since 1956.

Yet in 1968 40 per cent of eligible voters stayed away from the polls —the highest abstention rate since 1956. Richard Nixon, elected by about 27 per cent of the eligible voters, faced a nation in which faith in government processes had been dangerously eroded.

It had been a violent year; inquiry into causes and possible remedies was the order of the day. The Milton Eisenhower commission, which had acquired the title National Commission on the Causes and Prevention of Violence, was studying television programming as a possible factor. Senator John A. Pastore, chairman of the Senate subcommittee on communications, was planning hearings on television violence and sex. The U. S. Public Health Service was studying mental health implications of television brutality.

All this induced continual agitation at networks and film studios during the 1968–69 season. Networks scrapped some of the violent cartoon series still crowding Saturday morning schedules. NBC installed a live animal series and a children's quiz. Scripts for telefilms continued to be pruned of expendable mayhem. Hollywood's two hundred stunt men, who lived by being hurled from balconies and automobiles, encountered a sudden decline in employment. Mass fights were reduced to a punch or two. The shoot-out on a dusty street that had opened *Gunsmoke* for years was scrapped for a quiet opening; the west was tamed again. Though the counting of acts of violence had never seemed a valid procedure to CBS, the network adopted it and announced a 30 per cent reduction "in the number of violent incidents in our prime-time programs" for the following season. Quieter forms of murder were now favored, like the "Spock pinch" on NBC-TV's *Star Trek,* devised by Leonard Nimoy, the actor playing Mr. Spock, a Vulcan. "I decided that the Vulcans knew so much about the human anatomy that they could knock out an enemy just by pinching a nerve in the neck and the shoulders." In commercials on children's programs, war toys were yielding to science toys, with emphasis on space science.[16]

Soul-searching went with all this. Violence had always made for "easy" writing, said television writer Robert Sabaroff at a WGA-w discussion. "Violence is our obscenity." Instead of creating characters, it was always

16. *Variety,* September 25, 1968; New York *Times,* July 25, November 25, 1968; *Saturday Review,* July 12, 1969.

easier to "pick a label that evokes antagonism—cannibal, renegade, bandit, commie, nazi, Chinese Red. . . . We have all shared the corruption from the beginning."[17]

Many people thought it all ridiculous. "They tell me," said Efrem Zimbalist, Jr., star of *The FBI*, "we're going to do the show nonviolently. How they expect to do a show about violence without violence seems idiotic to me and the best way to drive the audience away."[18]

Some producers felt it would all blow over soon. They hoped so; they had other matters on their minds.

MOONSCAPE

Among the industry's concerns were the most extraordinary spectaculars ever conceived, and an anniversary. In 1970 it would be fifty years since KDKA had broadcast the Harding-Cox election returns and precipitated the epic adventure of commercial broadcasting. For some it had been a gold rush that had never stopped, intertwined with power and glory. While planning a new season—and fending off interference—the industry was also planning to spotlight the birthday of "the greatest communication system ever devised."

Meanwhile there were the spectaculars, which would take men outside their present habitat to a climax on the moon. From the start it was all planned as a series of television shows.

On-the-air tests of the Apollo spacecraft, which also functioned as television studio and transmitter, began late in 1968. During the first manned Apollo flight—163 times around the earth—television viewers shared the view of earth from orbit, and became acquainted with the euphoria that seemed to overtake men in weightlessness. Floating around their studio-capsule, Walter M. Schirra, Jr., and Donn F. Eisele clowned for the camera and held up a sign: "KEEP THOSE CARDS AND LETTERS COMING IN FOLKS."[1]

Late in December, during the first manned flight around the moon, viewers shared with astronauts the sight of its bleak and forbidding surface. "Looks like plaster of Paris or sort of grayish beach sand," said James A. Lovell, Jr., casually across space. Then on Christmas Eve, from the vicinity of the moon, the men staged a reading of Genesis. It was begun by William A. Anders.

17. *Variety*, August 27, 1968.
18. *TV Times*, May 18–24, 1969.
1. Apollo 7 telecast, October 14, 1968.

ANDERS: In the beginning God created the heaven and the earth. And the earth was without form, and void. And darkness was upon the face of the deep. . . .

After a few lines, James Lovell took up the reading.

LOVELL: And God called the light day, and the darkness He called night. . . .

Frank Borman ended it.

BORMAN: . . . and God saw that it was good.[2]

It was reminiscent of another Christmas Eve—1906—when wireless operators on ships in the Atlantic, accustomed to hearing dots and dashes on their earphones, were flabbergasted to hear, instead, a human voice. It was Reginald Fessenden reading a Bible passage in Brant Rock, Massachusetts. Pursuing his determination to transmit the human voice by wireless, he was testing his equipment with a Christmas message, and astonished his hearers as much as any voice from outer space would have done. The Bible reading from the astronauts served as reminder of how much had happened in those few decades.

For his six-year-old son, Anders mentioned from outer space that he had seen Santa Claus "and he was heading your way."

In March 1969 astronauts tested a spidery "bug" of the sort that would later take men to the moon's surface from an orbiting spaceship. During this flight Russell L. Schweickart also spent forty minutes outside a spacecraft taking pictures while doctors in Houston studied the curve of his heartbeat and other biological data sent by radio transmitters inside his space suit. In May astronauts took a bug to within eight miles of the moon, and viewers watched its departure from the spaceship via a color camera on the spaceship. The camera had been built by Westinghouse, the company that had launched and still owned KDKA.[3] As the men returned earthward, Eugene Cernan said, with a backward glance at the bleak surface behind them, "This moon of ours—had a rough beginning somewhere back there."[4]

But now it was July and time to make history. A bug named Eagle would take two men to the moon—to an area known as the Sea of Tran-

2. Apollo 8 telecast, December 24, 1968.
3. Ironically, the camera used the abandoned CBS color system with a scanning wheel, more compact than the RCA all-electronic system in use on earth.
4. Apollo 10 telecast, May 24, 1969.

quillity. With astonishing and frightening confidence, the planners released an advance scenario. A week before the event, viewers could read details in their newspapers.

> The greatest show in the history of television begins when Armstrong starts down the nine-rung ladder leading from Eagle's hatch to the surface. When he reaches the third rung from the top, the astronaut will reach out with his left arm and pull a D-shaped handle, opening a storage bay and exposing the lens of a black-and-white TV camera.
>
> In 1.3 seconds, the time it takes light to reach the earth, we will see Armstrong's legs carefully moving down the ladder.
>
> A moment later, men on earth will see man walking on the moon.[5]

Everything went by the script. Even words were ready beforehand. The occasion needed a touch of magic, like the words used by Alexander Morse to click the first telegraph message from Washington to Baltimore in 1844: "What hath God wrought?" But the NASA word experts were not up to its technicians. Neil A. Armstrong, standing on the moon, said awkwardly:

> ARMSTRONG: That's one small step for a man, one giant leap for mankind.

To viewers who had grown up on the radio serial *Jack Armstrong, All-American Boy* it seemed right that an Armstrong had done it. Later they saw Edwin Aldrin, Jr., join him. Together they set up a second television camera—these were both RCA-built—and viewers on earth saw the two men hop around, almost like kangaroos, gathering rocks. Then President Nixon appeared on a split-screen and became a telephone talker, from the Oval Room of the White House. The other half of the screen showed Armstrong and Aldrin in front of their spidery bug. The President said, "This certainly has to be the most historic phone call ever made." Then conversation gave way to statement.

> NIXON: Because of what you have done, the heavens have become a part of man's world. And as you talk to us from the Sea of Tranquillity, it requires us to redouble our efforts to bring peace and tranquillity to earth. For one priceless moment in the whole history of man, all the people on this earth are truly one—one in their pride in what you have done and one in our prayers that you will return safely to earth.

5. New York *Daily News*, July 15, 1969.

> ARMSTRONG: Thank you, Mr. President. It's a great honor and privilege for us to be here representing not only the United States but men of peace of all nations. . . .[6]

Later they planted an American flag made of nylon and stiffened at the top with an aluminum rod so that it would seem, on television, to blow in a nonexistent breeze.

When Armstrong and Aldrin, after two hours and twenty-one minutes, re-entered the bug to start the return trip, television and radio networks began picking up comments from points in the United States and abroad —including some behind the iron curtain. Most expressed awed, almost speechless wonder.

Comments from American scientists had a curious range. No one was quite certain why the feat had been undertaken. President Kennedy had said in 1961 that the nation "should commit itself" to put a man on the moon before the end of the decade, but Presidents make many declarations that are forgotten. Some said the attempt was necessary for the reason men climbed Mount Everest—"because it is there." Some said it was necessary to further the advance of knowledge in general, while others cited specific practical benefits, some of which seemed bizarre in the context of the moment; it was said that space travel was aiding development of heat-resistant paints and bath-tiles as well as world-communication instruments. One scientist considered it good to know that no matter what a mess mankind made of the earth, men would always be able to go elsewhere.

This was a chilling reminder of unsolved problems. Many scholars were stressing the precariousness of "spaceship Earth." Space flights had dramatized the importance of the life-supporting environment. Yet the earth's own "support system"—soil, water, air—was being despoiled at a staggering rate, in large part by industrial development abetted by the mass media. With population growth unchecked, some feared that the destruction of the support system would become irreversible within a few decades, dooming life on earth. Already much of mankind—including many in the United States—lived in a state of deprivation.

The taxpayer had spent $25 billion on the Apollo ventures, perhaps $50 billion on the space program as a whole—a sum comparable to the Marshall Plan funds that had helped Europe back on its feet after World

6. Apollo 11 telecast, July 20, 1969.

War II. How far could the sums have gone toward a livable world? "We are $50 billion further behind, and ten more years," thought Professor Bernard Baumrin of the City University of New York, "and thousands of men mistrained, and professionally narrow. To be an expert in a subject for which there is no basic need is to be a human frill. . . ." To many scientists the scientific justifications seemed flimsy. In the opinion of Nobel prize winner Harold C. Urey the moon landing was to be thought of not as a scientific venture but as "something like the building of the pyramids or the Parthenon."[7]

This suggested a main function of prestige—as President Kennedy had said, "to beat the Soviets. . . ."

Perhaps that was why it had to be a television show. The emphasis on television had been deplored by some among the colony of scientists at Houston, as drawing attention from scientific pursuits, while others had defended it as assuring more congressional appropriations. But for world impact it was essential.

In relation to that purpose, the television spectacular was an epoch-making success. A key element was the American willingness to allow live coverage of what might turn into disaster. For hundreds of millions of people throughout the world, watching via satellites, the unspoken possibility of disaster must have been ever-present. Against the background of this peril, the calm assurance had an epic quality. To such men, could anything be impossible?

It was important that the operation had international aspects. The Apollo program involved "cooperative arrangements" with eighty countries—in tracking, splash-down facilities, satellite schedules, ground-station operations. All over the world, the moon landing took the Vietnam war off the front pages and—at least momentarily—out of the minds of men.

Among the half-million who flocked to Cape Kennedy—formerly Cape Canaveral—to view the blast-off were newsmen, diplomats, and celebrities of many nations. It was a strange assemblage, arriving by airline, private plane and helicopter, railroad, limousine, trailer, bicycle, and on foot.

There was even a mule-cart procession of the Poor People's March led by Ralph Abernathy, who came to demonstrate in protest. But according to a *Time* report, as he saw the huge Apollo lift into the air as from a

7. New York *Times*, December 31, 1968; September 7, 1969.

thundering, fiery cauldron, he forgot about poverty and prayed for the safety of the men. So it may have been around the world.

If an empire needed bread and circuses, here was the greatest circus in history. For some this was its glory, and for others its fatal flaw.

Lewis Mumford said: "It is a symbolic act of war, and the slogan the astronauts will carry, proclaiming that it is for the benefit of mankind, is on the same level as the Air Force's monstrous hypocrisy—"Our Profession is Peace."[8]

In the *Saturday Review* critic Robert Lewis Shayon said it more gently, but memorably:

> Wherever explorers go in the future accompanied by television cameras, they will be actors, making their nebulous exits and entrances for the benefit of multi-planetary audiences. Nowhere will there ever again be pure events (if ever there were); everything hereafter will be stage-managed for cosmic Nielsens, in the interest of national or universal establishments.[9]

MILESTONE

The moon landing created a scientific euphoria, especially among broadcasters. They tended to feel, approaching their semi-centennial, that great changes were in store for their industry. They tended to visualize wired local and regional systems linked by domestic as well as international satellites, giving any viewer push-button command of scores of channels. These would be able to light up the walls of his rooms with countless items—events, plays, games, statistics, documents, lectures—from innumerable sources, near and far, live and recorded. The industry expected increasing influence in many ways. Broadcasters were absorbing major publishing companies and, in so doing, acquiring control of many copyrights, with the assumption that books and libraries would firmly enter the electronic age and, in one way or another, fit into a computerized push-button world of instant command.

The sheer multiplicity in these visions was not, in itself, reassuring to everyone. Those favoring access to diversity—civil liberties enthusiasts, for example—noted that satellites, cable systems, computers all created new control points. The viewer might be given numerous choices—but

8. *Ibid.*, July 21, 1969.
9. *Saturday Review*, August 9, 1969.

among pre-programmed items. Who would program the computers, command the satellites?

Within the industry the word was *private.* It saw its own story as a glorious chapter in the private-enterprise saga. Not only had it served other businesses, helping them toward vast new markets; it saw itself at the same time as a glittering example of private enterprise in action.

And so it had been. Its pioneers had been youths working in bedrooms, attics, and basements. And though the action had long ago moved elsewhere, into very different environments, the industry still clung to its "private" image—as did the industries it served.

But this had curious aspects. To apply the word "private" to an industry that had as its main resource the publicly owned airwaves and whose dominant units—RCA, General Electric, Westinghouse, and others—had the federal government (mainly military and space agencies) as their biggest customer, and whose revenue derived in large part from tax-financed research, was to stretch the word "private" to strange lengths. All this suggests changes enveloping broadcasting and the world of big business, particularly in their relations with government.

In the decade before World War II big business and government were generally at odds, and the broadcasting world was a battleground. But the later cold-war years saw the rise of a business-government partnership to which President Eisenhower gave the name "military-industrial complex." Television prime time, at home and abroad, became an expression of this complex in its dreams, themes, and taboos.

In spite of this, the broadcasting industry and its leading sponsors thought of themselves as private. What they did and wanted was always recommended in the name of private enterprise—just as the Soviet Union did everything in the name of socialism.

This illustrated another oddity. World conflicts had become a struggle between power-blocs which proclaimed different principles but increasingly resembled each other. Although differences were obvious enough, similarities were becoming more notable. In the Soviet military-industrial complex its powerful commissars of industry and armaments and communication seemed no different—even in style and appearance—from American tycoons, and were likewise served by privileged, suburban elites of technicians and prudent intellectuals.

On both sides the complexes were threatened by dissidents. In the United States the "security" specialists, such as the FBI, were still intent on pinning the label of communist sympathizer or collaborator on dissidents, but this was an increasingly fruitless obsession. To draft-resisters, anti-militarists, anti-Vietnam campaigners, and hippies, the Soviet Union was another military-industrial unit. Those whose heads were bloodied in Chicago were inclined to identify with those Czechs who, in the very same week, shook their fists at Soviet tanks—and not with the soldiers on the tanks.

"The day is not distant," said John Kenneth Galbraith at a Princeton seminar, "when the organization men on either side of the iron curtain will exchange sympathetic confidences on how to keep their respective poets and students in line."[1]

In both power-blocs, the individual increasingly suffered from a sense of powerlessness. The crisis overwhelming him was not so much a socialist-capitalist crisis—even some military men thought of this as a "cover story"—as it was a new crisis of the unfinished industrial revolution, which in the electronic and nuclear age had entered a new and devastating phase. The logic of industrial organization was leading to units demanding global elbowroom and resources. While their need for wide markets intensified, their dependence on reserves of labor was diminishing. They needed human beings as customers rather than as workers. The worker, an increasingly frustrated cog, was inclined to turn away from the reality of his position—toward the prime-time myth.

Such thoughts were seldom in the minds of broadcasting executives. They too found it prudent not to look too far beyond the job at hand. Millions were watching television, and that was enough. If people were interested, said Stanton, it was "in the public interest."

For defenders of the industry its statistics seemed a spectacular vote of confidence. Its leaders were courted by the White House; Senators and Representatives were its friends. Major film studios, recording studios, sports arenas, publishing empires had moved into its shelter. Its role in the business world, at home and abroad, was staggering. The industrial system, said Galbraith, "is profoundly dependent upon commercial television and could not exist in its present form without it." Television and radio, he wrote, had become the main instruments "for the management of consumer demand."[2]

To many broadcasters this was equivalent to saying "the greatest com-

1. New York *Times*, December 3, 1968.
2. Galbraith, *The New Industrial State*, p. 208.

munication system ever devised." But the "management of consumer demand" held for many people connotations of manipulation and control. And here the power of broadcasting became for many a cause for uneasiness rather than admiration. For the scope of the power had been dramatized by the history of television. That history had shown that the military-industrial alliance could to some extent influence and control the flow of news—through the creation of news events ranging from press conferences to moon probes, through control of transmitters throughout the world, through control of schedules, through restrictions on travel, through suppression of information on grounds of security, through occasional pressure on executives. But—perhaps more serious—the alliance had also come to control the whole myth-making machinery of broadcasting, which ultimately seemed to determine what would be regarded as good and admirable, and what, amid the swirling dust storms of news and information, people would notice or ignore, accept or reject, believe or disbelieve, remember or forget.

The news machinery of broadcasting, growing in maturity, was increasingly aware that its influence depended, in the long run, on the pervasive power of the myth-making machinery. And this, through the mechanism of sponsorship, had come under far tighter control. Together—so the history of the 1950's and 1960's clearly suggested—they could lead the nation into vast adventures in self-deception.

The anniversary, climaxing fifty years of stupendous growth and drama, was a time for celebration—and assessment.

5 / RECKONING

The messages wirelessed ten years ago have not reached some of the nearest stars.

GUGLIELMO MARCONI

I say we had best look our times and lands searchingly in the face.

WALT WHITMAN

The three volumes of this study, telling of an electronic tube and the fabulous genie that emerged from it—a creature of worldwide power—have sought mainly to provide a clear chronicle of events. It seems fitting at this point to add some comments on past events and future possibilities. In these comments, reference will be made to relevant passages in each of the volumes—*A Towel in Babel, The Golden Web,* and *The Image Empire.*

The "American system of broadcasting," as it has developed over the years, has been an extraordinary example of governmental laissez-faire. It has allowed private companies, almost without restraints, to set up tollgates across public highways of communication and to exact a toll from the traffic. Fortunes have been made from this privilege.

Meanwhile the tolls, levied substantially on a what-the-traffic-will-bear basis, have tended to eliminate some elements of society from the marketplace of ideas and to give dominance to others. Thus the private tollgates have caused a vast reshuffling of social influences—just as control-points relating to other media have done in the past. (*Babel,* pp. 3–5.)

The rationale for the private tollgates has been that the toll-keepers, in gratitude for their privilege, would handle the traffic with regard to the welfare of society as a whole. Some such obligation is enshrined in law in the phrase "public interest, convenience, and necessity," but its meaning has remained vague, and the toll-keepers have resisted governmental attempts to define it. Such attempts have in any case been only sporadic, and ineffective in the long run, partly because of ties between toll-keepers and governmental leaders—especially in the executive and legislative branches. (*Web*, pp. 227–36, *Empire*, pp. 68, 126.)

Most licensees have felt that their paramount obligation to the public interest was to "move goods" and keep the wheels of industry turning. Any other services they might render were considered subsidiary to this main function, which was also the source of profit. The conjunction of duty and profit was looked on as the beauty and strength of the system, and the reason for its certain triumph. (*Web*, p. 219.)

If such a system had been outlined in 1927 or 1934, when our basic broadcasting laws were written, it would certainly have been rejected. Legislative debates of those years reveal hardly a hint of the system that has developed, which has really been the product of a series of *faits accomplis*, gradually converting a public trust—defined in extremely idealistic terms in the early years—into a private power system firmly based on advertising. (*Babel*, pp. 199–200.)

That this has not always worked to the satisfaction of all concerned need not surprise us. "For it seems self-evident," wrote Leo Rosten, "that to strain the milk of life through the cheesecloth of advertising must curdle creativity and—more ominous—contaminate truth."[1]

What should perhaps surprise us is that it has worked as well as it has. It has worked at various times for various reasons.

During the 1930's broadcasting was already prosperous, but only about a third of the time was sold. The sold periods already reflected the beliefs and mythology of big business, but the unsold periods were a frontier area filled to some extent with the culture of the displaced—expressing itself through the Federal Theater, off-shoots like the Mercury Theater, and a burgeoning folk-song movement represented by such men as Woody Guthrie. Their unsponsored programming was often in sharp contrast with sponsored programming. Something resembling a "dialogue" existed. (*Web*, pp. 83–9, 118–25.)

The system also benefited from efforts of those FCC commissioners who interpreted the "public interest" in terms of diversity. The pressures of these men, from George Henry Payne to Nicholas Johnson, stimulated occasional flurries of program diversification. (*Web*, 63–70, 168–74, *Empire*, pp. 196–201, 204.)

The system worked well during World War II, partly because of a

1. *Harper's Magazine*, October 1957.

strong political consensus on the war, but also because of other factors. Many sponsors actually had no consumer goods to sell but continued advertising, being encouraged to do so by tax policies. Their commercials became brief institutional messages, and they backed programs of prestige rather than selling power. Although sponsorship was on the increase, the total effect was one of declining commercialism, and of declining sponsor control. The networks, amid rising prosperity, channeled much of their profit into magnificent, growing news systems, which during the war held large audiences and won wide acclaim. (*Web*, pp. 165–8.)

The postwar period and the rise of television brought sharp changes. Prosperity gradually eradicated the frontier of unsold time. A consumer-goods era put the accent first on big-money quizzes, then on action melodrama, which became an international phenomenon as business-military interests spread worldwide. These interests, often using the weapon of political blacklists, narrowed the zone of permitted conflict in ideas, and tended to thrust dissent into other media. The dominance of business-military interests became all-pervasive, leaving mainly news programming of various sorts, along with noncommercial broadcasting, as areas of comparative independence—used generally with caution, occasionally with boldness. Even here there were constant pressures to get into line. Networks, in choosing documentary topics, often leaned toward those that might be sponsorable; unsponsored topics tended to end up in fringe periods. In noncommercial broadcasting the influence of donors, boards of directors, and legislative bodies often had a conservative impact. (*Web*, pp. 253–303, *Empire*, pp. 294–5.)

Today the broadcasting industry rests on an economic base that is international. This has given it enormous wealth and power, and it has attracted to itself huge resources of talent and technical skill. To those touched by visions of the social leverage of broadcasting, the industry has been a powerful magnet.

Yet with all its assets, it faces built-in perils. Its muscle is derived largely from enclaves of prosperity within wastelands of poverty. Abroad, in much of the world, the enclaves are small. At home the area of prosperity is substantial, but even here the surrounding poverty and discontent make much of broadcasting seem irrelevant to the current scene.

During much of television history, its custodians of the advertising world have banned poverty—and the rancorous conflicts interlocked with it—from the home screen. They did so partly because it seemed sound merchandising strategy, or "upgrading"; partly because they believed, or wanted to believe, that prosperity would trickle down to less favored segments of society; and partly because, beguiled by the picture they themselves were holding up to the world, they scarcely believed in the reality of poverty. Perhaps it was, after all, just propaganda. (*Empire,* pp. 32–3, 304.)

But poverty has refused to remain invisible, even at home, and has pushed and shouted its way onto edges of the home screen.

Within the industry itself, discontents have mounted. Among writers and performers, the disillusioned are many. They range from the television news writer who said on a *PBL* program, "Nobody knows what isn't being said," to the army newscaster in Saigon who suddenly blurted out, at the end of a program: "In the military in Vietnam . . . a newscaster is not free to tell the truth."[2] They include Hollywood dramatists who feel chained by formula series, and newsmen who feel trapped by "objective" news policies.

But dissent has acquired a broader base than is suggested by such protests. Commercial broadcasting has been the voice of a "development" era when "growth rate" has been a magic phrase. Now this "development" has begun to connote the waste of irreplaceable resources and the pollution of our earthly environment. Concern has shifted to questions of the very survival of our "garden spot of the universe," as an astronaut fittingly called it. In November 1969 a San Francisco advertising man, Jerry Mander, told a UNESCO conference: "Beginning now, there should be national preparations toward a no-growth economy."

But an advertising-based system that has been the voice of laissez-faire development and the apostle of a spiraling "growth-rate" is scarcely the instrument to promote a no-growth economy. The needs conflict with the very nature of advertising as a brokerage operation. As our priorities shift, the spotlight must shift increasingly to a noncommercial or "public" system based on a different set of motives—through which Americans will, as the Carnegie commission envisioned it, "know themselves, their communities, and their world in richer ways." The commission has described the needed system as "a civilized voice in a civilized community."[3]

2. Al Kine, *PBL*, NET, February 16, 1969. Robert Lawrence, American Forces Vietnam Network, January 3, 1970; quoted, New York *Times,* January 4, 1970.
3. *Public Television,* p. 18.

There are those who appear to think that our commercial system can be reoriented by codes and review commissions, but the history of radio and television gives little reason, if any, for faith in these devices. Against such influences, built-in economic drives have always reasserted themselves without difficulty. The need for a supplementary system based on other motives is paramount and crucial.

With budgets of the sort suggested by the Carnegie commission—$100 million a year—there is no question that public broadcasting can become a dynamic element in our national life. The long association of American noncommercial broadcasting with panels, round-tables, and lectures was forced on it by abject poverty—also, to some extent, by its patrons in the commercial-broadcasting world who preferred that rival channels be committed to material not likely to stir undue excitement. In both Britain and Japan, adequately financed noncommercial systems have existed alongside commercial systems, and proved their ability to hold huge and growing audiences. Both have given some idea of "what isn't being said." In both, dramatists, poets, musicians, and pictorial artists were found ready and waiting.

The problem of funds, often cited as a stumbling block, is scarcely as overwhelming as it is made to appear. To suggest that the United States cannot afford what Britain and Japan can afford is, on the face of it, absurd. Various methods of finance have been suggested. The publicly owned airways have been used rent-free by commercial users to create vast fortunes; there is no reason, as Walter Lippmann suggested long ago, why commercial users should not pay a rental or royalty for such use of public property, to be earmarked for the needs of public broadcasting. The device suggested by the Ford Foundation—a royalty from domestic communication-satellite revenue—holds equal merit, since the technology has been almost entirely a product of tax funds, and the proposed royalty would be a reasonable "people's dividend." The Carnegie commission's proposal of a special tax on receiver sales is also logical, perhaps as a supplementary device.

That the problem is solvable, once we decide to solve it, is made clear by past and present government expenditures in broadcasting. During some years the taxpayer has apparently (though without having a say in the matter) supplied funds approaching $100 million for such activities as Radio Free Europe, Radio Liberation (later Radio Liberty), Radio Free Asia, Radio Swan, and other voices yet unidentified. When we decide that an investment in "a civilized voice in a civilized community" is worth more than an investment in cold war, we shall find that the financial problem is largely one of shifting our priorities.

The question is worth pursuing. An organization like Radio Liberty, originally conceived as an instrument for toppling the Soviet government but today more mellow in tone, seems a strange holdover from the Dulles era. That a crusade for getting truth through the iron curtain should in the first place have been launched under the patronage of Allen Dulles, who in each project—according to his memoirs—gave top priority to keeping facts from (1) the Soviets and (2) the American public, is surely ludicrous. One of the most important contributions Congress could make to our international credibility would be to divest the CIA of any role, financial or otherwise, in respect to broadcasting.

This raises the question of the continuing function of Radio Liberty, still calling itself "private" but still seemingly using public funds mysteriously transmitted. Much of what it does duplicates the Voice of America. Whatever on Radio Liberty is true and fine—and apparently there is much—could be done by the Voice of America; whatever is not had better not be done. If productive employment for well-informed refugees is a purpose, that can be done—and is done—at USIA.

To assess the values and liabilities of a project like Radio Liberty or Radio Free Europe, it may be useful to imagine a similar project in reverse. The Vietnam war caused many young Americans to move to Canada and elsewhere as refugees from militarism. Let us suppose that a colony of such emigres were suddenly provided with an array of superpower transmitters capable of reaching all parts of the United States. This group, announcing a round-the-clock truth crusade, could find much to talk about: administration pronouncements later proved false; statements by legislators, lawyers, and educators attacking the legality of the Vietnam war; reports from Americans in Vietnam about bombing errors, corruption, and napalm atrocities. Let us also suppose that this broadcasting venture was announced as "private" but clearly involved a multimillion-dollar investment by a foreign power. No doubt the project would delight some Americans but would anger others, including many opposed to the Vietnam involvement. They would feel that a truth crusade should not begin with a deception about auspices, and that the refugees had become pawns rather than free agents.

The notion that America needs an assortment of such false faces for different purposes is a cold-war heritage that could be usefully abandoned. The issue is relevant here because it has involved large investments of public funds in broadcasting ventures.

It is of course unlikely that any Russian-financed or Chinese-financed transmitters would ever be permitted on—let us say—Canadian soil; Canada, like the United States, does not allow foreign-controlled stations. This raises some questions about American propaganda transmitters in other parts of the world, which involve us in a double standard.

Our right to erect propaganda stations in West Germany was established by us as a conquering power. Most free countries do not readily yield such a privilege or right. It is significant that countries which have become important bases for American propaganda transmitters include Portugal, for Radio Free Europe; Spain and Taiwan, for Radio Liberty. In each case the right to erect our transmitters has been part of a complex relationship involving military aid buttressing a dictatorship. It is a sorry comment that our "free world" pronouncements have been powered by these dubious associations.

That the United States needs to communicate with the world through an agency like the USIA and its Voice of America seems self-evident. But several aspects of our communication activities are worth reconsidering.

One is the heavy reliance on *unattributed* material—productions which USIA does not announce as its work, though it may be prepared to acknowledge authorship if the issue is raised. In 1967 the New York *Times* reported that *most* USIA material appearing on foreign television stations—newsreel items, plays, documentaries, interviews—were not identified as USIA products.[4] Former USIA deputy director Thomas Sorensen has described this as good strategy because, he says, the USIA label brings a guarded reaction, which can be avoided by not announcing the USIA role.[5] But if the USIA label does bring such a reaction, is it perhaps because we have acquired a reputation for trickiness?

A factor in this situation appears to be that stations using a great deal of such material prefer not to proclaim their dependence on a foreign source, and will more readily use it if the source can be concealed.

Again, a way to assess the practice is to imagine its application in reverse. If an American television station or network were found to be

4. New York *Times*, March 29, 1967.
5. Sorensen, *The Word War*, p. 64.

using hours of material from a foreign government but not announced as such, we would no doubt consider it a suspect "infiltration," or worse. We may in the long run find it good strategy to treat people of other nations as we would prefer to be treated.

An even more serious problem is inherent in legal restrictions that make USIA films and broadcasts inaccessible to American audiences. In contrast to our practice, productions of the National Film Board of Canada—probably the most fruitful and intelligent government-sponsored film unit in the world—present the same face to Canadians as to the rest of the world. Government films seen by Canadians are available to others, and vice versa. United States citizens, on the other hand, are not permitted to know in what terms they are presented to the world, nor what pronouncements are made in their name.

The usual explanation—that USIA should not "propagandize the American people"—assumes the kind of material that, abroad as well as at home, can do us more harm than good.

Some of what is offered to the rest of the world on behalf of Americans would be of equal interest and value to Americans; under this heading one could list the Willis Conover radio series *Music USA,* such films as *The March,* and many others.

Other items, like *El Periodista!,* might get a less friendly domestic reaction. The corrective effect would be healthy.

The frequent USIA obsession with items for particular "target audiences" to meet specific problems is responsible for some of its least impressive products—sometimes produced on a crash basis and withdrawn a few months later when the situation has changed. On the other hand, material reflecting our long-range interests and concerns—including "cultural" material like *Music USA*—may in the long run have more potent political values.

The worst enemies of the USIA have been those who have conceived its task as one of merchandising something called "Americanism," to be sold with all the devices and stratagems of advertising. Fortunately their influence began to subside in the Murrow years, but the USIA is still perplexed and divided over its role. One reason may be that it operates in unhealthy isolation from the people it is supposed to represent. The question of *what to say* can only be approached intelligently within the context of larger questions: "Who am I? What do I represent?" And these are questions best not settled by information specialists huddling behind closed doors in Washington, but explored in full view and hearing of the American people.

This is to suggest that a nation already spending many millions on television, film, and radio projects does not have to search far for a solution to its needs for public broadcasting. The current sharp division between domestic and foreign "public" broadcasting is wasteful and poisonous. Must we really have a different face for each purpose? Has this perhaps contributed to our schizophrenia at home and shaky credibility abroad? A coalescing of the two activities, on a basis that provides secure income and therefore independence from week-by-week political pressures, could contribute to the health of both. Noncommercial stations are a natural place to exhibit the fruits of our culture as proclaimed abroad. And if our public television network is to be a "civilized voice in a civilized community," that is precisely what USIA should be transmitting to the world, rather than items especially concocted as an international façade.

These varied suggestions are advanced with awareness that they are not likely to find early acceptance. We are inclined to hang onto old formulas until disaster strikes—and for some time thereafter. The conflicts of interest built into our broadcast media probably guarantee that present arrangements will continue for some time. The advertising industry will not readily give up its custodianship of our cultural life, which it has purchased with good money. Our clandestine warfare agency will not readily surrender its purse strings over international transmitters. Major military contractors will not readily give up their central position in our principal communication medium. Congressmen will not readily surrender their right to accept benefits from an industry over which they legislate.

For all these reasons, the broadcasting industry will probably careen onward for some time with only minor changes. But it is meanwhile becoming increasingly conscious of shortcomings and problems.

It is becoming aware that the sales and mythology it has promoted so well are not solving the huge problems that surround it. In many lands the American broadcasting system has established itself with its promise of national development, but the reality has too often been a deepening dependence on the United States—while headache tablets, hair tonics, deodorants, lotions, stomach settlers, soft drinks, cigarettes, lipsticks, hair dyes, and television itself have been sold as symbols of the good life.

The products have tended to create little Americas throughout the world, walling themselves off from surrounding problems. In many cases they aggravate the problems, diverting energies and funds from dire necessities to props of a strange culture.

Each of these little Americas has its local toll-keepers, often believing strongly in their mission, and tending to dominate the traffic in ideas, while proclaiming law and order, but meanwhile building counter-pressures that will, in one way or another, sooner or later, expropriate the small screen—among the most precious of remaining resources, and key to the use of others.

APPENDIX A / CHRONOLOGY

1953 ABC and Paramount Theaters merge.
Noncommercial television begins in Houston.
Senator Joseph R. McCarthy directs purges and appointments at Voice of America and FCC.
Marty broadcast on *Goodyear Television Playhouse.*
Radio Liberation launched.
All networks broadcast *Dinner With the President.*
Formation of Aware, Inc.

1954 *See It Now* documentary on Senator McCarthy.
Army-McCarthy hearings carried live by ABC-TV and Dumont.
Increase in censorship disputes on anthology series.
Rise of filmed episodic series.
Disney and Warner Brothers contract to produce for ABC-TV.
Armed Forces Radio Service launches television outlets.

1955 Filming of Eisenhower press conferences permitted, subject to pre-release review.
The $64,000 Question precipitates big-money quiz boom.
Commercial television in Britain and Japan widens market for telefilms.
Robert Sarnoff takes over NBC presidency.

1956 John Henry Faulk files suit over Aware attack and blacklist conspiracy.
Cheyenne stirs avalanche of television westerns.
Release of hundreds of pre-1948 features brings decline in local television production.
Radio Free Europe criticized for role in Hungarian revolt.
Huntley-Brinkley team formed for NBC convention telecasts.
Eisenhower-Stevenson campaign features five-minute "hitch-hike" programs.

1957 Inauguration recorded for first time by videotape.
William Worthy broadcasts over CBS from Peking and is deprived of passport by State Department.
American Cancer Society report on smoking brings advertising emphasis on long and filter cigarettes.
Nikita Khrushchev on *Face the Nation* over CBS television and radio networks.
John C. Doerfer becomes FCC chairman.

1958 Bribery evidence brings resignation of FCC commissioner Richard A. Mack.
CBS ends *See It Now.*
Robert Kintner becomes NBC president as Robert Sarnoff assumes chairmanship.

1959 Guterma scandal at MBS.
Nixon-Khrushchev "kitchen debate" in Moscow.
Khrushchev tour of United States widely covered by television and radio.
Willis Conover of Voice of America visits Poland.
Revelations of quiz rigging bring rise in public-service budgets and debut of *CBS Reports.*
James T. Aubrey becomes CBS-TV president.

1960 Radio Swan launched by CIA to prepare for invasion of Cuba.
Time documentary *Primary* stirs *cinéma vérité* movement.
FCC chairman John C. Doerfer resigns on request over issue of fraternization with industry.
Networks ban news documentaries other than their own.
The Untouchables leads trend to violence in telefilms.
Khrushchev-Castro meeting on Harlem sidewalk televised.
Richard M. Nixon and John F. Kennedy meet in *The Great Debates.*

1961 Eisenhower, in television farewell, warns against military-industrial complex.
Kennedy inaugural signals policy of encouragement to arts.
Edward R. Murrow heads USIA.
Kennedy allows filming and televising of presidential press conferences without restriction.
Wide falsification of Bay of Pigs news by CIA fronts.
Senator Thomas J. Dodd studies television violence.
ABC-TV inaugurates 40-second station break for extra commercial, and other networks follow example.
FCC chairman Newton N. Minow finds a vast wasteland.

1962 Noncommercial television acquires New York channel 13.
John Henry Faulk wins unprecedented award over blacklist conspiracy.
Telstar I inaugurates satellite relays of television programs.
Growth of noncommercial television aided by federal grants for facilities.
Cuban missile crisis brings televised ultimatum.

NBC-TV broadcasts *The Tunnel.*
Rise of *The Beverly Hillbillies.*

1963 WRVR-FM, New York, records and broadcasts Alabama Ku Klux Klan meeting calling for action to halt integration.
E. William Henry succeeds Newton Minow as FCC chairman.
"I Have a Dream" speech by Martin Luther King climaxes March on Washington telecasts.
Westward move by baseball Giants and Dodgers linked to plans of Subscription Television, Inc.
Networks start half-hour evening newscasts.
Assassination and funeral of President Kennedy and murder of Lee Harvey Oswald rivet world attention through four days of coverage.

1964 United Church monitors television stations in Jackson, Mississippi.
Fred W. Friendly becomes CBS News president.
Instant replay adds new dimension to sport telecasts.
CBS buys New York Yankees baseball team.
Johnson-Goldwater campaign features girl-with-a-daisy and other spots.
Subscription Television, Inc., halted by California referendum.

1965 Former FCC commissioner Frederick W. Ford heads National Community Television Association.
Early Bird synchronous satellite makes debut.
Increased coverage of Vietnam as war escalates.
John A. Schneider succeeds James Aubrey as CBS-TV president.
Television network news shifting to color.
CBS occupies new skyscraper headquarters.

1966 Referendum outlawing pay-television declared unconstitutional by California Supreme Court.
Friendly resigns over CBS decision to halt coverage of Senate hearings on Vietnam war.
FCC assumes jurisdiction over cable television.
Rosel Hyde becomes FCC chairman on resignation of E. William Henry.
Spies and war dominate telefilms.
Asian tour by President Johnson receives wide coverage.

1967 Anti-war marches and demonstrations seen with increasing frequency on newscasts.
Occasional use of newsfilm from North Vietnam.
PBL series launched over NET stations.
Corporation for Public Broadcasting formed.

1968 NET schedules portion of Felix Greene film *Inside North Vietnam* and is assailed by congressmen.
Television coverage of New Hampshire primary indicates strong support for anti-war stand of Senator Eugene McCarthy.
Martin Luther King announces Poor People's March as war protest.
President Johnson broadcasts decision not to be a candidate for re-election.

Televised funeral of Martin Luther King, shot by sniper in Memphis.
CBS-TV broadcast of *Hunger in America* stirs wide response.
Televised funeral of Senator Robert Kennedy, shot in Los Angeles after primary victory.
Scenes of violence mark Democratic convention coverage.
Record sums spent on Nixon-Humphrey television campaign, but many voters abstain.
Television films pruned of violence.

1969 CBS cancels *Smothers Brothers Comedy Hour* after frequent censorship clashes.
Rise of *Rowan and Martin's Laugh-In.*
Apollo flights climaxed by televised moon landings.
Dean Burch becomes FCC chairman.

1970 Robert Sarnoff becomes RCA board chairman.
Industry marks fiftieth anniversary of KDKA election reports that launched broadcasting era.

APPENDIX B / LAWS

Central Intelligence Agency Act of 1949[1]

Public Law No. 110, June 20, 1949, 81st Congress, 1st session. An Act to provide for the administration of the Central Intelligence Agency, established pursuant to section 102, National Security Act of 1947, and for other purposes.

Be it enacted by the Senate and House of Representatives of the United States of America in Congress assembled,

DEFINITIONS

SEC. 1. That when used in this Act, the term—

(a) "Agency" means the Central Intelligence Agency;

(b) "Director" means the Director of Central Intelligence;

(c) "Government agency" means any executive department, commission, council, independent establishment, corporation wholly or partly owned by the United States which is an instrumentality of the United States, board, bureau, division, service, office, officer, authority, administration, or other establishment, in the executive branch of the Government; and

(d) "Continental United States" means the States and the District of Columbia.

[*Sections 2 and 3, dealing with* SEAL OF OFFICE *and* PROCUREMENT AUTHORITIES, *are here omitted.*]

1. This act provided the basis for CIA "action programs," including ventures in international broadcasting and film production and infiltration of organizations of many kinds, at home and abroad.

EDUCATION AND TRAINING

SEC. 4. (a) Any officer or employee of the Agency may be assigned or detailed for special instruction, research, or training, at or with domestic or foreign public or private institutions; trade, labor, agricultural, or scientific associations; courses or training programs under the National Military Establishment; or commercial firms.

(b) The Agency shall, under such regulations as the Director may prescribe, pay the tuition and other expenses of officers and employees of the Agency assigned or detailed in accordance with provisions of subsection (a) of this section, in addition to the pay and allowances to which such officers and employees may be otherwise entitled.

[*Section 5, dealing with* TRAVEL, ALLOWANCES, AND RELATED EXPENSES, *is here omitted.*]

GENERAL AUTHORITIES

SEC. 6. In the performance of its functions, the Central Intelligence Agency is authorized to—

(a) Transfer to and receive from other Government agencies such sums as may be approved by the Bureau of the Budget, for the performance of any of the functions or activities authorized under sections 102 and 303 of the National Security Act of 1947 (Public Law 253, Eightieth Congress), and any other Government agency is authorized to transfer to or receive from the Agency such sums without regard to any provisions of law limiting or prohibiting transfers between appropriations. Sums transferred to the Agency in accordance with this paragraph may be expended for the purposes and under the authority of this Act without regard to limitations of appropriations from which transferred;

(b) Exchange funds without regard to section 3651 Revised Statutes (31 U.S.C. 543);

(c) Reimburse other Government agencies for services of personnel assigned to the Agency, and such other Government agencies are hereby authorized, without regard to provisions of law to the contrary, so to assign or detail any officer or employee for duty with the Agency;

(d) Authorize couriers and guards designated by the Director to carry firearms when engaged in transportation of confidential documents and materials affecting the national defense and security;

(e) Make alterations, improvements, and repairs on premises rented by the Agency, and pay rent therefor without regard to limitations on expenditures contained in the Act of June 30, 1932, as amended: *Provided,* That in each case the Director shall certify that exception from such limitations is necessary to the successful performance of the Agency's functions or to the security of its activities.

SEC. 7. In the interests of the security of the foreign intelligence activities of the United States and in order further to implement the proviso of section 102 (d) (3) of the National Security Act of 1947 (Public Law 253, Eightieth Congress, first session) that the Director of Central Intelligence shall be responsible for protecting intelligence sources and methods from unauthorized disclosure, the Agency shall be exempted from the provisions of sections 1 and 2, chapter 795 of the Act of August 28, 1935 (49 Stat. 956, 957; 5 U.S.C. 654), and the provisions of any other law which require the publication or disclosure of the organization, functions, names, official titles, salaries, or numbers of personnel employed by the Agency:
Provided, That in furtherance of this section, the Director of the Bureau of the Budget shall make no reports to the Congress in connection with the Agency under section 607, title VI, chapter 212 of the Act of June 30, 1945, as amended (5 U.S.C. 947(b)).

SEC. 8. Whenever the Director, the Attorney General, and the Commissioner of Immigration shall determine that the entry of a particular alien into the United States for permanent residence is in the interest of national security or essential to the furtherance of the national intelligence mission, such alien and his immediate family shall be given entry into the United States for permanent residence without regard to their inadmissibility under the immigration or any other laws and regulations, or to the failure to comply with such laws and regulations pertaining to admissibility: *Provided,* That the number of aliens and members of their immediate families entering the United States under the authority of this section shall in no case exceed one hundred persons in any one fiscal year.

SEC. 9. The Director is authorized to establish and fix the compensation for not more than three positions in the professional and scientific field, within the Agency, each such position being established to effectuate those scientific intelligence functions relating to national security, which require the services of specially qualified scientific or professional personnel: *Provided,* That the rates of compensation for positions established pursuant to the provisions of this section shall not be less than $10,000 per anum nor more than $15,000 per annum, and shall be subject to the approval of the Civil Service Commission.

APPROPRIATIONS

SEC. 10. (a) Notwithstanding any other provisions of law, sums made available to the Agency by appropriation or otherwise may be expended for purposes necessary to carry out its functions, including—

(1) personal services, including personal services without regard to limitations on types of persons to be employed, and rent at the seat of government and elsewhere; health-service program as authorized by law (5 U.S.C. 150); rental of news-reporting services; purchase or rental and operation of photographic, reproduction, cryptographic, duplication and printing machines, equip-

ment and devices, and radio-receiving and radio-sending equipment and devices, including telegraph and teletype equipment; purchase, maintenance, operation, repair, and hire of passenger motor vehicles, and aircraft, and vessels of all kinds; subject to policies established by the Director, transportation of officers and employees of the Agency in Government-owned automotive equipment between their domiciles and places of employment, where such personnel are engaged in work which makes such transportation necessary, and transportation in such equipment, to and from school, of children of Agency personnel who have quarters for themselves and their families at isolated stations outside the continental United States where adequate public or private transportation is not available; printing and binding; purchase, maintenance, and cleaning of firearms, including purchase, storage, and maintenance of ammunition; subject to policies established by the Director, expenses of travel in connection with, and expenses incident to attendance at meetings of professional, technical, scientific, and other similar organizations when such attendance would be a benefit in the conduct of the work of the Agency; association and library dues; payment of premiums or costs of surety bonds for officers or employees without regard to the provisions of 61 Stat. 646; 6 U.S.C. 14; payment of claims pursuant to 28 U.S.C.; acquisition of necessary land and the clearing of such land; construction of buildings and facilities without regard to 36 Stat. 699; 40 U.S.C. 259, 267; repair, rental, operation, and maintenance of buildings, utilities, facilities, and appurtenances; and

(2) supplies, equipment, and personnel and contractual services otherwise authorized by law and regulations, when approved by the Director.

(b) The sums made available to the Agency may be expended without regard to the provisions of law and regulations relating to the expenditure of Government funds; and for objects of a confidential, extraordinary, or emergency nature, such expenditures to be accounted for solely on the certificate of the Director and every such certificate shall be deemed a sufficient voucher for the amount therein certified.

SEPARABILITY OF PROVISIONS

SEC. 11. If any provision of this Act, or the application of such provision to any person or circumstances, is held invalid, the remainder of this Act or the application of such provision to persons or circumstances other than those as to which it is held invalid, shall not be affected thereby.

SHORT TITLE

SEC. 12. This Act may be cited as the "Central Intelligence Agency Act of 1949."

Approved June 20, 1949.

Section 315 of the Communications Act of 1934

As amended 1959:[1]

FACILITIES FOR CANDIDATES FOR PUBLIC OFFICE

SEC. 315.

(a) If any licensee shall permit any person who is a legally qualified candidate for any public office to use a broadcasting station, he shall afford equal opportunities to all other such candidates for that office in the use of such broadcasting station: *Provided,* That such licensee shall have no power of censorship over the material broadcast under the provisions of this section. No obligation is imposed upon any licensee to allow the use of its station by any such candidate.[2] Appearance by a legally qualified candidate on any–

(1) bona fide newscast,

(2) bona fide news interview,

(3) bona fide news documentary (if the appearance of the candidate is incidental to the presentation of the subject or subjects covered by the news documentary), or

(4) on-the-spot coverage of bona fide news events (including but not limited to political conventions and activities incidental thereto),

shall not be deemed to be use of a broadcasting station within the meaning of this subsection. Nothing in the foregoing sentence shall be construed as relieving broadcasters, in connection with the presentation of newscasts, news interviews, news documentaries, and on-the-spot coverage of news events, from the obligation imposed upon them under this chapter to operate in the public interest and to afford reasonable opportunity for the discussion of conflicting views on issues of public importance.

(b) The charges made for the use of any broadcasting station for any of the purposes set forth in this section shall not exceed the charges made for comparable use of such station for other purposes.[3]

(c) The Commission shall prescribe appropriate rules and regulations to carry out the provisions of this section.

Amendment applying only to 1960:

Public Law 86–677, August 24, 1960, provided–

(1) That that part of section 315(a) of the Communications Act of 1934, as amended, which requires any licensee of a broadcast station who permits

1. Sec. 315 in its original form, as enacted 1934, consisted of paragraph (a) through the words "any such candidate."

2. The remainder of paragraph (a) and paragraph (c) were added September 14, 1959, by Public Law No. 274, 86th Congress, 1st session.

3. Paragraph (b) was added July 16, 1952, by Communication Act Amendments, Public Law No. 554, 82nd Congress, 2nd session.

any person who is a legally qualified candidate for any public office to use a broadcasting station to afford equal opportunities to all other such candidates for that office in the use of such broadcasting station, is suspended for the period of the 1960 presidential and vice presidential campaigns with respect to nominees for the offices of President and Vice President of the United States. Nothing in the foregoing shall be construed as relieving broadcasters from the obligation imposed upon them under this Act to operate in the public interest.

(2) The Federal Communications Commission shall make a report to the Congress, not later than March 1, 1961, with respect to the effect of the provisions of this joint resolution and any recommendations the Commission may have for amendments to the Communications Act of 1934 as a result of experience under the provisions of this joint resolution.

BIBLIOGRAPHY

Acknowledgements. The author wishes to thank the John Simon Guggenheim Memorial Foundation for a grant which made possible research in many parts of the United States; also, the many individuals who helped so generously with interviews, scripts, screenings, research materials, photographs—including personnel at networks, stations, advertising agencies, production companies, unions, libraries, publications, universities, government agencies.

Collections. Manuscript collections and oral history collections have proved of particular value. Collections are identified in the bibliography by letter, as follows:

(C) Columbia University Oral History Collection, New York.
(D) John Foster Dulles Collection in the Library of Princeton University, Princeton.
(M) Mass Communications History Center of the Wisconsin Historical Society, Madison.
(P) Broadcast Pioneers History Project, New York and Washington.
(S) University of Southern California Doheny Library, Los Angeles.

Abel, Elie. The Missile Crisis. New York, Bantam, 1966.
Action! Los Angeles, Directors Guild of America, bimonthly, 1966–
Advertising Age. Chicago, weekly, 1930–
AER Journal. Chicago, Association for Education by Radio, monthly, 1941–53; became AERT Journal, 1956–59. See also NAEB Journal.
AFRS Playback. Los Angeles, Armed Forces Radio Service, weekly, from 1944. Unpublished.
Allen, George V. Interview by Philip A. Crowl, 1965. Unpublished. (D)
Allen, Steve. The Funny Men. New York, Simon & Schuster, 1956.

The American Forces Network: serving American forces in Europe. U.S. Department of Defense memorandum, ca. 1943. Unpublished.

Anderson, Jack. See Pearson, Drew, and–

Ardrey, Robert. "Hollywood: the toll of the frenzied forties," The Reporter, March 21, 1957.

Arlen, Michael J. Living Room War. New York, Viking, 1969.

Armed Forces Radio Service. For Playback, see AFRS Playback.

Artime, Manuel. See Johnson, Haynes, with–

Averson, Richard. See White, David Manning, and–

Aware publication. New York, 1954–56.

Baker, Samm Sinclair. The Permissible Lie: the inside truth about advertising. Cleveland, World, 1968.

Baldwin, James. The Fire Next Time. New York, Delta, 1964.

Baldwin, James. Notes on a Native Son. New York, Bantam, 1964.

Barnouw, Erik. The Golden Web: a history of broadcasting in the United States, v. II–1933 to 1953. New York, Oxford University Press, 1968.

Barnouw, Erik. Mass Communication: television, radio, film, press. New York, Rinehart, 1956.

Barnouw, Erik. The Television Writer. New York, Hill & Wang, 1962.

Barnouw, Erik. A Tower in Babel: a history of broadcasting in the United States, v. I–to 1933. New York, Oxford University Press, 1966.

Barrett, Edward W. (ed.). Journalists in Action. Manhasset (N.Y.), Channel Press, 1963.

Barrett, Edward W. Truth Is Our Weapon. New York, Funk & Wagnalls, 1953.

Barrett, Marvin (ed.). Survey of Broadcast Journalism 1968–1969. New York, Grossett & Dunlap, 1969.

Barrow, Roscoe L. For "Barrow report" see Network Broadcasting.

Barrow, Roscoe L. "The Fairness Doctrine," Cincinnati Law Review, Summer 1968.

Barry, David W. See Parker, Everett C., and–

Batman: bat notes for bat writers. Hollywood, Greenway Productions, 1966. Unpublished.

Bellamy, Ralph. Reminiscences, 1958. Unpublished. (C)

Bentley, Eric. The Life of the Drama. New York, Atheneum, 1964.

Bernays, Edward L. Biography of an Idea: memoirs of a public relations counsel. New York, Simon & Schuster, 1965.

Bernstein, Victor, and Jesse Gordon. "The Press and the Bay of Pigs," Columbia University Forum, Fall 1967.

Berton, Pierre. Fast Fast Fast Relief. Toronto, McClelland & Stewart, 1962.

Billboard. Cincinnati, monthly, then weekly, 1894–

Bliss, Edward, Jr. (ed.). In Search of Light: the broadcasts of Edward R. Murrow 1938–1961. New York, Knopf, 1967.

Bluem, A. William. Documentary in American Television. New York, Hastings, 1965.

Blueprint for *Man Against Crime*. New York, William Esty agency, undated (ca. 1953). Unpublished.

Blum, Daniel C. Pictorial History of TV. Philadelphia, Chilton, 1958.

Boekemeier, Barbara. The Genesis of WNDT: a noncommercial station on a commercial channel. Master's essay. New York, Columbia University, 1963. Unpublished.

Bogart, Leo. The Age of Television. New York, Ungar, 1958.

Boorstin, Daniel J. The Image: a guide to pseudo-events in America. New York, Harper, 1961.

Boyd, James. Above the Law: the rise and fall of Senator Thomas J. Dodd. New York, New American Library, 1968.

Braddon, Russell. Roy Thomson of Fleet Street. London, Fontana, 1968.

Braden, Waldo W. See Pennybacker, John H., and–

Bricker, Harry. See Witty, Paul, and–

Briggs, Asa. "Broadcasting and Society," The Listener, November 22, 1962.

Britton, Florence (ed.). Best Television Plays 1957. New York, Ballantine, 1957.

Broadcast Advertisements: hearings before a subcommittee of the committee on interstate and foreign commerce, House of Representatives, 88th Congress, 1st session. Washington, Government Printing Office, 1963.

Broadcasting. Washington, semimonthly, then weekly, 1931–

Broadcasting and Government Regulation in a Free Society. Santa Barbara, Center for the Study of Democratic Institutions, 1959.

Broadcasting Yearbook. Washington, 1935, etc.

Brown, David, and W. Richard Bruner (eds.). How I Got That Story. New York, Dutton, 1967.

Brown, David, and W. Richard Bruner (eds.). I Can Tell It Now. New York, Dutton, 1964.

Brown, Donald E., and John Paul Jones. Radio and Television News. New York, Rinehart, 1954.

Bruner, W. Richard. See Brown, David, and–

Bryson, Lyman (ed.). The Communication of Ideas. New York, Harper, 1948.

Button, Robert E. Interview by Ed Dunham, 1967. Unpublished. (P)

Buxton, Frank, and Bill Owen. Radio's Golden Age: the programs and the personalities. New York, Easton Valley Press, 1966.

Cable, Charles P. Interview by Philip A. Crowl, 1965. Unpublished. (D)

Cassavetes, John. Reminiscences, 1959. Unpublished. (C)

Cassirer, Henry R. Television Teaching Today. Paris, UNESCO, 1960.

Cater, Douglass. "Every Congressman a Television Star," The Reporter, June 16, 1955.

Cater, Douglass, and Walter Pincus. "The Foreign Legion of U.S. Public Relations," The Reporter, December 22, 1960.

Chafee, Zechariah, Jr. Freedom of Speech and Press. New York, Freedom Agenda, 1955.

Chayefsky, Paddy. Papers. Unpublished. (M)

Chayefsky, Paddy. Television Plays. New York, Simon & Schuster, 1955.

Chester, Giraud, Garnet R. Garrison, and Edgar E. Willis. Television and Radio. New York, Appleton-Century-Crofts, 1963.

Chester, Lewis, Godfrey Hodgson, and Bruce Page. An American Melodrama: the presidential campaign of 1968. New York, Dell, 1969.

Childs, Marquis. Interview by Richard D. Challener, 1966. Unpublished. (D)

Clarke, Arthur C. "Extra-Terrestrial Relays," Wireless World, October 1945.

Clarke, Arthur C. "Faces from the Sky," Holiday, September 1959.

Coblentz, Gaston. See Drummond, Roscoe, and–

Codding, George A. Broadcasting Without Barriers. Paris, UNESCO, 1959.

Cogley, John. Report on Blacklisting, I: movies. Fund for the Republic, 1956.

Cogley, John. Report on Blacklisting, II: radio-television. Fund for the Republic, 1956.

Columbia Journalism Review. New York, Graduate School of Journalism, Columbia University, quarterly, 1962–

Comden, Betty, and Adolph Green. Reminiscences, 1959. Unpublished. (C)

Compton, Neil. "Camping in the Wasteland," Commentary, January 1966.

Compton, Neil. "Consensus Television," Commentary, October 1965.

Conant, Michael. Antitrust in the Motion Picture Industry. Berkeley, University of California Press, 1960.

Conniff, Frank. See Hearst, William Randolph, Jr., and–

Considine, Bob. See Hearst, William Randolph, Jr., and–

Coons, John E. (ed.). Freedom and Responsibility in Broadcasting. Evanston, Northwestern University Press, 1961.

Corwin, Norman. Overkill and Megalove. Cleveland, World, 1963.

Counterattack: the newsletter of facts on communism. New York, American Business Consultants, monthly, 1947–55.

Coverage of the Death of President Kennedy and the Subsequent Events by CBS News, Washington: an interview with William J. Small. Washington, CBS, 1963. Unpublished.

Cowan, Louis G. Reminiscences, 1968. Unpublished. (C)

Craig, Walter. Interview by Edwin L. Dunham recorded in Sarasota, Fla., 1965. Unpublished. (P)

Crankshaw, Edward. Khrushchev: a career. New York, Viking, 1966.

Cromwell, John. Reminiscences, 1958. Unpublished. (C)

Crosby, John. With Love and Loathing. New York, McGraw-Hill, 1963.

Crowther, Bosley. The Lion's Share: the story of an entertainment empire. New York, Dutton, 1957.

Davie, Michael. LBJ: a foreign observer's viewpoint. New York, Ballantine, 1967.

DeFleur, Melvin L. "Occupational Roles as Portrayed on Television," Public Opinion Quarterly, Spring 1964.

Dizard, Wilson P. Television: a world view. Syracuse, Syracuse University Press, 1966.

Douglas, Melvyn. Reminiscences, 1958. Unpublished. (C)

Draper, Theodore. Abuse of Power. New York, Viking, 1966.

Drummond, Roscoe, and Gaston Coblentz. Duel at the Brink: John Foster Dulles' command of American power. Garden City (N.Y.), Doubleday, 1960.

Dulles, Allen. The Craft of Intelligence. New York, Harper & Row, 1963.

Dulles, John Foster. Papers. Unpublished. (D)

Duncan, Hugh Dalziel. Symbols in Society. New York, Oxford University Press, 1968.

Edmerson, Estelle. A Descriptive Study of the American Negro in United States Professional Radio 1922–1953. Master's essay. Los Angeles, University of California, 1954.

Education on the Air: yearbook of the Institute for Education by Radio (later, "by Radio-Television"). Columbus, Ohio State University, 1930, etc.

The Eighth Art. Introduction by Robert Lewis Shayon. New York, Holt, Rinehart & Winston, 1962.

Eisenhower, Dwight D. Interview by Philip A. Crowl, 1964. Unpublished. (D)

Eisenhower, Dwight D. The White House Years: mandate for change 1953–1956. Garden City (N.Y.), Doubleday, 1963.

Emery, Walter B. Broadcasting and Government: responsibilities and regulation. East Lansing, Michigan State University Press, 1961.

Emery, Walter B. National and International Systems of Broadcasting: their history, operation, and control. East Lansing, Michigan State University Press, 1969.

Ernst, Harry W. The Primary That Made a President: West Virginia 1960. Eagleton Institute Cases in Practical Politics. New York, McGraw-Hill, 1962.

Evans, Rowland, and Robert Novak. Lyndon B. Johnson: the exercise of power. New York, Signet, 1968.

Ewers, Carolyn H. Sidney Poitier: the long journey. New York, New American Library, 1969.

Fadiman, Clifton. See Prize Plays of Television and Radio 1956.

Fair Trial vs. A Free Press. Santa Barbara, Center for the Study of Democratic Institutions, 1965.

Faulk, John Henry. Fear on Trial. New York, Simon & Schuster, 1964.

FCC Docket No. 12782 (study of network broodcasting). Hearings. Unpublished.

FCC Log: a chronology of events in the history of the Federal Communications Commission 1934–1956. Washington, Federal Communications Commission, 1956.

FCC Reports. Washington, Government Printing Office, annual, 1935–

Film Comment. New York, quarterly, 1962–

Film Daily. New York, 1918–

Film Daily Year Book of Motion Pictures. New York, 1919–
Films for Television. Evanston, Standard Rate and Data Service, 1955.
Fisher, Paul L., and Ralph L. Lowenstein (eds.). Race and the News Media. New York, Praeger, 1967.
Focal Encyclopedia of Film and Television: techniques. New York, Hastings, 1969.
A Forecast of USIA Television in 1962. Washington, U.S. Information Agency, 1958. Unpublished.
Four Days: the historical record of the death of President Kennedy, compiled by United Press International and American Heritage Magazine. New York, American Heritage, 1964.
Frank, Reuven. "Life With Brinkley," in Barrett, Edward W. (ed.), Journalists in Action. Manhasset (N.Y.), Channel Press, 1963.
Frank, Reuven. "The Making of *The Tunnel,*" Television Quarterly, Fall 1963.
Friendly, Fred W. Due to Circumstances Beyond Our Control . . . . New York, Random House, 1967.
Friendly, Fred W. Interview by Erik Barnouw, 1968. Unpublished. (C)
Frost, David, and Ned Sherrin (eds.). That Was The Week That Was. London, W. H. Allen, 1963.
Fulbright, J. William. The Arrogance of Power. New York, Random House, 1966.
Fulbright, J. William. Old Myths and New Realities. New York, Random House, 1964.
Galbraith, John Kenneth. The New Industrial State. Boston, Houghton Mifflin, 1967.
Gammons, Earl. The Twin Cities Story, 1964. Unpublished. (P)
Garrison, Garnet R. See Chester, Giraud, and–
Garry, Ralph, F. B. Rainsberry, and Charles Winick (eds.). For the Young Viewer: television programming for the young viewer–at the local level. New York, McGraw-Hill, 1962.
Geier, Leo. Ten Years With Television at Johns Hopkins. Baltimore, Johns Hopkins, 1958.
Gerassi, John. The Great Fear in Latin America. New York, Collier, 1965.
Gettleman, Marvin E. (ed.). Vietnam: history, documents, and opinions on a major world crisis. Greenwich (Conn.), Fawcett, 1965.
Goldenson, Leonard H. Responsible Broadcasting. Address delivered at Franklin Institute, January 17, 1964. New York, American Broadcasting Company, 1964.
Goldman, Eric F. The Tragedy of Lyndon Johnson. New York, Knopf, 1969.
Goldman, Eric F. "The White House and the Intellectuals," Harper's Magazine, January 1969.
Goodman, Paul. Growing Up Absurd: problems of youth in the organized society. New York, Vintage, 1960.
Goodman, Paul. People or Personnel: decentralizing and the mixed systems. New York, Random House, 1963.

Goodman, Walter. All Honorable Men: corruption and compromise in American life. Boston, Little, Brown, 1963.

Goralski, Robert S. Papers. Unpublished. (M)

Gordon, Jesse. See Bernstein, Victor, and—

Green, Adolph. See Comden, Betty, and—

Green, Gerald. "What Does the Monkey Do?" in Barrett, Edward W. (ed.), Journalists in Action. Manhasset (N.Y.), Channel Press, 1963.

Greene, Felix. A Curtain of Ignorance: how the American public has been misinformed about China. Garden City (N.Y.), Doubleday, 1964.

Greene, H. Carlton. The Broadcaster's Responsibility: speech delivered at Alfred I. Du Pont Awards Foundation Dinner in Washington, D.C., March 26, 1962. London, British Broadcasting Corporation, 1962.

Greene, Robert S. Television Writing: theory and technique. New York, Harper, 1956.

Gregory, Dick. Write Me In. New York, Bantam, 1968.

Gross, Ben. I Looked and Listened: informal recollections of radio and TV. New York, Random House, 1954.

Grove, Gene. "The CIA, FBI and CBS Bomb in Mission-Impossible," Scanlan's Monthly, March 1970.

Hackett, Albert. Reminiscences, 1958. Unpublished. (C)

Halberstam, David. The Unfinished Odyssey of Robert F. Kennedy. New York, Random House, 1969.

Hallerstein, Jerome R. Taxes, Loopholes, and Morals. New York, McGraw-Hill, 1963.

Hanser, Richard F. Reminiscences, 1967. Unpublished. (C)

Harrington, Michael. The Other America: poverty in the United States. Baltimore, Penguin, 1963.

Harvey, Frank. Air War—Vietnam. New York, Bantam, 1967.

Hazard, Patrick D. (ed.). TV as Art: some essays in criticism. Champaign (Ill.), National Council of Teachers of English, 1966.

Head, Sydney W. Broadcasting in America: a survey of television and radio. Boston, Houghton Mifflin, 1956.

Hearings of Antitrust Subcommittee of the Committee on the Judiciary, House of Representatives, 84th Congress, 2nd session. Washington, Government Printing Office, 1956.

Hearst, William Randolph, Jr., Bob Considine, and Frank Conniff. Khrushchev and the Russian Challenge. New York, Avon, 1961.

Hentoff, Nat. "Participatory Television," Evergreen Review, October 1969.

Hill, John W. The Making of a Public Relations Man. New York, David McKay, 1963.

Hill, Ruane B. See Koenig, Allen E., and—

Hilliard, Robert L. Understanding Television: an introduction to broadcasting. New York, Hastings, 1964.

Hilsman, Roger. To Move a Nation: the politics of foreign policy in the administration of John F. Kennedy. New York, Delta, 1968.

History of Communications-Electronics in the United States Navy. Prepared by Captain L. S. Howeth, USN (Ret.) under the auspices of the Bureau of Ships and Office of Naval History. Washington, Government Printing Office, 1963.

Hodgson, Godfrey. See Chester, Lewis, and–

Hofstadter, Richard. Anti-intellectualism in American Life. New York, Knopf, 1963.

Hofstadter, Richard. The Paranoid Style in American Politics. New York, Random House, 1965.

Hohenberg, John. Foreign Correspondence: the great reporters and their times. New York, Columbia University Press, 1964.

Hohenberg, John. The News Media: a journalist looks at his profession. New York, Holt, Rinehart & Winston, 1968.

Hollywood Reporter. Hollywood, daily, 1930–

Holt, Robert T. Radio Free Europe. Minneapolis, University of Minnesota Press, 1958.

Horton, Robert W. To Pay or Not to Pay: a report on subscription television. Santa Barbara, Center for the Study of Democratic Institutions, 1960.

Houghton, Neal D. (ed.). Struggle Against History: U.S. foreign policy in an age of revolution. New York, Washington Square Press, 1968.

How Television Stations Use Business Sponsored Films. New York, Modern Talking Picture Service, 1958.

Howe, Quincy. Reminiscences, 1962. Unpublished. (C)

Howeth, L. S. See History of Communications-Electronics in the United States Navy.

Hughes, Emmet John. The Ordeal of Power: a political memoir of the Eisenhower years. New York, Atheneum, 1963.

Hughes, Robert (ed.). Film Book 2: films of peace and war. New York, Grove, 1962.

Images of Peace: a television chronicle of a turning point in history. New York, CBS-TV, 1960.

In Defense of Fairness. New York, United Church of Christ, 1969.

Innis, Harold A. The Bias of Communication. Toronto, University of Toronto Press, 1951.

Innis, Harold A. Empire and Communications. Oxford, Clarendon Press, 1950.

International Television Manual. New York, Motion Picture Export Association, 1968.

Jaffe, Louis L. "The Scandal of TV Licensing," Harper's Magazine, September 1957.

Jahoda, Marie. "Anti-Communism and Employment Practices in Radio and Television," in Cogley, John, Report on Blacklisting, II: radio-television. Fund for the Republic, 1956.

Jenkins, Clive. Power Behind the Screen: ownership, control and motivations in British commercial television. London, MacGibbon and Kee, 1961.

Johnson, Haynes, with Manuel Artime, José Peréz San Román, Erneido Oliva, and Enrique Ruiz-Williams. The Bay of Pigs: the leaders' story of brigade 2506. New York, Norton, 1964.

Johnson, Nicholas. How to Talk Back to Your Television Set. Boston, Little, Brown, 1970.

Johnson, Nicholas. Television and Violence: perspectives and proposals. A statement . . . prepared at the invitation of the National Commission on the Causes and Prevention of Violence, for presentation Thursday, December 19, 1968. Unpublished.

Jones, John Paul. See Brown, Donald E., and—

Journal of Broadcasting. Association for Professional Broadcasting Education, quarterly, 1956—

Journal of the Producers Guild of America. Successor to Journal of the Screen Producers Guild. Quarterly, 1967—

Journal of the Screen Producers Guild. Succeeded 1967 by Journal of the Producers Guild of America.

Juvenile Delinquency: hearings before the subcommittee to investigate juvenile delinquency, committee on the judiciary, U.S. Senate, 87th Congress. Part 10: effects on young people of violence and crime portrayed on television. Washington, Government Printing Office, 1963.

Kahn, Frank J. (ed.). Documents of American Broadcasting. New York, Appleton-Century-Crofts, 1968.

Kalb, Marvin. "What Is Power Doing to the Pentagon?" Saturday Review, March 16, 1968.

Kaltenborn, H. V. Papers. Unpublished. (M)

Kaufman, William I. (ed.). How to Write for Television. New York, Hastings, 1955.

Kelley, Stanley, Jr. Professional Public Relations and Political Power. Baltimore, Johns Hopkins Press, 1966.

Kelly, Frank K. Who Owns the Air? Santa Barbara, Center for the Study of Democratic Institutions, 1960.

Kendrick, Alexander. Prime Time: the life of Edward R. Murrow. Boston, Little, Brown, 1969.

Kennedy, Ken. Interview, 1963. Unpublished. (C)

Kennedy, Robert F. Thirteen Days: a memoir of the Cuban missile crisis. New York, Signet, 1969.

Kieran, John. Not Under Oath: recollections and reflections. Boston, Houghton Mifflin, 1964.

Klapp, Orrin E. Symbolic Leaders: public dramas and public men. Chicago, Aldene, 1964.

Knight, Arthur. The Liveliest Art: a panoramic history of the movies. New York, Macmillan, 1957.

Knoll, Erwin. See McGaffin, William, and—

KNX-CBS Radio: continuity, growth and creativity. Los Angeles, KNX, 1961. Unpublished.

Koenig, Allen E., and Ruane B. Hill (eds.). The Farther Vision: educational television today. Madison, University of Wisconsin Press, 1967.

Kracauer, Siegfried. From Caligari to Hitler. Princeton (N.J.), Princeton University Press, 1947.

Lacy, Dan. Freedom and Communications. Urbana, University of Illinois Press, 1961.

Lamb, Edward. No Lamb for Slaughter. New York, Harcourt, Brace & World, 1963.

Lamb, Edward. Trial By Battle. Santa Barbara, Center for the Study of Democratic Institutions, 1964.

Lang, Kurt, and Gladys Engel Lang. Politics and Television. Chicago, Quadrangle Books, 1968.

Lederer, William J. A Nation of Sheep. Greenwich (Conn.), Fawcett, 1961.

Legislative History of the Fairness Doctrine: staff study for the committee on interstate and foreign commerce, House of Representatives, 90th Congress, 2nd session. Washington, Government Printing Office, 1968.

Levitan, Eli L. Animation Art in the Commercial Film. New York, Reinhold, 1960.

Levy, David. The Chameleons. New York, Dodd, Mead, 1964.

Lewis, Anthony ("and the New York *Times*"). Portrait of a Decade: the second American revolution. New York, Bantam, 1965.

Leyda, Jay. Films Beget Films. New York, Hill & Wang, 1964.

Lichty, Lawrence W. The Nation's Station: a history of radio station WLW. Ph.D. dissertation. Columbus, Ohio State University, 1964. Unpublished.

Lichty, Lawrence W. A Study of the Careers and Qualifications of Members of the Federal Radio Commission and Federal Communications Commission, 1927–1961. Master's essay. Columbus, Ohio State University, 1961.

The Listener. London, British Broadcasting Corporation, weekly, 1930–

Loewenstein, Ralph L. See Fisher, Paul L., and–

Lomax, Louis E. Thailand: the war that is, the war that will be. New York, Random House, 1967.

Lukacs, John. A New History of the Cold War. Garden City (N.Y.), Doubleday, 1966.

Lyle, Jack. See Schramm, Wilbur, and–

Lyons, Eugene. David Sarnoff: a biography. New York, Harper & Row, 1966.

Mabley, Jack. What Educational TV Offers You. Public Affairs Pamphlet. New York, Public Affairs Committee, 1954.

MacCann, Richard Dyer. Hollywood in Transition. Boston, Houghton Mifflin, 1962.

MacLeish, Archibald. Poetry and Journalism. Minneapolis, University of Minnesota, 1958.

MacNeil, Robert. The People Machine. New York, Harper & Row, 1968.

Madsen, Roy. Animated Film: concepts, methods, uses. New York, Pitman, 1969.

Mahoney, Haynes R. Broadcasting Communism: a survey of radio in the "German Democratic Republic." Unpublished, 1967.

Mailer, Norman. Miami and the Siege of Chicago: an informal history of the Republican and Democratic conventions of 1968. New York, Signet, 1968.

Manchester, William. The Death of a President: November 20 – November 25, 1963. New York, Harper & Row, 1967.

Mannes, Marya. "The Long Vigil," The Reporter, December 19, 1963.

Mannes, Marya. More in Anger. Philadelphia and New York, J. B. Lippincott, 1958.

Mayer, Martin. Madison Avenue, U.S.A. New York, Harper, 1958.

Mazo, Earl. Richard Nixon: a political and personal portrait. New York, Harper, 1959.

McCarthy, Eugene. The Limits of Power. New York, Dell, 1968.

McGaffin, William, and Erwin Knoll. Anything But the Truth: the credibility gap—how the news is managed in Washington. New York, Putnam, 1968.

McGinniss, Joe. The Selling of the President 1968. New York, Trident, 1969.

McKinney, Eleanor (ed.). The Exacting Ear: the story of listener-sponsored radio. New York, Pantheon, 1966.

McLuhan, Marshall. The Gutenberg Galaxy. Toronto, University of Toronto Press, 1962.

McLuhan, Marshall. Understanding Media: the extensions of man. New York, McGraw-Hill, 1965.

McMahon, Robert S. For the "McMahon Report," see Regulation of Broadcasting.

Mehling, Harold. The Great Time-Killer. Cleveland, World, 1962.

Mehta, Gaganvihari Lallubhai. Interview by Louis L. Gerson, Bombay, 1966. Unpublished. (D)

Memorandum Submitted to the American Public Relations Association: in the Silver Anvil competition. New York, Anti-Defamation League of B'nai B'rith, 1953.

Merson, Martin. "My Education in Government," The Reporter, October 7, 1954.

Merson, Martin. The Private Diary of a Public Servant. New York, Macmillan, 1955.

Meyer, Karl E. See Szulc, Tad, and—

Michie, Allan A. Voices Through the Iron Curtain: the Radio Free Europe story. New York, Dodd, Mead, 1963.

Miller, Arthur. Reminiscences, 1959. Unpublished. (C)

Miller, Merle. The Judges and the Judged. Garden City (N.Y.), Doubleday, 1952.

Miller, Merle. Only You, Dick Daring: or how to write one television script and make $50,000,000. New York, Sloane, 1964.

Miller, William Burke. Interview by Edwin Dunham, undated (ca. 1964). Unpublished. (P)

Montgomery, Robert. Open Letter from a Television Viewer. New York, Heineman, 1968.

Morgan, Edward P. Clearing the Air. Washington, Luce, 1963.

Morse, Carlton. Interview, 1963. Unpublished. (C)

Motivations. Motivational Publications, Inc., Croton-on-Hudson, N.Y.

Murrow, Edward R. See Bliss, Edward, Jr. (ed.), In Search of Light: the broadcasts of Edward R. Murrow 1938–1961.

Myrdal, Gunnar. Asian Drama: an inquiry into the poverty of nations. New York, Random House, 1968.

NAEB Journal. Bimonthly, successor to AERT Journal. Urbana (Ill.), National Association of Educational Broadcasters, 1957–

NAEB Monitoring Studies. See Smythe, Dallas W., Three Years of New York Television 1951–1953.

Network Broadcasting: report of the committee on interstate and foreign commerce, House of Representatives, 85th Congress, 2nd session (The "Barrow report"). Washington, Government Printing Office, 1958.

Neuman, E. Jack. Papers. Unpublished. (M)

Newfield, Jack. Robert Kennedy: a memoir. New York, Dutton, 1969.

Newman, Paul. Reminiscences, 1959. Unpublished. (C)

Nixon, Richard M. Six Crises. Garden City (N.Y.), Doubleday, 1962.

Nizer, Louis. The Jury Returns. Garden City (N.Y.), Doubleday, 1966.

Novak, Robert. See Evans, Rowland, and–

Ogilvy, David. Confessions of an Advertising Man. New York, Dell, 1964.

Oliva, Erneido. See Johnson, Haynes, with–

One Week of Educational Television. Waltham (Mass.), Brandeis University, 1964. Unpublished.

Opotowsky, Stan. TV–the Big Picture. New York, Collier, 1962.

Opt: the journal of American Women in Radio and Television. New York, 3 times per year, 1967–

Owen, Bill. See Buxton, Frank, and–

Packard, Vance. The Waste Makers. New York, McKay, 1960.

Page, Bruce. See Chester, Lewis, and—

Panofsky, Erwin. "Style and Medium in the Motion Pictures," in Talbot, Daniel (ed.), Film: an anthology. Berkeley, University of California Press, 1966.

Parker, Edwin. See Schramm, Wilbur, and–

Parker, Everett C., David W. Barry, and Dallas W. Smythe. The Television-Radio Audience and Religion. New York, Harper, 1955.

Pastore, John O. The Story of Communications: from beacon light to Telstar. New York, Macfadden-Bartell, 1964.

Paulu, Burton. Radio and Television on the European Continent. Minneapolis, University of Minnesota Press, 1967.
Pearson, Drew, and Jack Anderson. The Case Against Congress. New York, Pocket Books, 1969.
Peck, Ira. The Life and Words of Martin Luther King, Jr. New York, Scholastic, 1966.
Pennybacker, John H., and Waldo W. Braden (eds.). Broadcasting and the Public Interest. New York, Random House, 1969.
Petrie, Dan. Reminiscences, 1968. Unpublished. (C)
Pincus, Walter. See Cater, Douglass, and—
Political Broadcast Catechism and the Fairness Doctrine (6th edition). Washington, National Association of Broadcasters, 1968.
Potter, David M. People of Plenty: economic abundance and the American character. Chicago, University of Chicago Press, 1954.
Powell, John Walker. Channels of Learning: the story of educational television. Washington, Public Affairs Press, 1962.
Pratt, Robert L. Interview, 1963. Unpublished. (C)
Prize Plays of Television and Radio 1956. Foreword by Clifton Fadiman. New York, Random House, 1957.
Public Television: a program for action. The report and recommendations of the Carnegie commission on educational television. New York, Bantam, 1967.
Quarterly of Film, Radio and Television. Berkeley, University of California Press, quarterly, 1951–57.
Radio Annual. New York, Radio Daily, annual, 1938–64.
Radio Daily. New York, daily, 1937–65.
Radio Free Europe Policy Handbook. Undated.
Radio Liberty. Papers, 1960–61. Unpublished. (M)
Radio Round-Up. Hollywood, Armed Forces Radio and Television Service, weekly, 1942–
Rafetto, Michael. Interview, 1963. Unpublished. (C)
Rainsberry, F. B. See Garry, Ralph, and —
Rasky, Harry. Interview by Erik Barnouw, 1969. Unpublished. (C)
Rasky, Harry. Letters to Holly. United Features Syndicate, series of articles, 1965.
RCA Annual Report. New York, annual, 1921–
Red Channels: the report of communist influence in radio and television. New York, American Business Cosultants, 1950.
Regulation of Broadcasting: half a century of government regulation of broadcasting and the need for further legislative action. Study for the committee on interstate and foreign commerce, House of Representatives, 85th Congress, 2nd session (The "McMahon Report"). Washington, Government Printing Office, 1958.
Reisman, Philip H. Interview by Erik Barnouw, 1968. Unpublished. (C)

The Relation of the Writer to Television. Santa Barbara, Center for the Study of Democratic Institutions, 1960.

Report of the Committee on Broadcasting, 1960. London, Her Majesty's Stationery Office, 1962.

Report of the Committee on Broadcasting ("Fowler committee"). Ottawa, Queen's Printer, 1965.

Report on the Morale Effectiveness of AFL-TV. Loring Air Force Base, January 1956. Unpublished.

Report of the National Advisory Commission on Civil Disorders. New York, Bantam, 1968.

Reston, James. Interview by Philip A. Crowl, 1965. Unpublished. (D)

Rice, Elmer. "The Biography of a Play," Theatre Arts, November 1959.

Robinson, Otis E. Interview, 1963. Unpublished. (C)

Robson, Shirley Vaughn. Advertising and Politics: a case study of the relationship between Doyle Dane Bernbach, Inc., and the Democratic National Committee during the 1964 presidential campaign. Master's essay. Washington, American University, 1966. Unpublished.

Robson, William N. Reminiscences, 1966. Unpublished. (C)

Roddenberry, Gene. Papers. Unpublished. (S)

Roe, Yale. The Television Dilemma: search for a solution. New York, Hastings, 1962.

Rogers, Edward A. Face to Face. New York, Morrow, 1962.

Rose, Reginald. Six Television Plays. New York, Simon & Schuster, 1956.

Rosenthal, Raymond (ed.). McLuhan: pro and con. New York, Penguin, 1968.

Ross, Thomas B. See Wise, David, and—

Rovere, Richard H. Senator Joe McCarthy. Cleveland, World, 1960.

Ruiz-Williams, Enrique. See Johnson, Haynes, with—

Russell, Bertrand. War Crimes in Vietnam. New York, Monthly Review Press, 1967.

Ryan, J. Harold. Reminiscences, 1960. Unpublished. (C)

San Román, José Peréz. See Johnson, Haynes, with—

Sargeant, Howland H. Communicating With the People Behind the Iron Curtain: lecture at the New School, New York, April 4, 1957. Radio Liberty, Papers. (M)

Scheer, Robert. How the United States Got Involved in Vietnam. Santa Barbara, Center for the Study of Democratic Institutions, 1965.

Scher, Saul N. Voice of the City: the history of WNYC, New York City's municipal radio station, 1924–1962. Ph.D. dissertation. New York University, 1965. Unpublished.

Schickel, Richard. The Disney Version: the life, times, art and commerce of Walt Disney. New York, Simon & Schuster, 1968.

Schiller, Herbert I. Mass Communications and American Empire. New York, Kelley, 1969.

Schlesinger, Arthur M., Jr. The Bitter Heritage: Vietnam and American democracy: 1941–1966. New York, Fawcett, 1967.

Schlesinger, Arthur M., Jr. A Thousand Days: John F. Kennedy in the White House. Boston, Houghton Mifflin, 1965.

Schoenbrun, David. Interview by Philip A. Crowl, 1964. Unpublished. (D)

Schramm, Wilbur (ed.). The Effects of Television on Children and Adolescents: an annotated bibliography with an introductory overview of research results. Paris, UNESCO, 1964.

Schramm, Wilbur (ed.). The Impact of Educational Television. Urbana, University of Illinois Press, 1960.

Schramm, Wilbur (ed.). Mass Communications. Urbana, University of Illinois Press, 1949, 1960.

Schramm, Wilbur. Mass Media and National Development. Palo Alto, Stanford University Press, 1964.

Schramm, Wilbur (ed.). The Process and Effects of Mass Communications. Urbana, University of Illinois Press, 1954.

Schramm, Wilbur, Jack Lyle, and Edwin Parker. Television in the Lives of Our Children. Palo Alto, Stanford University Press, 1961.

Schumach, Murray. The Face on the Cutting Room Floor: the story of movie and television censorship. New York, Morrow, 1964.

Schurmann, Franz, Peter Dale Scott, and Reginald Zelnik. The Politics of Escalation in Vietnam. Greenwich (Conn.), Fawcett, 1966.

Schwartz, Bernard. The Professor and the Commissions. New York, Knopf, 1959.

Scott, Peter Dale. See Schurmann, Franz, and—

Seldes, Gilbert. The New Mass Media: challenge to a free society. Washington, Public Affairs Press, 1968.

Seldes, Gilbert. The Public Arts. New York, Simon & Schuster, 1956.

Serling, Rod. Papers. Unpublished. (M)

Serling, Rod. Patterns. New York, Simon & Schuster, 1957.

Servan-Schreiber, J. J. The American Challenge. New York, Atheneum, 1969.

Settel, Irving. A Pictorial History of Radio. New York, Citadel, 1960.

Sevareid, Eric. Politics and the Press: an address delivered before a joint session of the Massachusetts State Legislature, Boston, Mass., January 24, 1967. New York, Columbia Broadcasting System, 1967.

Sevareid, Eric. This Is Eric Sevareid. New York, McGraw-Hill, 1964.

Shayon, Robert Lewis. See The Eighth Art.

Shayon, Robert Lewis. Reminiscences, 1967. Unpublished. (C)

Sheldon, Sidney. Papers. Unpublished. (M)

Sherrill, Robert. The Accidental President. New York, Pyramid, 1968.

Sherrin, Ned. See Frost, David, and—

Siepmann, Charles A. Radio, Television, and Society. New York, Oxford University Press, 1950.

Siepmann, Charles A. Television and Education in the United States. Paris, UNESCO, 1952.

Skornia, Harry J. Television and the News: a critical appraisal. Palo Alto, Pacific Books, 1968.

Skornia, Harry J. Television and Society: an inquest and agenda for improvement. New York, McGraw-Hill, 1965.

Slaton, John L. Interview, 1963. Unpublished. (C)

Small, William J. See Coverage of the Death of President Kennedy and the Subsequent Events by CBS News, Washington.

Smiley, Robert. Interview, 1963. Unpublished. (P)

Smythe, Dallas W. Three Years of New York Television 1951–1953: monitoring study No. 6. Urbana (Ill.), National Association of Educational Broadcasters, 1953.

Smythe, Dallas W. See also Parker, Everett C., and –

Sopkin, Charles. Seven Glorious Days, Seven Fun-filled Nights: one man's struggle to survive a week watching commercial television in America. New York, Simon & Schuster, 1968.

Sorensen, Theodore C. Kennedy. New York, Harper, 1965.

Sorensen, Thomas C. The Word War: the story of American propaganda. New York, Harper, 1968.

Sponsor. New York, biweekly, then weekly, 1948–

Stanley, Kim. Reminiscences, 1959. Unpublished. (C)

Stanton, Frank. Books and Television. New York, New York Public Library, 1963.

Stearns, Marshall W. The Story of Jazz. New York, Oxford University Press, 1958.

Steel, Ronald. Pax Americana. New York, Viking, 1967.

Steiger, Rod. Reminiscences, 1959. Unpublished. (C)

Steinberg, Charles S. (ed.). Mass Media and Communication. New York, Hastings, 1966.

Steiner, Gary A. The People Look at Television: a study of audience attitudes. New York, Knopf, 1963.

Stephenson, Ralph. Animation in the Cinema. London, Zwemmer, 1967.

Sterling, Christopher Hastings. Second Service: a history of commercial FM broadcasting. Ph.D. dissertation. Madison, University of Wisconsin, 1969. Unpublished.

Storer, George B. The Storer Story. Memorandum, 1964. Unpublished. (P)

Subversive Infiltration of Radio, Television and the Entertainment Industry: hearings before the subcommittee to investigate the administration of the internal security act and other internal security laws of the committee on the judiciary, U.S. Senate, 82nd Congress. Washington, Government Printing Office, 1952.

Summers, Harrison B. (ed.). A Thirty-Year History of Programs Carried on

National Radio Networks in the United States 1926–1956. Columbus, Ohio State University, 1958.

Summers, Harrison B. See also Summers, Robert E., and —

Summers, Robert E., and Harrison B. Summers. Broadcasting and the Public. Belmont (Calif.), Wadsworth, 1966.

Swing, Raymond. Good Evening! a professional memoir. New York, Harcourt, Brace & World, 1964.

Szulc, Tad. "The New York *Times* and the Bay of Pigs," in Brown and Bruner (eds.), How I Got That Story. New York, Dutton, 1967.

Szulc, Tad, and Karl E. Meyer. The Cuban Invasion: the chronicle of a disaster. New York, Ballantine, 1962.

Talbot, Daniel. Film: an anthology. Berkeley, University of California Press, 1966.

Tales of Seven Cities: and of video that teaches. Washington, National Citizens Commission for Educational Television, 1954.

Taylor, Davidson. Reminiscences, 1967. Unpublished. (C)

Taylor, Telford. Grand Inquest: the story of congressional investigations. New York, Simon & Schuster, 1955.

Tebbel, John. David Sarnoff: putting electrons to work. Chicago, New York, Encyclopaedia Britannica Press, 1963.

Telefilm. Hollywood, monthly, 1956–

Teletips. New York, Armed Forces Press, Radio and Television Service, weekly, 1955–

Television. New York, monthly, 1944–67.

Television for Children. Boston, Foundation for Character Education, undated.

Television Digest. Washington, Triangle, weekly, 1945–

Television in Education: a summary report . . . of the educational television program institute held at Pennsylvania State College April 20–24, 1952. Washington, American Council on Education, 1952.

Television Factbook. Supplements to Television Digest. Radnor (Pa.), Triangle, 1945–

Television and Juvenile Delinquency: interim report of the subcommittee to investigate juvenile delinquency, committee on the judiciary, U.S. Senate, 88th Congress, 2nd session. Unpublished.

Television Market List. Hollywood, Writers Guild of America, West, annual, 1955–

Television Network Program Procurement: report of the committee on interstate and foreign commerce, House of Representatives, 88th Congress, 1st session. Washington, Government Printing Office, 1963.

Television Quarterly: journal of the National Academy of Television Arts and Sciences, 1962–

Television and the Wired City: a study of the implications of a change in the mode of transmission. Washington, National Association of Broadcasters, 1968.

Terkel, Studs. Division Street. New York, Pantheon, 1967.
Thomas, Bob. King Cohn: the life and times of Harry Cohn. New York, Putnam, 1967.
Turner, C. Nigel. American Influence in British Broadcasting Since World War II. Master's essay. New York, Columbia University, 1968. Unpublished.
TV Communications: the professional journal of the cable television industry. (Original title, 1964–65: TV & Communications.) Oklahoma City, monthly, 1964–
TV Guide. Philadelphia, Triangle, weekly, 1948–
United States House of Representatives, committee on interstate and foreign commerce. See following titles: Broadcast Advertisements; Legislative History of the Fairness Doctrine; Network Broadcasting; Regulation of Broadcasting; Television Network Program Procurement.
United States House of Representatives, committee on the judiciary. See Hearings of the Antitrust Subcommittee.
United States Senate, committee on the judiciary. See following titles: Juvenile Delinquency; Subversive Infiltration of Radio, Television and the Entertainment Industry; Television and Juvenile Delinquency.
Variety. New York, weekly, 1905–
Vidal, Gore (ed.). Best Television Plays. New York, Ballantine, 1956.
Vidal, Gore. Sex, Death and Money. Boston, Little, Brown, 1968.
Wade, Robert J. Designing for TV. New York, Pelligrini & Cudahy, 1952.
Wald, Jerry. Reminiscences, 1959. Unpublished. (C)
Walton, Richard. "The Last 170 Days of Adlai Stevenson," Esquire, September 1968.
Waters, David. Interview by Philip A. Crowl, 1966. Unpublished. (D)
Weinberg, Meyer. TV in America: the morality of hard cash. New York, Ballantine, 1962.
Weir, E. Austin. The Struggle for National Broadcasting in Canada. Toronto, McClelland & Stewart, 1965.
Weiss, Peter. The Investigation (play). English version by Jon Swan and Ulu Grosbard. New York, Atheneum, 1966.
WGA-e Newsletter. New York, Writers Guild of America, East, monthly, 1967–
WGA-w Newsletter. Hollywood, Writers Guild of America, West, monthly, 1965–
Whalen, Richard J. The Founding Father: the story of Joseph P. Kennedy. New York, New American Library, 1964.
Where Was Radio When the Lights Went Out? Washington, National Association of Broadcasters, 1966.
White, David Manning, and Richard Averson (eds.). Sight, Sound, and Society: motion pictures and television in America. Boston, Beacon, 1968.
White, Theodore H. The Making of the President 1960. New York, Atheneum, 1961.

White, Theodore H. The Making of the President 1964. New York, Atheneum, 1965.

White, Theodore H. The Making of the President 1968. New York, Atheneum, 1969.

White, William S. The Professional: Lyndon B. Johnson. Greenwich (Conn.), Fawcett. 1964.

Whitfield, Stephen E. The Making of *Star Trek*. New York, Ballantine, 1968.

Whitworth, William. "An Accident of Casting" (Huntley-Brinkley profile), New Yorker, August 3, 1968.

Wiggins, James Russell. Freedom or Secrecy. New York, Oxford University Press, 1964.

Wiland, Harry A. The Pirate Radio Invasion of Great Britain. Master's Essay. New York, Columbia University, 1968. Unpublished.

Wilford, John Noble. We Reach the Moon: the New York *Times* story of man's greatest adventure. New York, Bantam, 1969.

Willis, Edgar E. Foundations in Broadcasting. New York, Oxford University Press, 1951.

Willis, Edgar E. See also Chester, Giraud, Garnet R. Garrison, and–

Wilson, H. H. Pressure Group: the campaign for commercial television in England. New Brunswick (N.J.), Rutgers University Press, 1961.

Wilson, Richard. "The Ring Around McCarthy," Look, December 1, 1953.

Winick, Charles. Taste and the Censor in Television. New York, Fund for the Republic, 1959.

Winick, Charles. See also Garry, Ralph, and–

Wise, David, and Thomas B. Ross. The Invisible Government. New York, Random House, 1964.

Witcover, Jules. 85 Days: the last campaign of Robert F. Kennedy. New York, Ace, 1969.

Witty, Paul, and Harry Bricker. Your Child and Radio, TV, Comics and Movies. Chicago, Science Research Associates, 1952.

Wood, Robin. Arthur Penn. London, Studio Vista, 1967.

Wood, William A. Electronic Journalism. New York, Columbia University Press, 1967.

Woodward, Joanne. Reminiscences, 1959. Unpublished. (C)

Woolley, Thomas Russell, Jr. A Rhetorical Study: the radio speaking of Edward R. Murrow. Ph.D. dissertation. Evanston (Ill.), Northwestern University, 1957. Unpublished.

World Radio-TV Handbook. Hellerup (Denmark), annual, 1947–

Writing *Mission: Impossible*. Hollywood, Geller, undated (ca. 1966). Unpublished.

Wyckoff, Gene. The Image Candidates: American politics in the age of television. New York, Macmillan, 1968.

Zelnik, Reginald. See Schurmann, Franz, and–

INDEX